Progress in Medicinal Chemistry 47

Editors:

G. LAWTON, B.SC., PH.D., C.CHEM.

Garden Fields
Stevenage Road
St. Ippolyts
Herts SG4 7PE
United Kingdom

and

D. R. WITTY, B.A., M.A., D.PHIL., C.CHEM., F.R.S.C.

GlaxoSmithKline
New Frontiers Science Park (North)
Third Avenue
Harlow, Essex CM19 5AW
United Kingdom

ELSEVIER

AMSTERDAM · BOSTON · HEIDELBERG · LONDON · NEW YORK · OXFORD
PARIS · SAN DIEGO · SAN FRANCISCO · SINGAPORE · SYDNEY · TOKYO

Elsevier
Radarweg 29, PO Box 211, 1000 AE Amsterdam, The Netherlands
Linacre House, Jordan Hill, Oxford OX2 8DP, UK

First edition 2009

Copyright © 2009 Elsevier B.V. All rights reserved

No part of this publication may be reproduced, stored in a retrieval system
or transmitted in any form or by any means electronic, mechanical, photocopying,
recording or otherwise without the prior written permission of the publisher

Permissions may be sought directly from Elsevier's Science & Technology Rights
Department in Oxford, UK: phone (+44) (0) 1865 843830; fax (+44) (0) 1865 853333;
email: permissions@elsevier.com. Alternatively you can submit your request online by
visiting the Elsevier web site at http://www.elsevier.com/locate/permissions, and selecting
Obtaining permission to use Elsevier material

Notice
No responsibility is assumed by the publisher for any injury and/or damage to persons
or property as a matter of products liability, negligence or otherwise, or from any use
or operation of any methods, products, instructions or ideas contained in the material
herein. Because of rapid advances in the medical sciences, in particular, independent
verification of diagnoses and drug dosages should be made

British Library Cataloguing-in-Publication Data
A catalogue record for this book is available from the British Library

Library of Congress Cataloging-in-Publication Data
A catalog record for this book is available from the Library of Congress

ISBN: 978-0-444-53300-5
ISSN: 0079-6468

For information on all Elsevier publications
visit our website at elsevierdirect.com

Printed and bound in Hungary

09 10 11 12 13   10 9 8 7 6 5 4 3 2 1

Working together to grow
libraries in developing countries

www.elsevier.com | www.bookaid.org | www.sabre.org

ELSEVIER   BOOK AID International   Sabre Foundation

Progress in
Medicinal Chemistry 47

Books are to be returned on or before
the last date below.

LIVERPOOL
JOHN MOORES UNIVERSITY
AVRIL ROBARTS LRC
TITHEBARN STREET
LIVERPOOL L2 2ER
TEL 0151 231 4022

Preface

This volume provides reviews of six topics demonstrating the breadth of the field and recent successes in medicinal chemistry. Each of the first five chapters takes an important biochemical target as its theme and provides an insight into current progress in drug design. The last chapter focuses on the vital subject of pharmacokinetics and the great strides that have been made in this discipline during the past decade. All chapters provide an insight into the skills required of the modern medicinal chemist, in particular, the use of an appropriate selection of the wide range of tools now available to solve key scientific problems.

Migraine headaches are still not adequately treated in a significant proportion of patients, despite the success of the tryptan family. This is partly because of side effects caused by cardiovascular liabilities in currently available therapies. It is therefore very exciting that recently developed small molecule Calcitonin Gene-Related Peptide (CGRP) receptor antagonists have been reported to show efficacy in late Phase clinical trials as a novel treatment for migraine, which potentially offer a safer profile. Chapter 1 describes the challenges encountered in designing and optimising effective potential small molecule CGRP antagonists starting from the peptidic lead and includes discussion of the discovery of telcagepant.

Phosphodiesterase-4 (PDE4) inhibition has been an anti-inflammatory target for drug designers for many years, however, a selective inhibitor has yet to be marketed. Recent developments in understanding the positive and negative effects of PDE4 inhibition, as well as the structural and pharmacokinetic properties required of potential compounds, have clearly brought that day closer. Chapter 2 analyses current medicinal chemistry research into modulators of this pleiotropic enzyme, including subtype selective agents. It describes the potential for topical, inhaled and systemic drugs as well as dual therapy opportunities with complementary or synergistic agents.

The advent of the proton pump inhibitors had a significant impact on the lives of the 20% of the population who suffer with varying degrees of severity from gastro-intestinal disease. Recent developments in this 20-year old field are reviewed in Chapter 3.

Cardiovascular diseases, neuroprotection and modulation of lipid metabolism have all been proposed as therapeutic opportunities for compounds which agonise or modulate the adenosine A1 G-protein coupled receptor. The limitations imposed by the chemistry and properties of typical adenosine analogues have led to slow progress and, so far, no successful drugs have been developed which target this mechanism. Chapter 4 reviews recent work in the area and describes the discovery of a promising drug-like potent and selective series of non-purinergic adenosine A1 receptor agonists with potential to provide an effective treatment for pain.

The endothelin receptor represents one of the clear successes of the target-based approach to new therapeutics and it took only a relatively short time from discovery of the target to launch of the first drug. Endothelin was first characterised in 1988 and three antagonists were launched during the subsequent 20 years. Chapter 5 reviews recent developments in the field of endothelin antagonists including the search for receptor subtype-selective compounds.

One of the major successes of medicinal chemistry in the past decade has been the significant reduction in project failures for pharmacokinetic reasons. This was achieved in large part through increased understanding of metabolic liabilities. Extensive use of metabolic profiling during lead optimisation (and earlier) is now a routine part of drug discovery projects. Barry Jones and colleagues in Chapter 6 describe the strategies and experimental procedures used to develop structure–activity relationships in cytochrome p450 inhibition. They give several case histories and discuss challenges remaining to the rational design of metabolically stable compounds.

September 2008   G. Lawton
                                                                                D. R. Witty

Contents

Preface		v
List of Contributors		ix
1	**Calcitonin Gene-Related Peptide Receptor Antagonists for the Treatment of Migraine** Theresa M. Williams, Christopher S. Burgey and Christopher A. Salvatore	1
2	**PDE4 Inhibitors – A Review of the Current Field** Neil J. Press and Katharine H. Banner	37
3	**H^+/K^+ ATPase Inhibitors in the Treatment of Acid-Related Disorders** Mark Bamford	75
4	**The Adenosine A_1 Receptor and its Ligands** Peter G. Nell and Barbara Albrecht-Küpper	163
5	**Endothelin Receptor Antagonists: Status and Learning 20 Years On** Michael J. Palmer	203
6	**Cytochrome P450 Metabolism and Inhibition: Analysis for Drug Discovery** Barry C. Jones, Donald S. Middleton and Kuresh Youdim	239
Subject Index		265
Cumulative Index of Authors for Volumes 1–47		273
Cumulative Index of Subjects for Volumes 1–47		281

Christopher A. Salvatore
Department of Pain Research, Merck Research Laboratories, West Point, PA 19486, USA

Theresa M. Williams
Department of Medicinal Chemistry, Merck Research Laboratories, West Point, PA 19486, USA

Kuresh Youdim
Pfizer Global Research and Development, Ramsgate Road, Sandwich, Kent CT13 9NJ, UK

Contents

Preface		v
List of Contributors		ix
1	**Calcitonin Gene-Related Peptide Receptor Antagonists for the Treatment of Migraine** Theresa M. Williams, Christopher S. Burgey and Christopher A. Salvatore	1
2	**PDE4 Inhibitors – A Review of the Current Field** Neil J. Press and Katharine H. Banner	37
3	**H^+/K^+ ATPase Inhibitors in the Treatment of Acid-Related Disorders** Mark Bamford	75
4	**The Adenosine A_1 Receptor and its Ligands** Peter G. Nell and Barbara Albrecht-Küpper	163
5	**Endothelin Receptor Antagonists: Status and Learning 20 Years On** Michael J. Palmer	203
6	**Cytochrome P450 Metabolism and Inhibition: Analysis for Drug Discovery** Barry C. Jones, Donald S. Middleton and Kuresh Youdim	239
Subject Index		265
Cumulative Index of Authors for Volumes 1–47		273
Cumulative Index of Subjects for Volumes 1–47		281

List of Contributors

Barbara Albrecht-Küpper
Global Drug Discovery – Department of Cardiology Research,
Bayer HealthCare AG, Bayer Schering Pharma, Aprather Weg 18a,
42096 Wuppertal, Germany

Mark Bamford
GlaxoSmithKline, Medicines Research Centre, Gunnels Wood Road,
Stevenage, Herts SG1 2NY, UK

Katharine H. Banner
Novartis Institutes for Biomedical Research, Horsham,
West Sussex RH12 5AB, UK

Christopher S. Burgey
Department of Medicinal Chemistry, Merck Research Laboratories,
West Point, PA 19486, USA

Barry C. Jones
Pfizer Global Research and Development, Ramsgate Road,
Sandwich, Kent CT13 9NJ, UK

Donald S. Middleton
Pfizer Global Research and Development, Ramsgate Road,
Sandwich, Kent CT13 9NJ, UK

Peter G. Nell
Global Drug Discovery – Operations, Bayer HealthCare AG,
Bayer Schering Pharma, Müllerstraße 178, 13353 Berlin, Germany

Michael J. Palmer
Sandwich Discovery Chemistry, Pfizer Global Research and Development,
Sandwich Laboratories, Ramsgate Road, Sandwich, Kent CT13 9NJ, UK

Neil J. Press
Novartis Institutes for Biomedical Research, Horsham,
West Sussex RH12 5AB, UK

Christopher A. Salvatore
Department of Pain Research, Merck Research Laboratories, West Point, PA 19486, USA

Theresa M. Williams
Department of Medicinal Chemistry, Merck Research Laboratories, West Point, PA 19486, USA

Kuresh Youdim
Pfizer Global Research and Development, Ramsgate Road, Sandwich, Kent CT13 9NJ, UK

1 Calcitonin Gene-Related Peptide Receptor Antagonists for the Treatment of Migraine

THERESA M. WILLIAMS[1], CHRISTOPHER S. BURGEY[1] and CHRISTOPHER A. SALVATORE[2]

[1]*Department of Medicinal Chemistry, Merck Research Laboratories, West Point, PA 19486, USA*

[2]*Department of Pain Research, Merck Research Laboratories, West Point, PA 19486, USA*

INTRODUCTION	2
MIGRAINE AND A ROLE FOR CGRP	4
Epidemiology and Clinical Manifestations of Migraine	4
CGRP and Migraine	4
PHARMACOLOGICAL CHARACTERIZATION OF THE CGRP RECEPTOR	5
The CGRP Receptor	5
In Vitro Assays	7
Small Molecule Antagonists can Display Species Selectivity	8
Selectivity Versus the Adrenomedullin Receptors	9
Additional Determinants of Antagonist Affinity	9
Pharmacodynamic Assay to Assess *In Vivo* Potency	10
DISCOVERY OF CGRP RECEPTOR ANTAGONISTS	11
Early Antagonists and the Discovery of BIBN4096BS	11
Proof of Concept in Human Clinical Trials with BIBN4096BS	14
Search for Orally Active CGRP Receptor Antagonists	15
Spirohydantoin-Based CGRP Receptor Antagonists	16
Pyridinone as Replacement for the MK-0974 Caprolactam	17
Additional Exploration of HTS Hit (15): Spirohydantoin Replacements	19
Caprolactam as a Benzodiazepinone Replacement	22
Caprolactam Optimization	24

Practical Synthesis of MK-0974   28
In Vitro and In Vivo Characterization of MK-0974   29
Clinical Trial Results for MK-0974: Human Pharmacokinetics   30
Capsaicin-Induced Dermal Blood Flow   30
Clinical Efficacy   31

CONCLUSION   31

ACKNOWLEDGEMENTS   32

REFERENCES   32

INTRODUCTION

Migraine headache is a disabling condition with characteristic symptoms that can last for several days. The earliest written descriptions of migraine date from ancient Sumeria *circa* 3000–4000 BC [1]. Primitive treatment included trepanation, in which a hole was drilled in the skull in an attempt to alleviate head pain.

Caffeine was probably the first pharmacological agent to treat migraine, and it is likely that the availability and prevalence of coffee in 17th-century England helped to popularize this remedy. Two hundred years later, ergot extracts were discovered to be useful in treating some headaches [2]. More widespread use followed when pure, crystalline ergotamine tartrate was obtained in 1918, and in 1925 the successful treatment of migraine with subcutaneous ergotamine (ET) was achieved [3]. Hydrogenation of the 9,10 double bond produced dihydroergotamine (DHT), which was better tolerated than ET and effective in treating migraine. Both ET and DHT bind with high affinity to multiple serotonin, noradrenaline and dopamine receptors. Since rare but potentially serious adverse cardiac events have been reported, the use of ET and DHT is contraindicated in patients with cardiovascular disease.

The discovery of subtype-selective serotonin agonists known as triptans significantly improved treatment options. First introduced in 1993, this class of drugs revolutionized migraine therapy and is now considered the "gold standard". Triptans bind and activate 5-HT_{1B} and 5-HT_{1D} receptors to promote vasoconstriction of cranial blood vessels, and to block inflammatory peptide release in sensory neurons [4]. Unlike ET and DHT, triptans are orally bioavailable, convenient to dose and generally well tolerated. These drugs are safe when used appropriately but, because they cause vasoconstriction, triptans are also contraindicated in patients with cardiovascular disease.

After the successful introduction of the triptans, evidence began to accumulate linking calcitonin gene-related peptide (CGRP) to migraine pathophysiology. CGRP had been identified as a vasoactive neuropeptide largely expressed in sensory neurons [5]. It was observed that plasma levels of CGRP were elevated during migraine, and that successful treatment with a triptan returned it to basal levels [6, 7]. This observation, along with other evidence, suggested that CGRP receptor antagonists might be useful antimigraine drugs. Because CGRP receptor antagonists are not direct vasoconstrictors, it was suggested that they would not have the cardiovascular liabilities associated with triptans, and thus could possess a therapeutic advantage. Clinical proof of concept was obtained with the potent CGRP receptor antagonist BIBN4096BS (olcegepant) (1) [8]. Following iv administration, BIBN4096BS proved efficacious in treating acute migraine headache in human clinical trials. A practical goal was to identify a CGRP receptor antagonist that was orally bioavailable and thus convenient for patients to take. This was the focus of the medicinal chemistry effort that led to the discovery of MK-0974 (telcagepant) (2), an oral CGRP receptor antagonist [9, 10].

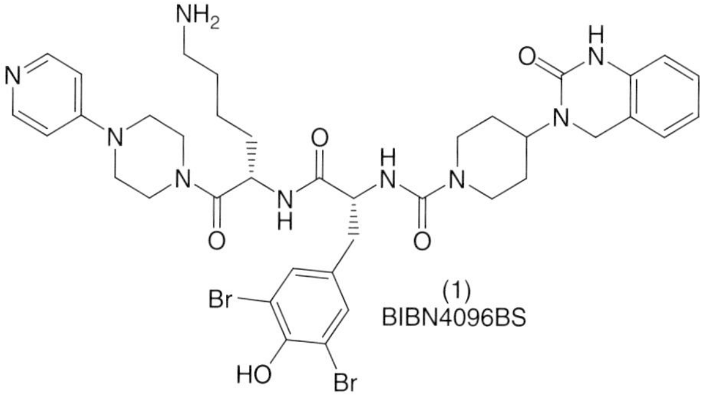

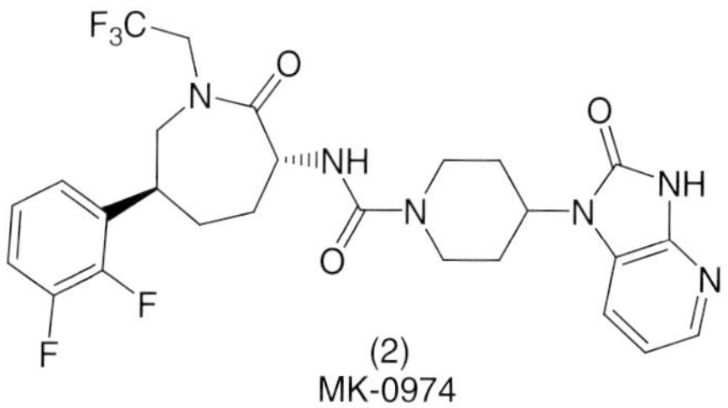

This review seeks to summarize key aspects of research in the discovery and characterization of CGRP receptor antagonists for therapeutic application to the treatment of migraine.

MIGRAINE AND A ROLE FOR CGRP

EPIDEMIOLOGY AND CLINICAL MANIFESTATIONS OF MIGRAINE

Migraine is a common and disabling neurological disorder estimated to afflict 11% of the global adult population [11]. Prevalence ranges from 6 to 9% among men and 15 to 17% among women [12]. After gender, the second biggest factor associated with migraine is age. Migraine prevalence increases with age, peaking in the middle years (40–50 years of age) and decreasing thereafter [13]. The severity and high prevalence of migraine headache is estimated to have a societal cost of $16.6 billion in the United States [14]. A variety of indirect costs, including absenteeism and reduced productivity at work, account for a large portion of this cost. Migraine is generally agreed to be under-diagnosed. A population-based survey identified individuals with severe headache that met the criteria for migraine, but about 50% of these migraineurs had not been diagnosed by a physician [15]. Therefore there is a significant opportunity for improvement in the diagnosis and management of migraine [14, 15].

Migraine can be separated into five phases: the prodrome, aura, headache, resolution and postdrome phases [12]. The most common prodrome (premonitory) symptoms reported by approximately 60% of migraineurs are fatigue, mood changes and gastrointestinal symptoms [16, 17]. The large proportion of migraineurs experiencing prodrome symptoms underscores the potential for preventing the headache phase with early pharmacological intervention. An acute migraine attack does not always include an aura phase. When present, aura typically occurs just before or simultaneously with the headache. Aura is characterized by one or more transient, neurologic visual, sensory, motor, aphasic or basilar symptoms, typically lasting less than 60 min [12, 18]. In adults, untreated migraine attacks without aura have typically unilateral head pain, last between 4 and 72 h and are accompanied by one or more of the following: nausea, vomiting, sensitivity to sound and sensitivity to light [19].

CGRP AND MIGRAINE

CGRP is a 37 amino acid neuropeptide belonging to the calcitonin family of peptides which includes calcitonin, adrenomedullin and amylin. CGRP exists in two forms, α- and β-CGRP, which differ from each other by three amino acids. α-CGRP (3) is produced by tissue specific alternative mRNA splicing of the calcitonin gene [20], whereas β-CGRP is encoded by a discrete gene which does not produce calcitonin [21, 22]. Of the two forms, it is considered that α-CGRP is the more abundant and it is found in discrete areas of the central and peripheral nervous system [23]. Although there are proposed differences in the receptor-mediated effects of α- and β-CGRP, throughout this review it is the effects of α-CGRP (3) which will be discussed.

ACDTATCVTHRLAGLLSRSGGVVKNNFVPTNVGSKAF-NH$_2$

(3)
CGRP

CGRP is widely distributed in the peripheral and central nervous system [24] and exhibits a wide range of biological effects on tissues, the most pronounced being vasodilation [25]. Cerebral blood vessels are densely innervated by CGRP-expressing trigeminal nerve endings [26] and the vasoactive effects of CGRP have been demonstrated in a variety of blood vessels, including those of the cerebral vasculature [27]. The above observations are aligned with the early theories proposed over 60 years ago by Wolff [28] that migraine is related to dilation of the extracerebral vasculature. It has been demonstrated that stimulation of the trigeminal ganglia in both cats and humans results in the elevation of CGRP in the cranial circulation [29] and analogously, that CGRP levels are increased in the cranial circulation during severe migraine attacks [30, 31]. Intravenous administration of CGRP to migraineurs induced a delayed migrainous headache in some patients [32], and successful treatment of migraine headache with sumatriptan resulted in the normalization of CGRP levels [33].

The prevailing consensus view is that migraine is a neurological disorder, where the primary site of dysfunction resides in the brain [34]. This is supported by the following observations: (1) positron emission tomography showed evidence of brainstem activation during a migrainous attack [35]; (2) intravenous administration of the potent CGRP receptor antagonist BIBN4096BS reduced spontaneous and thermally evoked activity in the spinal trigeminal nucleus of rats [36], and inhibited trigeminocervical superior sagittal sinus-evoked activity in the cat [37]; (3) peripheral application of CGRP to the meningeal dura mater in rats caused an increase in blood flow, but did not cause the sensitization of meningeal nociceptors that are observed during migraine [38] and (4) intravenous infusion of vasoactive intestinal peptide (VIP) to migraineurs did not induce migraine despite causing a marked dilation of cranial arteries [39].

PHARMACOLOGICAL CHARACTERIZATION OF THE CGRP RECEPTOR

THE CGRP RECEPTOR

The CGRP receptor is heterodimeric (Figure 1.1A). One component is a G-protein coupled receptor (GPCR) identified as the calcitonin-receptor-like

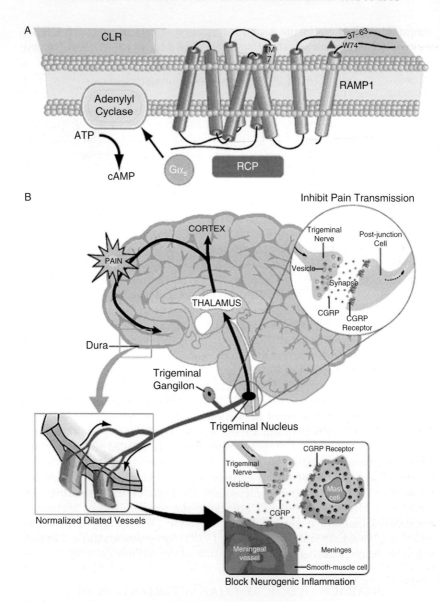

Fig. 1.1 (A) The heterodimeric CGRP receptor. Regions implicated in the interaction of small molecule antagonists and the CGRP receptor are indicated and (B) potential sites of action for CGRP receptor antagonists.

receptor (CLR). The second component is a single membrane-spanning protein designated receptor activity-modifying protein 1 (RAMP1) [40]. An additional intracellular protein, receptor component protein (RCP), is required for G-protein signal transduction [41]. Activation of the CGRP receptor causes dissociation of the G-protein (Gα_s) leading to an increase in the second messenger cAMP. The CL receptor (CLR) is in the class B GPCR family, whose members are characterized by large N-terminal extracellular domains that interact with various regulatory peptides. The RAMPs comprise a group of three single transmembrane-spanning proteins (RAMP1, RAMP2 and RAMP3). RAMPs participate in trafficking of CLR to the cell surface and determine the glycosylation state and pharmacological properties of CLR. A functional CGRP receptor requires co-expression of CLR and RAMP1. When CLR is co-expressed with RAMP2 or RAMP3, a receptor with high affinity for adrenomedullin is produced [40].

The distribution of CGRP receptors in the trigeminovascular system is consistent with a role in the pathophysiology of migraine (Figure 1.1B). CGRP receptor antagonists could act through numerous mechanisms as highlighted by the broad distribution of CGRP receptors in the trigeminovascular system. CGRP receptor antagonism at peripheral receptors on blood vessels and mast cells could block neurogenic inflammation and normalize dilated blood vessels whereas blockade of central receptors in the brainstem may inhibit pain transmission. Immunohistochemical localization of CLR and RAMP1 in the human cerebral vasculature was first demonstrated by Oliver *et al.* [42]. Specifically, it was demonstrated that both CLR and RAMP1 were expressed in human meningeal blood vessels that are thought to be involved in migraine pain. More recently in the rat, CLR and RAMP1 have been localized in the cranial dura mater in blood vessels and mast cells [43], supporting a role for CGRP in mast cell degranulation and the subsequent neurogenic inflammation [44]. CGRP and CGRP receptors are widely expressed in many parts of the brain including periaqueductal gray, parabrachial nucleus, nucleus solitaris, cerebellum, hippocampus and amygdala [45–47]. These areas may play a role in the initiation of migraine and are potential sites of action for CGRP receptor antagonists. CLR and RAMP1 are also located centrally in the spinal trigeminal nucleus of rat [43] and are co-localized in the pre-synaptic terminals of the spinal cord dorsal horn [48] indicating a potential role for CGRP in pain transmission.

IN VITRO ASSAYS

To support the drug discovery process, stable cell lines expressing the human CGRP and human adrenomedullin receptors were generated in human embryonic kidney (HEK) cells. To confirm that the human recombinant CGRP receptor stably expressed in HEK293 cells exhibited physiologically relevant pharmacology, comparisons were made to the native human receptor found in SK-N-MC cells [49]. Radioligand binding assays utilized membranes from these cell lines to monitor affinity (K_i) for the CGRP receptor. Additionally, binding assays were used to determine affinity for CGRP receptors from other species, and to track selectivity versus the related adrenomedullin receptors.

As antagonism of the CGRP receptor inhibits agonist-induced cAMP production, this provides an easy method of monitoring *in vitro* functional antagonism. The functional monitoring of antagonist potency (IC_{50}) was conducted both with and without 50% serum to assess the effect of plasma protein binding on functional activity. Antagonist potency in the "serum-shift" assay correlated well with *in vivo* potency in the rhesus pharmacodynamic (PD) assay, described in detail later in this chapter. The "serum shift" assay provided a robust method to prioritize compounds to be evaluated in the PD model. Details on stable cell line generation and *in vitro* assays have been described [50].

SMALL MOLECULE ANTAGONISTS CAN DISPLAY SPECIES SELECTIVITY

Class B GPCRs are attractive drug targets with the potential to treat a wide variety of biological activities. They are notoriously difficult targets for drug development, in part because the natural peptide ligands have extensive binding sites. As a consequence, compounds in development or clinical practice tend to be peptidic in nature. Another factor complicating small molecule drug discovery is the pronounced species-selective pharmacology displayed by many antagonists of the CGRP receptor [51–53]. For example, BIBN4096BS demonstrated very high affinity for the human CGRP receptor ($K_i = 14$ pM) and the marmoset receptor, but had > 100-fold lower affinity for the rat, rabbit, dog and guinea pig receptors [51]. The CGRP receptor antagonist MK-0974 displayed approximately 1,000-fold higher affinity for the human ($K_i = 0.77$ nM) and rhesus ($K_i = 1.2$ nM) CGRP receptors compared to the dog and rat CGRP receptors [50].

The observation that small molecule antagonists of the CGRP receptor display species-selective pharmacology provided an opportunity to map the determinants of receptor affinity exhibited by BIBN4096BS and related analogues (Figure 1.1A). For the heterodimeric CGRP receptor, high affinity binding of small molecule antagonists could be driven by either CLR or RAMP1, or both. In the case of the human and rat CGRP receptors, CLR is 91% homologous and RAMP1 is 71% homologous. Mixed species co-transfection showed that co-expression of human CLR with rat RAMP1 resulted in rat-like CGRP receptor pharmacology for BIBN4096BS. Conversely, rat CLR co-expressed with human RAMP1 produced human CGRP receptor pharmacology. This implicated RAMP1 as the primary driver for high affinity binding of BIBN4096BS [54]. Chimeric rat/human RAMP1 receptors and subsequent site-directed mutagenesis identified a single amino acid (residue 74, Figure 1.1A) in RAMP1 as being responsible for the species differences between rat and human receptor binding of BIBN4096BS-like compounds. In human RAMP1 this amino acid residue is a tryptophan, and in rat it is a lysine.

SELECTIVITY VERSUS THE ADRENOMEDULLIN RECEPTORS

CLR can heterodimerize with RAMP2 (AM_1 receptor) and RAMP3 (AM_2 receptor) to produce high affinity adrenomedullin (AM) receptors. Since both CGRP and adrenomedullin receptors contain CLR, initially it was thought that developing highly selective CGRP receptor antagonists would be challenging. In practice, the RAMP1-dependence displayed by many small molecule CGRP receptor antagonists also provided selectivity for the CGRP receptor over the AM receptor. The residue analogous to Trp74 in RAMP1 is Glu74 in RAMP2 and RAMP3 [55]. BIBN4096BS displayed no significant activity (up to 10 µM) on transiently expressed human AM_1 and AM_2 receptors measured via inhibition of adrenomedullin-induced cAMP production [56]. Likewise, MK-0974 was highly selective for the CGRP receptor as measured by the ability to compete with ^{125}I-human adrenomedullin, with K_i values of >100 and 29 µM on AM_1 and AM_2 receptors, respectively [50].

ADDITIONAL DETERMINANTS OF ANTAGONIST AFFINITY

Not all small molecule antagonists of the CGRP receptor display species and receptor subtype selectivity. During the drug discovery process (4) was identified, which had K_i values of 3.5 and 6.8 µM on the human and rat CGRP receptors and similar affinity for the human AM_1 receptor [57]. The observation that (4) exhibited similar affinity for these receptors suggested that this compound was binding differently compared to BIBN4096BS and MK-0974, in possibly a RAMP-independent manner. A related GPCR, the calcitonin receptor (CTR), which has also been shown to interact with RAMP1 and shares approximately 57% identity on the amino acid level with CLR, was utilized in a chimeric receptor approach to identify key domains involved in binding CGRP receptor antagonists. Chimeric receptors were generated by exchanging regions of CLR with the corresponding regions of CTR followed by co-expression with RAMP1 and determination of antagonist affinities. By using this approach two distinct regions were identified within CLR that were responsible for conferring high-affinity antagonist interaction. The RAMP1-dependent compounds were found to interact with the extracellular domain (aminoterminus) of CLR; specifically with residues 37–63 (Figure 1.1A). In contrast, (4) was shown to interact with transmembrane domain 7 in CLR, which is identical in human CGRP and AM receptors, and highly homologous across species. The finding that BIBN4096BS interacts with both CLR and RAMP1 for high affinity binding to the CGRP receptor further highlights the complexity of targeting this unique receptor.

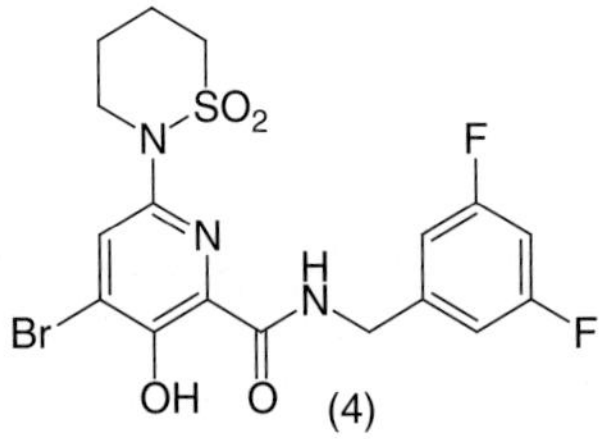

PHARMACODYNAMIC ASSAY TO ASSESS *IN VIVO* POTENCY

As a consequence of the pronounced species selectivity exhibited by CGRP receptor antagonists, *in vivo* potency has to be monitored in a non-human primate. Researchers at Boehringer Ingelheim employed a marmoset model of CGRP-mediated neurogenic vasodilation to test the effect of BIBN4096BS. In this model, antidromic electrical stimulation of the trigeminal ganglion releases neuropeptides, including CGRP, causing neurogenic vasodilation and a subsequent increase in facial blood flow. BIBN4096BS completely inhibited the neurogenic vasodilation, indicating CGRP plays a significant role in this marmoset model [51]. Unfortunately, this is an invasive and terminal procedure which limits its utility as a preclinical study.

A major advancement was the development of a non-invasive pharmacodynamic assay in rhesus monkeys, based on topical capsaicin-induced dermal vasodilation (CIDV) [58]. This assay was developed to determine the *in vivo* activity of potential drug development candidates in a pharmacologically sensitive species, with the additional potential for translation to the clinical setting. Topical application of capsaicin to the ventral side of the rhesus forearm resulted in an increase in microvascular blood flow that could be quantified by laser Doppler imaging. Increased blood flow is a direct result of endogenous CGRP release via capsaicin activation of TRPV1 receptors [59]. The ability of a CGRP receptor antagonist to block the CGRP-mediated increase in microvascular blood flow provides pharmacodynamic evidence of in vivo receptor blockade. Infusion of increasing doses of MK-0974 produced a concentration-dependent inhibition of capsaicin-induced blood flow in the rhesus forearm, affording EC_{50} and EC_{90} values of 127 and 994 nM, respectively [50]. This assay provided a simple and rapid method to evaluate novel drug candidates in vivo.

DISCOVERY OF CGRP RECEPTOR ANTAGONISTS

EARLY ANTAGONISTS AND THE DISCOVERY OF BIBN4096BS

The first non-peptide, small molecule CGRP receptor antagonist was reported by the group at SmithKline Beecham (now GlaxoSmithKline). A high-throughput screening (HTS) using porcine lung membranes identified quinine analogue 2′-(4-chlorophenyl)dihydroquinine (5) as having weak potency on the porcine receptor ($IC_{50} = 6.0\,\mu M$) [60]. Binding activity was confirmed with cells expressing the native human receptor, where (5) inhibited [^{125}I]-CGRP binding to SK-N-MC cell membranes with an $IC_{50} = 5.9\,\mu M$. Functional activity was demonstrated in the same cell line, where (5) blocked CGRP-stimulated cAMP production with an $IC_{50} = 26\,\mu M$. Structure activity studies established that the dihydroquinine 2′-phenyl substituent was critical for activity. While a number of different 2′-aryl substitutions were tolerated, none improved on the weak activity reported for (5).

HTS also identified benzamide (6) (SB-211973) as modestly potent in binding and functional assays ($K_i = 1.9\,\mu M$, $IC_{50} = 0.9\,\mu M$, SK-N-MC cells) [61]. Optimization of the benzamide series produced chiral sulfoxide (7) ((+)-SB-273779), which had binding $K_i = 310\,nM$ and functional $IC_{50} = 390\,nM$. Further studies suggested (7) bound irreversibly to the receptor. Cells pre-treated with (7) for greater than 30 min and thoroughly washed showed a significant decrease in cAMP activity upon stimulation with CGRP. The effect was selective for the CGRP receptor, as cAMP activity elicited by other receptor agonists remained unaffected.

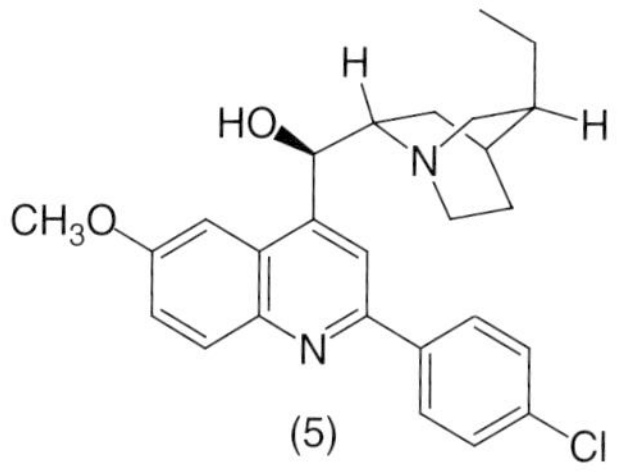

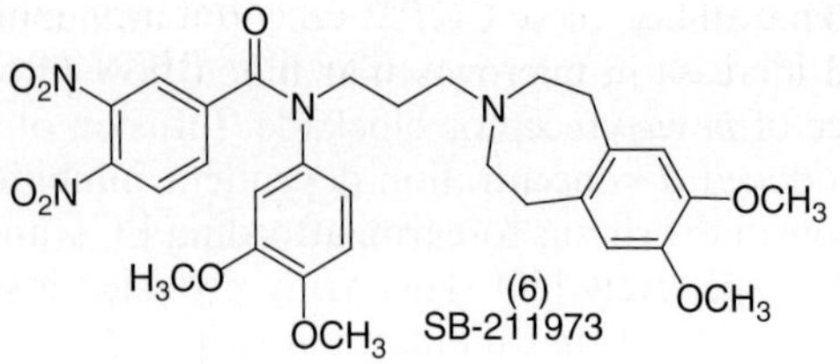

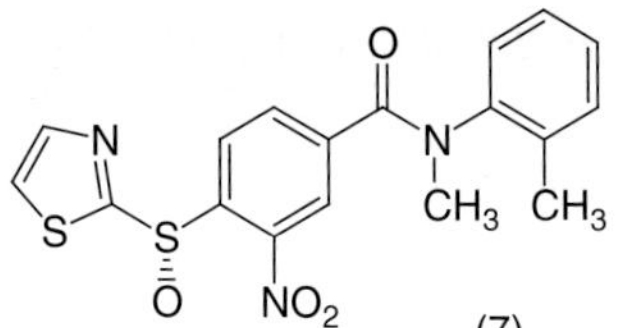

A major breakthrough in potency came from research at Boehringer Ingelheim. Initial hits from HTS included weakly active (*R*)-Tyr-(*S*)-Lys dipeptide amides such as (8), which had micromolar affinity for the CGRP receptor (binding $IC_{50} = 17\,\mu M$, SK-N-MC cells) and contained the unusual (*R*)-dibromotyrosine residue [62]. The absolute stereochemistry of the dipeptide amide was important since the synthesis of the other three stereoisomers produced inactive compounds ($IC_{50} > 300\,\mu M$). Assuming additive SAR, the N-terminus, C-terminus and the two amino acids were each varied while keeping the rest of the structure constant. Using this strategy, *N*-(4-piperidinyl)piperazine was identified as the optimal C-terminal amide (9) ($IC_{50} = 10.3\,\mu M$). Variations to the N-terminal amide strongly suggested the importance of maintaining an aromatic group a certain distance from the (*R*)-dibromotyrosine residue. Significant potency enhancement accompanied rigidifying the phenethyl amide component by replacing it with a urea group containing *N*-(2-methyoxyphenyl)piperidine (10) ($IC_{50} = 250\,nM$). A key discovery was that incorporation of a hydrogen bond donor (and/or acceptor) in this region of the molecule had a profound effect on potency (carboxamide (11) $IC_{50} = 4.5\,nM$). Constraining the aryl and hydrogen bond donor/acceptor group into a bicyclic heterocycle produced compounds with subnanomolar potency: (12) ($IC_{50} = 0.2\,nM$) and BIBN4096BS (1) ($IC_{50} = 0.03\,nM$). Alternatively, aryl-substituted heterocycles with the requisite hydrogen bonding group were also quite active (13) ($IC_{50} = 0.05\,nM$). Among these analogues, BIBN4096BS was the most potent antagonist of CGRP-mediated cAMP production in SK-N-MC cells, with a pA$_2$ of 11.1.[1] Overall, binding potency increased over five orders of magnitude from the screening lead to BIBN4096BS.

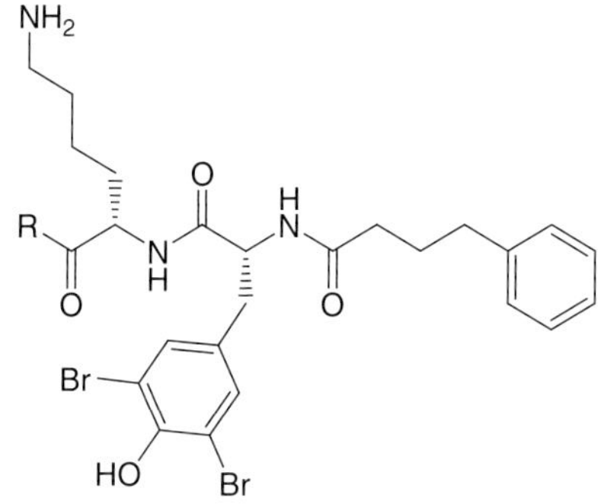

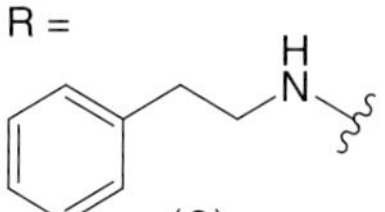

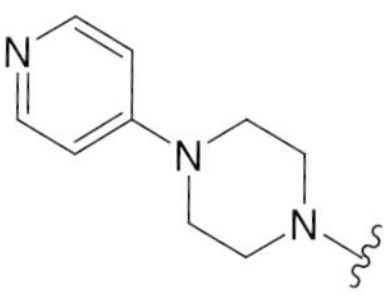

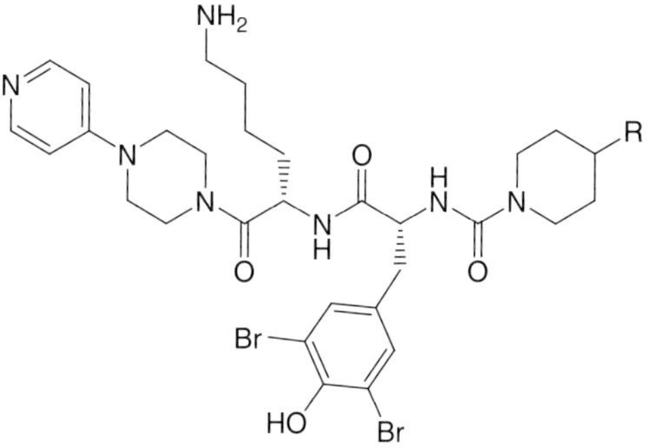

---

[1] pA$_2$ represents the functional potency of a competitive antagonist. It is the negative log$_{10}$ of the antagonist concentration, expressed in molar units, that produces a 2-fold shift to the right of the agonist concentration-response curve.

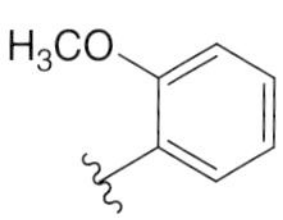

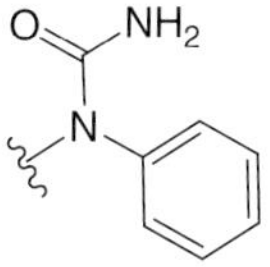

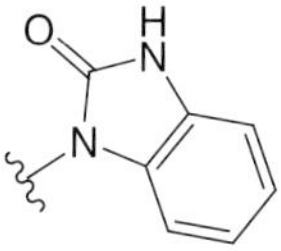

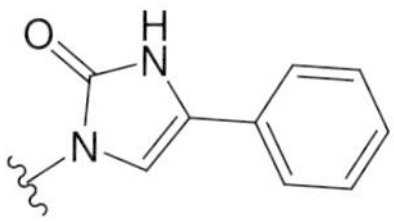

Detailed pharmacological studies of BIBN4096BS showed that it was a pure antagonist, with no intrinsic agonist activity [51]. Affinity for related receptors was low: $IC_{50} > 1,000$ nM was reported for calcitonin, amylin and rat adrenomedullin receptors. Affinity for primate CGRP receptors was much greater compared to rat, dog, guinea pig and rabbit receptors. For example, it binds to the marmoset CGRP receptor with an IC_{50} of 0.06 nM, and with an IC_{50} of 6.4 nM for the rat receptor [62]. As mentioned earlier, the first reported *in vivo* study assessed the ability of BIBN4096BS to block effects on facial blood flow following stimulation of the trigeminal ganglion and endogenous CGRP release [51]. Because of the profound species selectivity, marmoset monkeys were used in this study. Following iv administration of 3 µg/kg, 50% of the increase in facial blood flow was inhibited from antidromic stimulation of the trigeminal ganglion, and at 30 µg/kg, 100% inhibition was obtained. Importantly, no effect on basal heart rate or blood pressure was noted up to an iv dose of 1 mg/kg. Of note, triptans were also effective in this assay, presumably through pre-synaptic inhibition of CGRP release [51].

PROOF OF CONCEPT IN HUMAN CLINICAL TRIALS WITH BIBN4096BS

While BIBN4096BS had good potency and selectivity, the size (MW = 870), high polar surface area (180 Å^2), and dipeptide core resulted in low oral bioavailability ($F<1\%$ rat, dog) [62]. Therefore, iv administration of BIBN4096BS was required in human clinical trials to establish proof of concept for efficacy in treating migraine. A double-blind, placebo-controlled study was conducted in human volunteers experiencing an acute migraine attack. The primary endpoint was either absence of headache or the presence of a mild headache 2 h after treatment with BIBN4096BS. Following iv administration of 2.5 mg over a 10 min period, 66% of 32 patients experienced a response at 2 h, compared to 27% receiving placebo [8]. Secondary endpoints to assess effects on nausea, photo- and phonophobia, and ability to function also improved with treatment. Thus BIBN4096BS established proof of concept for treating acute migraine headache with a CGRP receptor antagonist. The rate of headache recurrence was lower in the treatment group (19%) compared to placebo (46%). Mild paresthesias were the most frequently reported adverse event. In other studies, BIBN4096BS had no effect on middle cerebral artery blood flow or global or regional cerebral blood flow [63]. No effect was observed on the diameter of temporal or radial arteries. Systemic hemodynamics such as blood pressure and heart rate were unaffected. From these observations together with results from animal studies [64] it is reasonable to conclude that BIBN4096BS is not a vasoconstrictor. However, additional studies with more patients are required to determine if safety is improved relative to the triptans.

SEARCH FOR ORALLY ACTIVE CGRP RECEPTOR ANTAGONISTS

Although acceptable for proof of concept studies, intravenous infusion would likely limit use of a drug to the clinical setting. For drugs lacking oral bioavailability, alternative routes of delivery include subcutaneous injection, inhaled aerosol and intranasal delivery. BIBN4096BS is suitably potent to accommodate doses compatible with these routes, and a patent application claimed an inhaled aerosol formulation [65].

An orally active CGRP receptor antagonist would be ideal in terms of patient compliance and convenience, but represented a significant challenge. Class B GPCR peptide agonists occupy extensive binding sites, and orthosteric antagonists would have to compete with CGRP binding to these sites. The probability of obtaining an orally bioavailable drug decreases with increasing molecular weight and peptidic character. In the Merck programme, relatively potent dipeptide leads such as (14) were not pursued owing to the difficult nature of extracting oral bioavailability from these templates [66]. HTS identified weaker, non-peptidic antagonists that were more attractive in terms of the ultimate goal of achieving oral bioavailability.

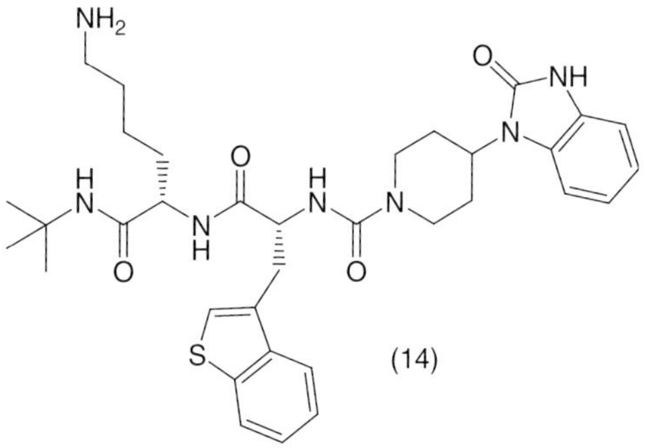

(14)

The HTS hit (15) exhibited modest binding to native human CGRP receptors ($K_i = 4.8\,\mu M$, SK-N-MC cell membranes) and antagonized CGRP-stimulated cAMP production ($IC_{50} = 6\,\mu M$, SK-N-MC cells) [67]. It incorporated two heterocycles, benzodiazepinone and hydantoin, each found in orally active drugs. The strategy for optimization took advantage of the modular nature of the lead. In one series, an optimized spirohydantoin fragment was held constant, and used to search for benzodiazepinone replacements. In a complementary approach, the benzodiazepinone fragment was used to identify spirohydantoin surrogates. The benzodiazepinone in the latter series was re-engineered as a caprolactam to produce the clinical candidate MK-0974 (2). Before reviewing the discovery of MK-0974, the spirohydantoin series derived from (15) will be discussed, along with compounds in the pyridinone series that were prepared as analogues of MK-0974.

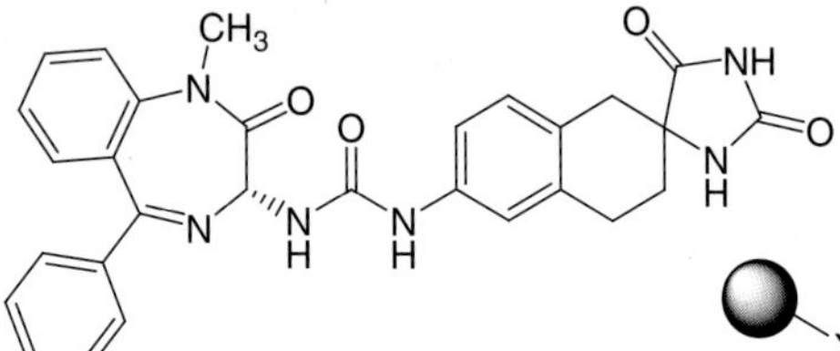

(15)

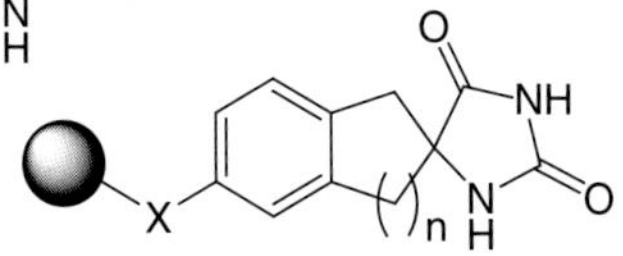

(16)
n = 1, 2
X = linker

⬤ = benzodiazepinone replacement

SPIROHYDANTOIN-BASED CGRP RECEPTOR ANTAGONISTS

A rapid analogue strategy was pursued to identify replacements to the benzodiazepinone in (15) [68]. Utilizing several templates (tetralins and indanes) and linkers (X) a variety of analogues (16) were investigated. In this way a wide array of trajectories and structures linked to the hydantoin core were rapidly explored.

A number of structurally diverse antagonists were identified using this strategy ($K_i < 100\,\mu M$), with the relatively simple benzimidazolinone (17) being selected for further studies ($K_i = 280\,nM$, cAMP $IC_{50} = 690\,nM$). Resolution of the spirohydantoin revealed that the major component of potency was due to the (R)-enantiomer, and methylation resulted in the potent compound (18) ($K_i = 43\,nM$, cAMP $IC_{50} = 120\,nM$). The amide nitrogen of the benzimidazolinone was exploited as a point at which the physicochemical and pharmacokinetic properties could be modulated. Production of the 2-pyridyl-substituted compound (19) afforded a CGRP receptor antagonist with good PK across rat ($F = 36\%$), dog ($F = 83\%$), and rhesus ($F = 29\%$). This spirohydantoin study revealed the intriguing inverse correlation between calculated polar surface area (PSA) and oral bioavailability. For this series, it appears that the probability of attaining good oral bioavailability is better for compounds with PSA < 130 Å^2 than for those compounds with PSA > 140 Å^2, consistent with published data [69].

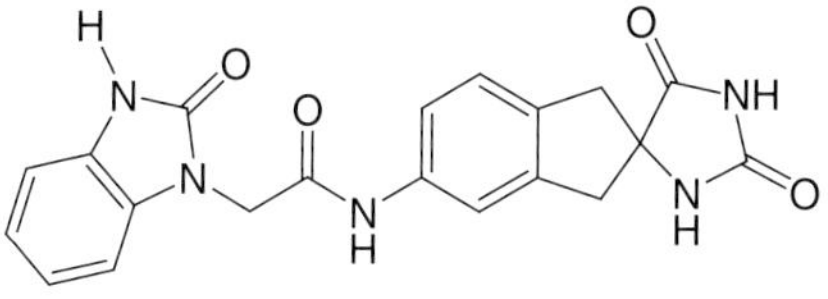

(17)
K_i = 280 nM
cAMP IC$_{50}$ = 690 nM

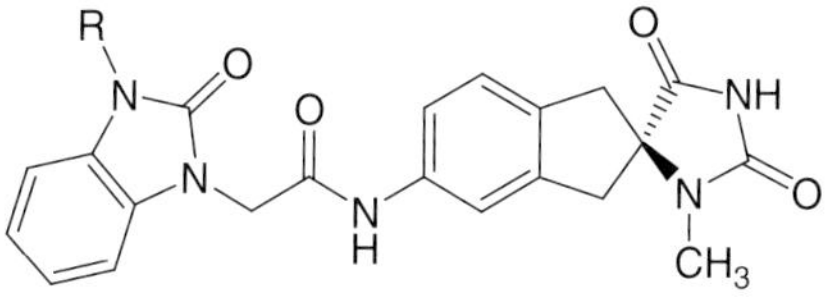

(18)
R = H
K_i = 43 nM
cAMP IC$_{50}$ = 120 nM

(19)
R = 2-pyridyl
K_i = 21 nM
cAMP IC$_{50}$ = 78 nM
Rat F = 36%
$t_{1/2}$ = 1.6 h

PYRIDINONE AS REPLACEMENT FOR THE MK-0974 CAPROLACTAM

Alternative templates for the chiral, substituted caprolactam of MK-0974 were explored (20, 21). The pyridinone moiety was proposed as a planar surrogate that would allow for a similar spatial orientation of the critical substituents required for potency [70]. Additional potential advantages proposed were: the achiral nature of the pyridinone core, which would greatly simplify synthesis and thus facilitate SAR studies; the inherent rigidity and the degree of unsaturation in the pyridinone heterocycle which could lead to good pharmacokinetic profiles.

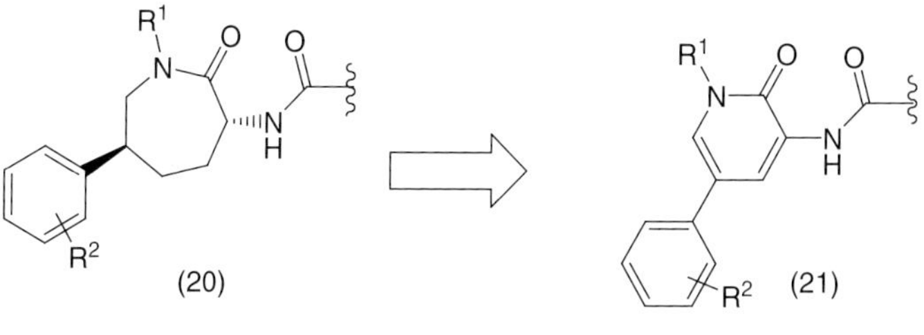

(20)   (21)

Indeed this strategy delivered potent CGRP receptor antagonists, represented by the *N*-trifluoroethyl analog (22), with a phenyl group at C-5 position of the pyridinone. Polar aryl substituents and heterocycles were tolerated on the C-5 aryl group, as exemplified by (23) and (24). As described later, this was in contrast to the equivalent substitution on the caprolactams. Nevertheless, the direct comparator to MK-0974, (25), containing the *N*-trifluoroethyl amide side-chain and the 2,3-difluorphenyl C-5 moiety, was the optimal analogue from this series. This compound was about 9-fold less potent than MK-0974 in the functional cAMP assay, but had a larger shift in the presence of human serum and was correspondingly 80-fold less potent. The oral bioavailability was similar to MK-0974, and half-lives in rats and dogs were longer.

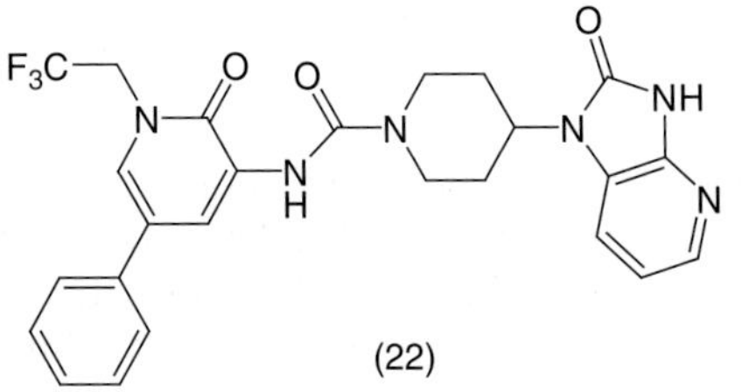

(22)
K_i = 95 nM
cAMP IC_{50} = 209 nM
Rat F = 37%, $t_{1/2}$ = 0.5 h

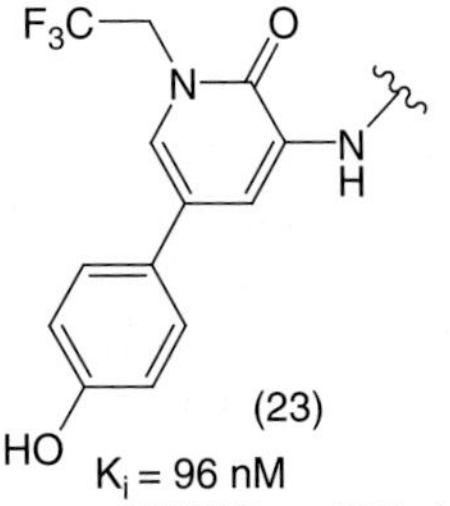

(23)
K_i = 96 nM
cAMP IC_{50} = 282 nM
cAMP+HS IC_{50} = 971 nM

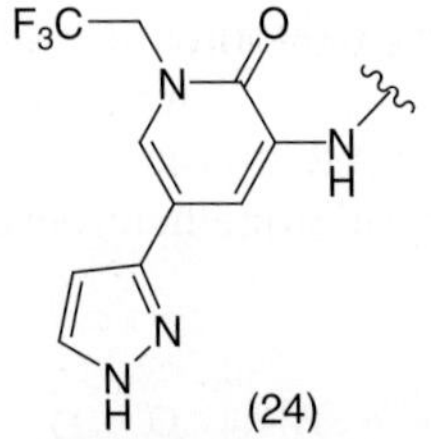

(24)
K_i = 205 nM
cAMP IC_{50} = 550 nM
cAMP+HS IC_{50} = 1421 nM

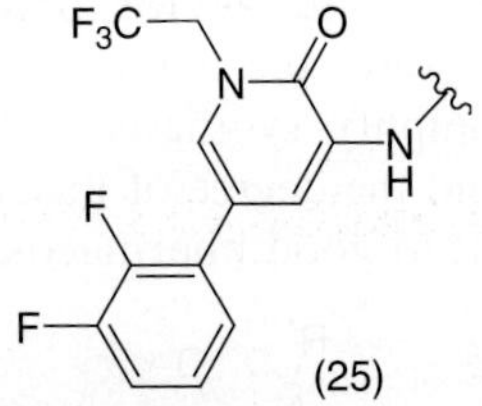

(25)
K_i = 17 nM
cAMP IC_{50} = 18 nM
cAMP+HS IC_{50} = 811 nM
Rat F = 34%, $t_{1/2}$ = 6.3 h
Dog F = 65%, $t_{1/2}$ = 3.3 h

ADDITIONAL EXPLORATION OF HTS HIT (15): SPIROHYDANTOIN REPLACEMENTS

Although the heterocyclic nature of (15) made it attractive for the ultimate goal of bioavailability, it had some marginal characteristics compared to orally bioavailable compounds [69, 71]. While the $c\log P$ value of 2.3 was similar to that of known oral drugs, less desirable characteristics included molecular weight over 500 (MW = 522), 9 H-bond donors/acceptors, and high calculated polar surface area (PSA = 147 Å^2). Therefore optimization focused on increasing both potency and drug-like properties.

At the time, the known potent CGRP receptor antagonists contained either piperidinylbenzimidazolone, eg, (12) and (14), piperidinyldihydroquinazolinone (BIBN4096BS), or phenylimidazolinone (13). One common feature was a C(O)NH hydrogen bond donor-acceptor pharmacophore (26) embedded in the benzimidazolinone, dihydroquinazolinone or phenylimidazolinone heterocycles. It is interesting to note that a hydrogen bond donor–acceptor pharmacophore was also present in the peptide agonist CGRP and the truncated antagonist CGRP(8-37) (27) in the C-terminal phenylalanine amide. In CGRP(8-37), both the phenylalanine benzyl side chain and carboxamide were essential for activity [72]. Thus, it is possible that the dihydroquinazolinone substructure (29) in BIBN4096BS functioned as a conformationally constrained version of the C-terminal dipeptide amide Ala-Phe-NH$_2$ (28) in CGRP(8-37) (Figure 1.2).

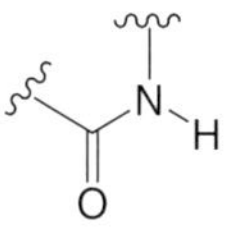

(26)

VTHRLAGLLSRSGGVVKNNFVPTNVGSKAF-NH$_2$
(27)
CGRP(8-37)

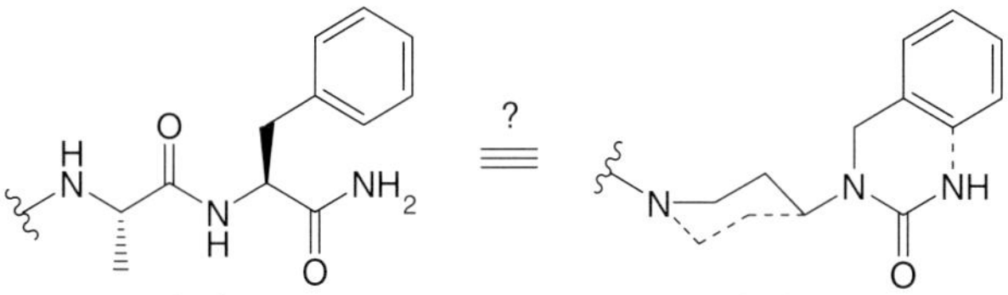

(28) (29)

Fig. 1.2 Schematic representation of structural similarities between CGRP C-terminal dipeptide (28) and piperidinyl dihydroquinazolinone (29).

Fig. 1.3 Overlay of energy-minimized piperidinyl dihydroquinazolinone (dark grey) and tetralin spirohydantoin (light grey).

Furthermore, the hydantoin in (15) had two C(O)NH hydrogen bond donor–acceptor pairs embedded in the heterocycle. This led us to speculate that BIBN4096BS and (15) shared a common hydrogen bond donor–acceptor pharmacophore. Modelling studies showed good alignment between carbonyl amide oxygen–NH pairs in an overlay of energy minimized piperidinyldihydroquinazolinone (30) and one tetralin spirohydantoin enantiomer (31) (Figure 1.3).

Hybrid structures (32) and (33) were prepared to test the pharmacophore hypothesis, and whether the original hydantoin substructure in (15) could be replaced with either the piperidinylbenzimidazolinone or piperidinyldihydroquinazolinone heterocycles. Although benzimidazolinone (32) was inactive at the concentrations tested, racemic dihydroquinazolinone (33) had low micromolar activity, and the more active enantiomer (34) was approximately five-fold more potent than the original lead (15) (Table 1.1) [67]. The hybrid molecule (34) had a molecular weight similar to (15), but other properties showed a trend towards increasing potential for oral bioavailability: fewer H-bond donor/acceptors (6 vs. 9) and significantly lower PSA (95 Å^2 vs. 147 Å^2). In support of the hydrogen bond

Table 1.1 SUMMARY OF BENZODIAZEPINONE SAR

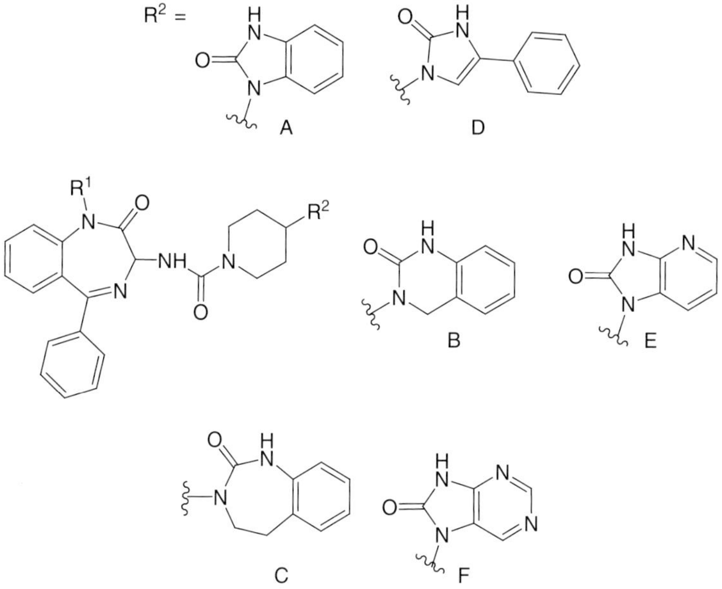

Compound	R^1	C-3	R^2	K_i (nM)	IC_{50} (nM)
(32)	CH_3	R,S	A	>13000	–
(33)	CH_3	R,S	B	1400	–
(34)	CH_3	R	B	2200	3,640
(35)	CF_3CH_2	R	B	44	38
(36)	CF_3CH_2	R	A	2370	1580
(37)	CF_3CH_2	R	C	61	43
(38)	CF_3CH_2	R	D	11	14
(39)	CF_3CH_2	R	E	23	49
(40)	CF_3CH_2	R	F	51	430

donor–acceptor pharmacophore model, N-alkylation of the dihydroquinazolinone nitrogen was highly detrimental to activity (although steric effects between the ligand and receptor cannot be ruled out as the cause of the poor activity). SAR studies showed that relatively small, lipophilic N-1 benzodiazepinone substituents could increase activity; for example trifluoroethyl (35) increased potency 50-fold. The pharmacokinetic profile of (35) was determined in dogs and rats to assess oral bioavailability. In dogs, (35) was not detected in plasma after an oral dose of 2 mg/kg (1% aqueous methocel suspension). However in rats, (35) showed low but consistent oral bioavailability when dosed as a suspension in 1% aqueous methocel ($F = 10\%$, $C_{max} = 240$ nM, 10 mg/kg). The iv half life and clearance values were reasonable ($t_{1/2} = 1.7$ h, Cl $= 20$ mL/min/kg). These modest results in one species were encouraging.

Unfortunately, the dihydroquinazoline benzylic methylene group in (35) underwent oxidation at ambient laboratory conditions. Stable alternatives to this critical group thus became a high priority. From both literature and in-house studies, potential alternatives included the ring-contracted benzimidazolinone, the ring-expanded benzodiazepine-2-one and the unfused phenylimidazolinone. Incorporating these into the lead structure gave (36), (37) and (38), respectively. While (36) lost 40-fold in potency compared to (35), both (37) and (38) retained or improved activity (37: $K_i = 61$ nM, 38: $K_i = 11$ nM). Both benzimidazolinone (36) and 1,3-benzodiazepin-2-one (37) were air-stable. Phenylimidazolinone (38) underwent decomposition under ambient conditions, most likely through oxidation of the electron-rich imidazolidinone ring.

As described earlier, the functional assay could be run in the presence of 50% human serum to mimic *in vivo* effects of plasma protein binding on activity. Adding human serum to the functional assay raised the IC$_{50}$ for (35) 36-fold, and for (38) 70-fold. Further SAR studies sought to increase intrinsic potency and decrease protein-binding effects on functional activity. For compound (36), it was discovered that adding a nitrogen at the 4-position of benzimidazolinone increased potency. Thus, aza-analogue (39) improved potency 100-fold in the binding assay, and 30-fold in the functional assay [73]. In addition, the serum shift decreased to 15-fold. This novel "aza binding site" was also evident in the 4,6-diaza analogue (40). While (40) was somewhat less active than (39), the serum shift was further reduced to 5-fold. Curiously, the "aza binding site" was uniquely realized in the benzimidazolinone substructure. Incorporating the same aza modification in the dihydroquinazolinone (35), or phenylimidazolinone (37) (to give 2-pyridylimidazolinone) decreased potency relative to the parent compound [73].

CAPROLACTAM AS A BENZODIAZEPINONE REPLACEMENT

While the benzodiazepinone-based CGRP receptor antagonists possessed very good potency and functional antagonism, these derivatives in general displayed poor pharmacokinetic profiles; for example, (41) $K_i = 55$ nM, rat $F = 4\%$. In an effort to improve physicochemical properties to potentially confer better pharmacokinetics and oral bioavailabilities, alternate scaffolds for the benzodiazepinone core were investigated.

Deconstruction of the benzodiazepinone core by successive removal of the fused and pendant aryl groups in (41) revealed that the simplified 7-membered ring lactam (43) (via 42) retained both binding and functional activity [74]. Subsequent SAR demonstrated that the corresponding 5- and 6-membered lactams were less potent than the caprolactams. In an attempt to regain the high levels of potency of the benzodiazepinones, re-incorporation of the phenyl substituent around the caprolactam ring was undertaken. Installation of the phenyl group at the C-6 position was found to be optimal, specifically the trans-(3R,6S)-diastereomer (44) was established as the most potent by several fold ($K_i = 25$ nM, cAMP IC$_{50} = 36$ nM). Incorporation of the phenylimidazolinone, a privileged structure shown previously to confer high levels of potency, delivered (45). This compound was an exceptionally potent antagonist ($K_i = 2$ nM, cAMP IC$_{50} = 4$ nM) and, importantly, possessed good oral bioavailability with moderate clearance in rats ($F = 27\%$, 36 mL/min/kg); thereby paving the way for the development of a new caprolactam-based series of orally bioavailable CGRP receptor antagonists.

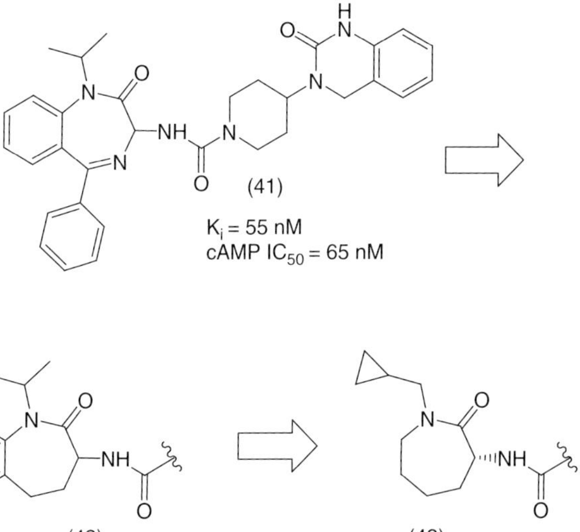

(41)
$K_i = 55$ nM
cAMP IC$_{50} = 65$ nM

(42)
$K_i = 3.7$ μM
cAMP IC$_{50} = 3.1$ μM

(43)
$K_i = 7.8$ μM
cAMP IC$_{50} = 2.4$ μM

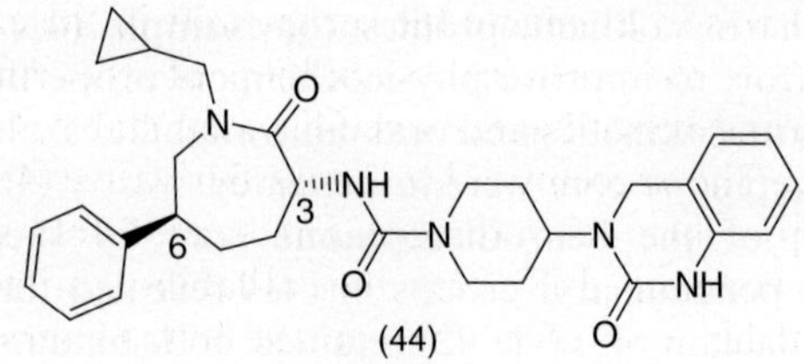

(44)

$K_i = 25$ nM
cAMP $IC_{50} = 36$ nM

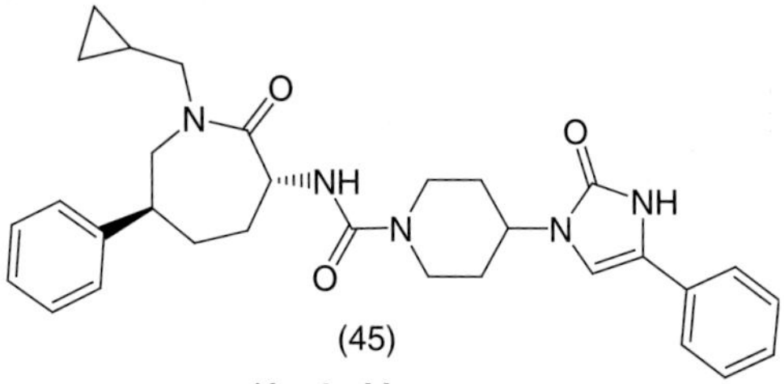

(45)

$K_i = 2$ nM
cAMP $IC_{50} = 4$ nM
serum shift = 28 fold

CAPROLACTAM OPTIMIZATION

The (3R)-amino-(6S)-phenyl caprolactam linked through the urea to the piperidinyl phenylimidazolinone (45) showed good potency and rat pharmacokinetics but oral bioavailability in dogs was low ($F = 6\%$) [9]. Further evaluation revealed chemical instability associated with this structure, specifically ambient oxidation of the imidazolinone ring. Another issue identified was the 28-fold shift in potency when the functional cAMP production assay was performed in the presence of 50% human serum; indicative of a significant degree of plasma protein binding. The use of the cAMP + 50% human serum assay to rapidly assess the impact of serum protein binding on *in vitro* efficacy was critical to the optimization of this series of compounds (*vide infra*).

The various alternative GPCR piperidine-based privileged structures that evolved throughout the course of the chemical research of CGRP receptor antagonists were evaluated as replacements for phenylimidazolinone [74].

The primary goal was to attenuate the serum shift in the cAMP functional assay and improve the chemical stabilities of the final compounds. Triazolinone (46) and benzodiazepinone (47) were stable, but retained high serum shift values. The azabenzimidazolinone derivative (48), containing an embedded basic pyridine, demonstrated a reduced 9-fold serum shift and, importantly, showed improved oral bioavailability in dogs ($F = 41\%$). Compounds with this level of potency in the human serum cAMP assay ($IC_{50} = 339$ nM) required high plasma levels of antagonist to demonstrate an effect in the rhesus pharmacodynamic model. Therefore, an SAR study of both the C-6 aryl substituent and the N-1 amide side chain moiety was undertaken to improve potency while maintaining good pharmacokinetics.

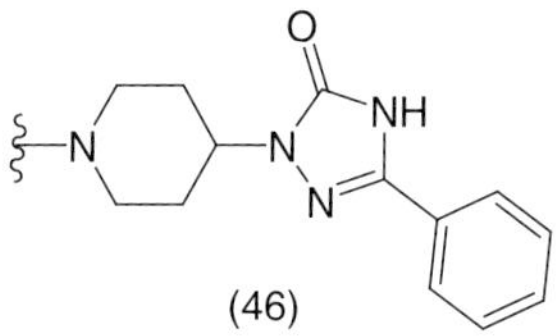

(46)
K_i = 30 nM
cAMP IC_{50} = 144 nM
serum shift > 42 fold

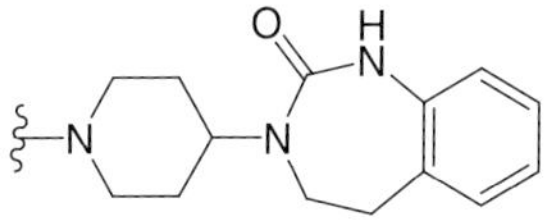

(47)
K_i = 29 nM
cAMP IC_{50} = 94 nM
serum shift = 39 fold

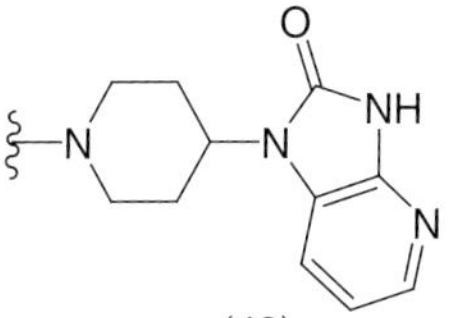

(48)
K_i = 11 nM
cAMP IC_{50} = 38 nM
serum shift = 9 fold

In order to undertake this SAR study a new synthetic route was developed in which the C-6 aryl substituent and the N-1 amide side chain moiety could be installed readily and at a late stage (Scheme 1.1) [9]. The key reaction was an unprecedented vinyl bromide ring-closing metathesis of (49), allowing rapid production of intermediate (50). Suzuki couplings with substituted aryl boronic acids effected introduction of the C-6 substituents, to yield (51). Hydrogenation, amide alkylation, deprotection and urea

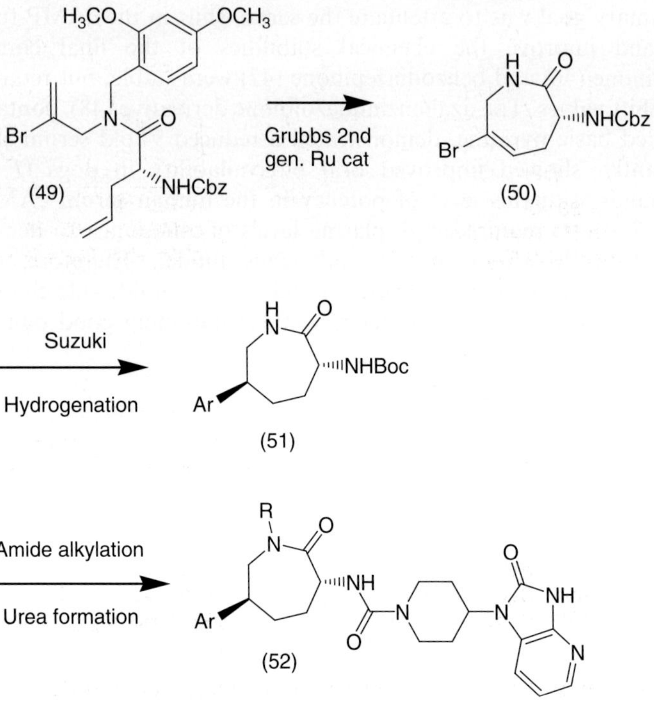

Scheme 1.1 Synthesis of caprolactam analogs for SAR studies.

coupling with the azabenzimidazolinone piperidine provided the final targets (52).

This chemistry enabled a range of C-6 groups to be investigated. Many of these analogues were designed to keep the serum shifts and solubility in check (e.g., pyridines); however, most of these derivatives suffered large potency losses compared to the phenyl analogue (53). From this rather extensive study, the fluorinated analogues were the most promising, including the 2-fluorophenyl (54, $K_i = 22$ nM) and 3-fluorophenyl (55, $K_i = 51$ nM) derivatives. The combination 2,3-difluorophenyl analogue (56), resulted in a greater than additive 20-fold increase in affinity ($K_i = 3.6$ nM) compared with the parent phenyl compound (53). Potency in the serum shift functional assay was also greatly improved ($IC_{50} = 22$ nM). Overall the 2,3-difluorophenyl group provided large potency enhancements and low serum shifts with only a modest increase in molecular weight and lipophilicity.

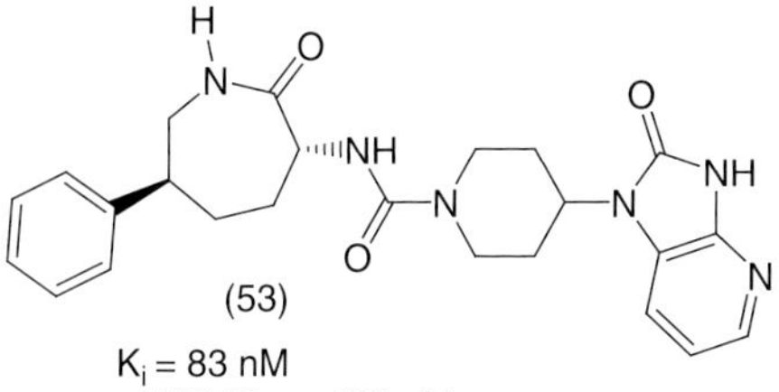

(53)
K_i = 83 nM
cAMP IC_{50} = 520 nM
cAMP+HS IC_{50} = 700 nM

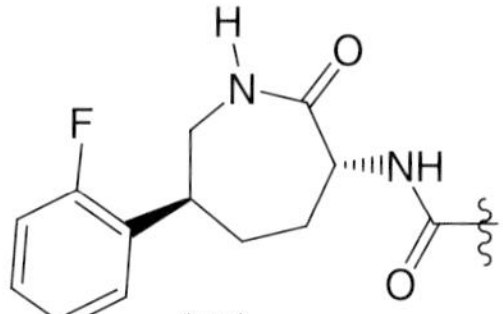

(54)
K_i = 22 nM
cAMP IC_{50} = 65 nM
cAMP+HS IC_{50} = 120 nM

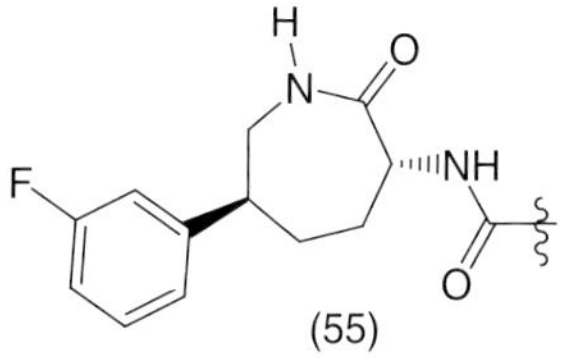

(55)
K_i = 51 nM
cAMP IC_{50} = 220 nM
cAMP+HS IC_{50} = 250 nM

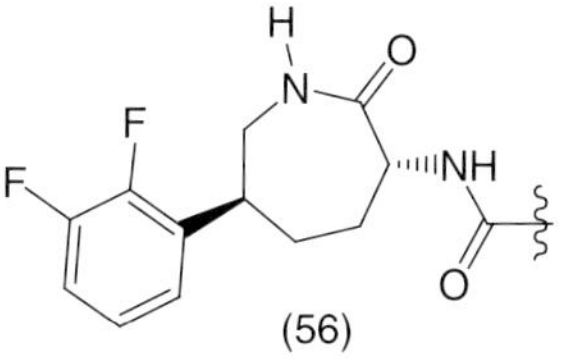

(56)
K_i = 3.6 nM
cAMP IC_{50} = 14 nM
cAMP+HS IC_{50} = 22 nM

Exploration of the amide substituent was undertaken in order to optimize PK and potency. Compounds with polar side chains (e.g., hydroxy, alkoxy and heteroaryl derivatives) maintained good efficacy in the cAMP functional assay in the presence of 50% human serum, but oral bioavailability was poor in both rats and dogs. Small alkyl groups, however, provided potent compounds with high Caco-2 permeabilities and good pharmacokinetic profiles. For example, the methyl and ethyl analogues (57) and (58) showed good dog PK profiles, with low clearance (4–8 mL/min/kg) and high oral bioavailability (Table 1.2) [9]. Introduction of the trifluoroethyl side-chain afforded a further improvement in intrinsic potency (2, K_i = 0.8 nM) accompanied by a low serum shift (IC_{50} = 11 nM). This compound also displayed good levels of oral bioavailability and low to moderate clearance in both rats (%F = 20, Cl = 9.4 mL/min/kg) and dogs (%F = 35, Cl = 17 mL/min/kg).

The lead compounds (57), (58) and (2) demonstrated a robust effect in the rhesus *in vivo* pharmacodynamic model of capsaicin-induced dermal blood flow, with compound (2) (EC_{50} = 127 nM, EC_{90} = 994 nM) the most potent by > 3.5-fold. Additionally, (2) demonstrated good rhesus pharmacokinetics and high selectivity and was selected as the development candidate MK-0974.

Table 1.2 SUMMARY OF CAPROLACTAM *IN VITRO* AND *IN VIVO* POTENCY, AND ORAL BIOAVAILABILITY

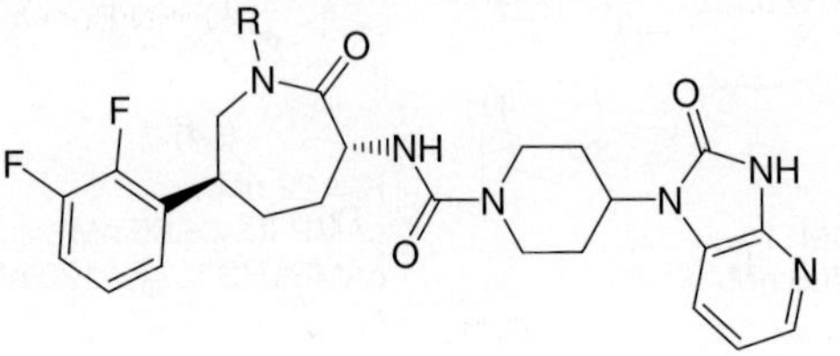

R = polar sidechain poor %F

Compound	R	K_i (nM)	cAMP – IC_{50} (nM)	cAMP+HS – IC_{50} (nM)	%F (Rat)	%F (Dog)	Rhesus PD – EC_{90} (μM)
(57)	CH_3CH_2	2	6	30	30	61	3.6
(58)	CH_3	3	8	22	12	60	4.0
(2)	CF_3CH_2	0.8	2	11	20	35	1.0

PRACTICAL SYNTHESIS OF MK-0974

In support of comprehensive preclinical evaluation of MK-0974, a scaleable, stereoselective synthesis of (3R,6S)-3-amino-6-(2,3-difluorophenyl)-1-(2,2,2-trifluoroethyl)-azepan-2-one was developed [75]. The preceding route, utilized to deliver limited quantities of MK-0974 for preliminary characterization, relied on an inefficient Ru-catalyzed metathesis reaction followed by a non-diastereoselective reduction to produce the C-6 stereocenter. A second-generation route was developed in which a diastereoselective Hayashi-Miyaura Rh-catalyzed arylboronic acid addition [76, 77] to a nitroalkene installed the key C-6 stereocenter (Scheme 1.2).

The commercially available D-glutamic acid derivative (59) was elaborated to give the nitro alkene (60), upon which the key Hayashi-Miyaura Rh-catalyzed arylboronic acid addition could be undertaken. Since application of the standard literature conditions led to rapid hydrolysis of the 2,3-difluorophenylboronic acid, the generation of more active catalysts was explored as a means to accelerate this reaction. Reaction optimization studies revealed that $NaHCO_3$ (0.5 equiv) was an efficacious additive, leading to an improved procedure that reproducibly delivered the desired product (61) in high yield and diastereoselectivity (96%, 93:7) on a scale of up to 2 kg. Straightforward transformations through intermediate (62) were then used to deliver MK-0974.

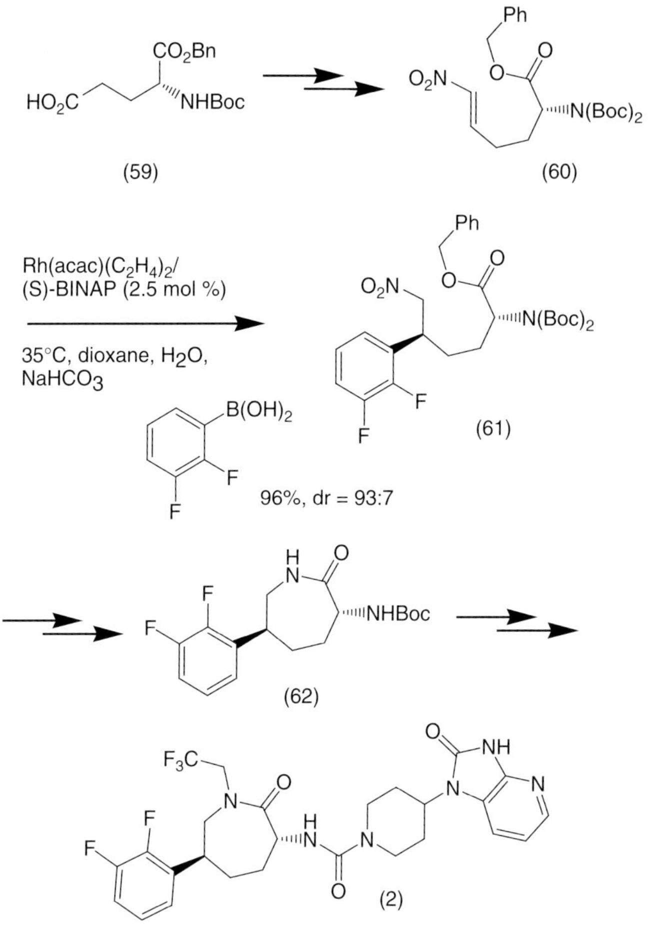

Scheme 1.2 A practical synthesis of MK-0974 (2).

IN VITRO AND *IN VIVO* CHARACTERIZATION OF MK-0974

Competitive binding experiments were carried out to determine the relative affinity of MK-0974 for rhesus, rat and dog CGRP receptors as measured by the ability to compete with ^{125}I-hCGRP binding [50]. MK-0974 displayed a similar affinity (K_i) for the rhesus receptor (1.2 nM) as for human (0.77 nM), but displayed >1,000-fold lower affinity for the dog and rat receptors with values of 1204 and 1192 nM, respectively. MK-0974 displayed little to no affinity for the human adrenomedullin receptors as measured by the ability to compete with ^{125}I-human adrenomedullin with K_i values of $>100\,\mu M$ and $29\,\mu M$ on CLR/RAMP2 and CLR/RAMP3, respectively.

Consistent with the binding data, MK-0974 potently blocked human CGRP-stimulated cAMP responses in cells expressing the human CGRP receptor. MK-0974 had an IC_{50} of $2.2\,nM$ and $IC_{50} = 10.9\,nM$ in the absence and presence of 50% human serum, respectively. Schild plot analysis revealed that increasing concentrations of MK-0974 caused a dose-dependent rightward shift in the CGRP dose response-curve (no reduction in the maximal agonist response) and $K_B = 1.1\,nM$.

Infusion of increasing doses of MK-0974 produced a concentration-dependent inhibition of capsaicin-induced blood flow in the rhesus forearm, affording EC_{50} and EC_{90} values of 127 and $994\,nM$, respectively. This pharmacodynamic response was also shown to be time-dependent, correlating well with plasma levels, and specific, as demonstrated by lack of direct TRPV1 antagonism.

CLINICAL TRIAL RESULTS FOR MK-0974: HUMAN PHARMACOKINETICS

A randomized, double-blind, placebo-controlled clinical study was performed to assess the safety, tolerability and pharmacokinetics of MK-0974 during and between migraine attacks [78]. Gastrointestinal symptoms (e.g., stasis) are part of a migraine attack and could potentially have an impact on the oral absorption of drugs. For example, the oral migraine drug zolmitriptan demonstrates slower absorption during a migraine than between attacks [79]. In this study, 15 patients received 300 mg of MK-0974 within 2 h of the onset of a migraine attack, and also during a period free of migraine attacks. No serious adverse experiences occurred in either period and treatment was generally well tolerated. The pharmacokinetics for the two treatments were similar: C_{max}(during) $= 2.60\,\mu M$, C_{max}(between) $= 2.93\,\mu M$; T_{max}(during) $= 1.5\,h$, T_{max}(between) $= 1.0\,h$; AUC(during) $= 11.98\,\mu M^*h$, AUC(between) $= 11.21\,\mu M^*h$; $t_{1/2}$(during) $= 8.5\,h$, $t_{1/2}$(between) $= 10.6\,h$. This data suggests that absorption of MK-0974 was not significantly affected by migraine-associated gastrointestinal symptoms.

CAPSAICIN-INDUCED DERMAL BLOOD FLOW

The capsaicin-induced dermal blood flow (CIDV) model, initially developed in the rhesus monkey, was translated to the clinical setting. This enabled the generation of a pharmacokinetic/pharmacodynamic (PK/PD) relationship between plasma exposure and inhibition of capsaicin-induced dermal microvascular blood flow (DBF) in man [80, 81]. A three-period crossover clinical study was designed to evaluate the response to MK-0974. Each subject received a single oral dose of MK-0974 300 mg, MK-0974 800 mg or placebo followed by topical application of capsaicin on the volar surface of the subject's forearm. A PK/PD relationship was observed 1 and 4 h postdose with maximal inhibition of capsaicin-induced DBF at plasma concentrations similar to those required in the rhesus CIDV study. These results provided evidence that engagement of the CGRP receptor had been achieved and served as a valuable tool for dose selection for a Phase II dose finding study in the acute treatment of migraine.

CLINICAL EFFICACY

A randomized, double-blind, placebo- and active-controlled (rizatriptan 10 mg) outpatient clinical study was conducted to evaluate the efficacy and tolerability of the novel, oral CGRP receptor antagonist, MK-0974, in patients with an acute migraine attack [10]. Headache severity was recorded using a 4-grade scale (no pain, mild pain, moderate pain, severe pain) at baseline (0 h – time of taking study medication) and 0.5, 1, 1.5, 2, 3, 4 and 24 h postdose. Presence or absence of associated symptoms (nausea, vomiting, photophobia or phonophobia) and rating of functional disability (4-grade scale – normal, mildly impaired, severely impaired, requires bed rest) were recorded at the same time points.

MK-0974 (300–600 mg) was effective in treating moderate or severe migraine attacks on the primary endpoint of pain relief (mild or no pain) at 2 h. The proportion of patients achieving pain relief at 2 h were 68.1%, 48.2% and 67.5% with 300 mg, 400 mg and 600 mg MK-0974, respectively; 69.5% with 10 mg rizatriptan, and 46.3% with placebo. The effects were mirrored at the other endpoints of pain freedom (no pain), improvement of associated symptoms and functional disability and use of additional medication. In most cases, the effective MK-0974 doses produced results that appeared comparable to rizatriptan. However, the study was not powered to detect differences between active treatments and will require larger comparative clinical trials.

MK-0974 was generally well tolerated. The incidence of patients reporting adverse experiences within 14 days appeared comparable between active treatment groups and placebo, with no evidence of an increase in adverse experiences with increasing dose.

CONCLUSION

The discovery of small molecule CGRP receptor antagonists has generally begun with weak, micromolar HTS leads being optimized to potent antagonists such as BIBN4096BS and MK-0974. A clinical trial using an iv infusion of BIBN4096BS established Proof of Concept for the effectiveness of CGRP receptor antagonists in treating acute migraine headache. Obtaining an orally active CGRP receptor antagonist represented a significant challenge. Lead optimization of the weakly active benzodiazepinone spirohydantoin (15) focused on improving functional CGRP receptor antagonism, first in cell culture with added human serum, then *in vivo* in a non-invasive rhesus monkey PD model using topical capsaicin-induced dermal vasodilation (CIDV). This assay was later transferred to the clinic to assist in dose selection for a Phase II trial. Along with functional activity, properties associated with good oral absorption were concurrently optimized. This led to the discovery of MK-0974 (telcagepant) [82], a potent orally bioavailable CGRP receptor antagonist that effectively treated acute migraine in a Phase II clinical trial. A preliminary report recently indicated that MK-0974 also showed efficacy in a Phase III clinical study [83].

CGRP receptor antagonists do not appear to have the cardiovascular liabilities associated with triptans. However, larger studies are needed to confirm these early observations. If borne out, CGRP receptor antagonists would present a significant therapeutic advantage over earlier migraine-specific treatments.

ACKNOWLEDGEMENTS

The authors gratefully acknowledge Ian Bell, Tony Ho and Dan Paone for critical review of this manuscript and helpful suggestions.

REFERENCES

[1] Unger, J. (2006) *Dis. Mon.* **52**, 367–384.
[2] Saper, J.R. and Siberstein, S. (2006) *Headache* **46**, S171–S181.
[3] Rothlin, E. (1955) *Int. Arch. Allergy Immunol.* **7**, 205–209.
[4] Goadsby, P.J. (2007) Handbook of Experimental Pharmacology, **177**(Analgesia); pp. 129–143, Springer GmbH, Berlin, Heidelberg.
[5] Poyner, D.R. (1992) *Pharmac. Ther.* **56**, 23–51.
[6] Goadsby, P.J. and Edvinsson, L. (1993) *Ann. Neurol.* **33**, 48–56.
[7] Sarchielli, P., Pini, L.A., Zanchin, G., Alberti, A., Maggioni, F., Rossi, C., Floridi, A. and Calabresi, P. (2006) *Cephalalgia* **26**, 257–265.
[8] Olesen, J., Diener, H.-C., Husstedt, I.W., Goadsby, P.J., Hall, D., Meier, U., Pollentier, S. and Lesko, L.M. (2004) *N. Engl. J. Med.* **350**, 1104–1110.
[9] Paone, D.V., Shaw, A.W., Nguyen, D.N., Burgey, C.S., Deng, J.Z., Kane, S.A., et al. (2007) *J. Med. Chem.* **50**(23), 5564–5567.

[10] Ho, T.W., Mannix, L.K., Fan, X., Assaid, C., Furtek, C., Jones, C.J., Lines, C.R. and Rapoport, A.M. (2008) *Neurology* **70**(16), 1304–1312.
[11] Stovner, L.J., Hagen, K., Jensen, R., Katsarava, Z., Lipton, R.B., Scher, A.I., Steiner, T.J. and Zwart, J.-A. (2007) *Cephalalgia* **27**, 193, 210.
[12] Linde, M. (2006) *Acta Neurol. Scand.* **114**, 71–83.
[13] Rasmussen, B.K. (2006) *In* "The Headaches". Olesen, J., Goadsby, P.J., Ramadan, N.M., Tfelt-Hansen, P. and Welch, K.M.A. (eds), (3rd ed.), pp. 235–242. Lippincott Williams & Wilkins, Philadelphia.
[14] Berg, J. and Ramadan, N.M. (2006) *In* "The Headaches". Olesen, J., Goadsby, P.J., Ramadan, N.M., Tfelt-Hansen, P. and Welch, K.M.A. (eds), (3rd ed.), pp. 35–42. Lippincott Williams & Wilkins, Philadelphia.
[15] Lipton, R.B., Diamond, S., Reed, M., Diamond, M.L. and Stewart, W.F. (2001) *Headache* **41**, 638–645.
[16] Kelman, L. (2004) *Headache* **44**, 865–872.
[17] Dahlof, C. and Linde, M. (2001) *Cephalalgia* **21**, 664–671.
[18] Eriksen, M.K., Thomsen, L.L., Anderson, I., Nazim, F. and Olesen, J. (2004) *Cephalalgia* **24**, 564–575.
[19] Headache Classification Subcommittee of the International Headache Society. (2004) The international classification of headache disorders.*Cephalalgia* **24**(Suppl. 1), 1–160.
[20] Amara, S.G., Jonas, V., Rosenfeld, M.G., Ong, E.S. and Evans, R.M. (1982) *Nature (London)* **298**, 240–244.
[21] Steenbergh, P.H., Hoppener, J.W.M., Zandberg, J., Lips, C.J.M. and Jansz, H.S. (1985) *FEBS Lett.* **183**, 403–407.
[22] Alevizaki, M., Shiraishi, A., Rassool, F.V., Ferrier, G.J.M., MacIntyre, I. and Legon, S. (1986) *FEBS Lett.* **206**, 47–52.
[23] Brain, S.D. and Grant, A.D. (2004) *Physiol. Rev.* **84**, 903–934.
[24] van Rossum, D., Hanisch, U.-K. and Quirion, R. (1997) *Neurosci. Biobehav. Rev.* **21**, 649–678.
[25] Brain, S.D., Williams, T.J., Tippins, J.R., Morris, H.R. and MacIntyre, I. (1985) *Nature (London)* **313**, 54–56.
[26] Uddman, R., Edvinsson, L., Ekman, R., Kingman, T. and McCulloch, J. (1985) *Neurosci. Lett.* **62**, 131–136.
[27] Jansen, I., Uddman, R., Ekman, R., Olesen, J., Ottosson, A. and Edvinsson, L. (1992) *Peptides* **13**, 527–536.
[28] Ray, B.S. and Wolff, H.G. (1940) *Arch. Surgery* **41**, 813–856.
[29] Goadsby, P.J., Edvinsson, L. and Ekman, R. (1988) *Ann. Neurol.* **23**, 193–196.
[30] Goadsby, P.J., Edvinsson, L. and Ekman, R. (1990) *Ann. Neurol.* **28**, 183–187.
[31] Gallai, V., Sarchielli, P., Floridi, A., Franceschini, M., Codini, M., Glioti, G., Trequattrini, A. and Palumbo, R. (1995) *Cephalalgia* **15**, 384–390.
[32] Lassen, L.H., Jacobsen, V.B., Petersen, P., Sperling, B., Iversen, H.K. and Olesen, J. (1998) *Eur. J. Neurol.* **5**, S63.
[33] Goadsby, P.J. and Edvinsson, L. (1993) *Ann. Neurol.* **23**, 193–196.
[34] Durham, P.L. (2006) *Headache* **46**, S3–S8.
[35] Afridi, S.K., Giffin, N.J., Kaube, H., Friston, K.J., Ward, N.S., Frackowiak, R.S.J. and Goadsby, P.J. (2005) *Arch. Neurol.* **62**, 1270–1275.
[36] Fischer, M.J.M., Koulchitsky, S. and Messlinger, K. (2005) *J. Neurosci.* **25**, 5877–5883.
[37] Storer, R.J., Akerman, S. and Goadsby, P.J. (2004) *Br. J. Pharmacol.* **142**, 1171–1181.

[38] Levy, D., Burstein, R. and Strassman, A.M. (2005) *Ann. Neurol.* **58**, 698–705.
[39] Rahmann, A., Wienecke, T., Hansen, J.M., Fahrenkrug, J., Olesen, J. and Ashina, M. (2008) *Cephalalgia* **28**, 226–236.
[40] McLatchie, L.M., Fraser, N.J., Main, M.J., Wise, A., Brown, J., Thompson, N., Solari, R., Lee, M.G. and Foord, S.M. (1998) *Nature (London)* **393**, 333–339.
[41] Evans, B.N., Rosenblatt, M.I., Mnayer, L.O., Oliver, K.R. and Dickerson, I.M. (2000) *J. Biol. Chem.* **275**, 31438–31443.
[42] Oliver, K.R., Wainwright, A., Edvinsson, L., Pickard, J.D. and Hill, R.G. (2002) *J. Cereb. Blood Flow Metab.* **22**, 620–629.
[43] Lennerz, J.K., Ruhle, V., Ceppa, E.P., Neuhuber, W.L., Bunnett, N.W., Grady, E.F. and Messlinger, K.J. (2008) *Comp. Neurol.* **507**, 1277–1299.
[44] Williamson, D.J. and Hargreaves, R.J. (2001) *Microsc. Res. Tech.* **53**, 167–178.
[45] Sexton, P.M., McKenzie, J.S., Mason, R.T., Moseley, J.M., Martin, T.J. and Mendelsohn, F.A.O. (1986) *Neuroscience* **19**, 1235–1245.
[46] Sexton, P.M. (1991) *Mol. Neurobiol.* **5**(2–4), 251–273.
[47] Christopoulos, G., Paxinos, G., Huang, X.F., Beaumont, K., Toga, A.W. and Sexton, P.M. (1995) *Can. J. Physiol. Pharmacol.* **73**(7), 1037–1041.
[48] Marvizon, J.C.G., Perez, O.A., Song, B., Chen, W., Bunnett, N.W., Grady, E.F. and Todd, A.J. (2007) *Neuroscience* **148**, 250–265.
[49] Semark, J.E., Middlemiss, D.N. and Hutson, P.H. (1992) *Mol. Neuropharmacol.* **2**, 311–317.
[50] Salvatore, C.A., Hershey, J.C., Corcoran, H.A., Fay, J.F., Johnston, V.K., Moore, E.L., et al. (2008) *J. Pharmacol. Exp. Ther.* **324**, 416–421.
[51] Doods, H., Hallermayer, G., Wu, D., Entzeroth, M., Rudolf, K., Engel, W. and Eberlein, W. (2000) *Br. J. Pharmacol.* **129**, 420–423.
[52] Edvinsson, L., Sams, A., Jansen-Olesen, I., Tajti, J., Kane, S.A., Rutledge, R.Z., Koblan, K.S., Hill, R.G. and Longmore, J. (2001) *Eur. J. Pharmacol.* **415**, 39–44.
[53] Hasbak, P., Sams, A., Schifter, S., Longmore, J. and Edvinsson, L. (2001) *Br. J. Pharmacol.* **133**, 1405–1413.
[54] Mallee, J.J., Salvatore, C.A., LeBourdelles, B., Oliver, K.R., Longmore, J., Koblan, K.S. and Kane, S.A. (2002) *J. Biol. Chem.* **277**, 14294, 14298.
[55] Hay, D.L., Howitt, S.G., Conner, A.C., Schindler, M., Smith, D.M. and Poyner, D.R. (2003) *Br. J. Pharmacol.* **140**, 477, 486.
[56] Hay, D.L., Christopoulos, G., Christopoulos, A. and Sexton, P.M. (2006) *Mol. Pharmacol.* **70**, 1984–1991.
[57] Salvatore, C.A., Mallee, J.J., Bell, I.M., Zartman, C.B., Williams, T.M., Koblan, K.S. and Kane, S.A. (2006) *Biochemistry* **45**, 1881–1887.
[58] Hershey, J.C., Corcoran, H.A., Baskin, E.P., Salvatore, C.A., Mosser, S., Williams, T.M., Koblan, K.S., Hargreaves, R.J. and Kane, S.A. (2005) *Regul. Pept.* **127**, 71–77.
[59] Akerman, S., Kaube, H. and Goadsby, P.J. (2003) *Br. J. Pharmacol.* **140**, 718–724.
[60] Daines, R.A., Sham, K.K.C., Taggart, J.J., Kingsbury, W.D., Chan, J., Breen, A., Disa, J. and Aiyar, N. (1997) *Bioorg. Med. Chem. Lett.* **7**(20), 2673–2676.
[61] Aiyar, N., Daines, R.A., Disa, J., Chambers, P.A., Sauermelch, C.F., Quiniou, M.-J., Khandoudi, N., Gout, B., Douglas, S.A. and Willette, R.N. (2001) *J. Pharmacol. Exp. Ther.* **296**(3), 768–775.
[62] Rudolf, K., Eberlein, W., Engel, W., Pieper, H., Entzeroth, M., Hallermayer, G. and Doods, H. (2005) *J. Med. Chem.* **48**(19), 5921–5931.
[63] Petersen, K.A., Birk, S., Lassen, L.H., Krusse, C., Jonassen, O., Lesko, L. and Olesen, J. (2004) *Cephalalgia* **25**, 139–147.

[64] Kapoor, K., Arulmani, U., Heiligers, J.P.C., Willems, E.W., Doods, H., Villalon, C.M. and Saxena, P.R. (2003) *Eur. J. Pharmacol.* **475**, 69–77.
[65] Mueller, S.G., Rudolf, K., Lustenberger, P., Stenkamp, D., Doods, H., Arndt, K. and Schaenzle, G. (2007) *PCT Int. Appl.*, WO 2007028812, CAplus, via STN AnaVist, version 1.1; Chemical Abstracts Service: Columbus, OH, 2006; AN 2007:284303.
[66] Patchett, A.A., Hill, R.G. and Yang, L. (2003) US 6,552,043, April 22.
[67] Williams, T.M., Stump, C.A., Nguyen, D.N., Quigley, A.G., Bell, I.M., Gallicchio, S.N., et al. (2006) *Bioorg. Med. Chem. Lett.* **16**, 2595–2598.
[68] Bell, I.M., Bednar, R.A., Fay, J.F., Gallicchio, S.N., Hochman, J.H., McMasters, D.R., et al. (2006) *Bioorg. Med. Chem. Lett.* **16**, 6165–6168.
[69] Clark, D.E. (1999) *J. Pharm. Sci.* **88**, 807.
[70] Nguyen, D.N., Paone, D.P., Shaw, A.W., Burgey, C.S., Mosser, S.D., Johnston, V.K., et al. (2008) *Bioorg. Med. Chem. Lett.* **18**, 755–758.
[71] Lipinski, C.A. (1997) *Adv. Drug Deliv. Rev.* **23**, 3.
[72] Smith, D.D., Saha, S., Fang, G., Schaffert, C., Waugh, D.J.J., Zeng, Z., Toth, G., Hulce, M. and Abel, P.W. (2003) *J. Med. Chem.* **46**, 2427–2435.
[73] Burgey, C.S., Stump, C.A., Nguyen, D.N., Deng, J.Z., Quigley, A.G., Norton, B.R., et al. (2006) *Bioorg. Med. Chem. Lett.* **16**, 5052–5056.
[74] Shaw, A.W., Paone, D.V., Nguyen, D.N., Stump, C.A., Burgey, C.S., Mosser, S.D., et al. (2007) *Bioorg. Med. Chem. Lett.* **17**, 4795–4798.
[75] Burgey, C.S., Paone, D.V., Shaw, A.W., Deng, Z.J., Nguyen, D.N., Potteiger, C.M., Graham, S.L., Vacca, J.P. and Williams, T.M. (2008) *Org. Lett.* **10**, 3235–3238.
[76] Hayashi, T. and Yamasaki, K. (2003) *Chem. Rev.* **103**, 2829–2844.
[77] Hayashi, T., Senda, T. and Ogasawara, M. (2000) *J. Am. Chem. Soc.* **122**, 10716–10717.
[78] Sinclair, S.R., Boyle, J.E., de Lepeleire, I., Kane, S.A., Blanchard, R., Willson, K., Xu, Y., Agrawal, N., Palcza, J. and Murphy, M.G. (2007) *Headache* **47**, 811.
[79] Thomsen, L.L., Dixon, L.H., Gibbens, M., Langemark, M., Bendtsen, L., Daugaard, D. and Olesen, J. (1996) *Cephalalgia* **16**, 270–275.
[80] Van der Schueren, B.J., de Hoon, J.N., Vanmolkot, F.H., Van Hecken, A., Depre, M., Kane, S.A., De Lepeleire, I. and Sinclair, S.R. (2007) *Br. J. Clin. Pharm.* **64**(5), 580–590.
[81] Sinclair, S.R., Kane, S.A., Xiao, A., Willson, K., Xu, Y., Hickey, L., Palcza, J., de Lepeleire, I., Vanmolkot, F., de Hoon, J. and Murphy, M.G. (2007) *Headache* **47**, 811.
[82] Wang, Y., Serradell, N., Rosa, E. and Bolos, J. (2008) *Drugs Fut.* **33**(2), 116–122.
[83] Ho, T.W., Ferrari, M.D., Dodick, D.W., Galet, V., Kost, J., Fan, X., et al. (2008) *Headache* **48**(s1), S7.

2 PDE4 Inhibitors – A Review of the Current Field

NEIL J. PRESS and KATHARINE H. BANNER

Novartis Institutes for Biomedical Research, Horsham, West Sussex RH12 5AB, UK

INTRODUCTION TO THE PDE FAMILY, FOCUSING ON PDE4	37
Molecular Cloning, Localisation and Regulation	38
PUBLISHED PDE4 INHIBITORS	48
Marketed Compounds	48
CONCLUSION AND OUTLOOK	64
REFERENCES	66

INTRODUCTION TO THE PDE FAMILY, FOCUSING ON PDE4

Cyclic adenosine monophosphate (cAMP) and cyclic guanosine monophosphate (cGMP) are second messengers that regulate a number of critical cellular processes such as metabolism, cell proliferation and differentiation, secretion, vascular and airway smooth muscle relaxation and the release of inflammatory mediators [1]. The phosphodiesterase (PDE) enzyme family hydrolyses cAMP and cGMP, to inactive 5′-AMP and 5′-GMP, respectively, and thus inhibition of PDEs represents a potential mechanism by which cellular processes can be modulated [2]. To date 11 major PDE gene families have been identified, denoted PDE1–11, which differ in primary structures, affinities for cAMP and cGMP, responses to specific effectors, sensitivities to specific inhibitors and mechanisms of regulation [3]. Each family contains at least one member, and in some cases the members are products of more than one gene. Inhibitors of a variety of specific PDE enzymes are in development for a range of indications [4]. This review will focus on PDE4 enzyme inhibitors and their potential therapeutic utility.

MOLECULAR CLONING, LOCALISATION AND REGULATION

The enzyme PDE4 is described as a low K_m ($\sim$ 1–10 µM) cAMP-specific PDE with only a weak affinity for cGMP (K_m > 50 µM). The PDE4 gene family comprises four genes (PDE4A, B, C, D), with each gene having multiple splice variants. To date more than 60 splice variants have been identified [3]. Each gene has a 'long form', as well as one or more short forms. This 'long' and 'short' form terminology is based on the length of the N-terminal portion. Each long form has an N-terminal domain, including two regions described as upstream conserved regions (UCR1 and UCR2), which are thought to play a regulatory role, and a highly conserved catalytic region in the C-terminal domain, which exhibits $\sim$ 75% sequence identity between PDE4 family members. The PDE4 variants arise due to differences in their N-termini. Given that the N-termini encode regulatory domains and phosphorylation sites, each splice variant will be subject to different regulatory mechanisms. While several PDE4 promoters have been characterised, a common finding is their regulation by cAMP via cyclicAMP response element (CRE)/CRE-binding protein.

Selected PDE4 variants are localised to subcellular domains through interactions with a variety of scaffolding/anchoring structures. These include proline-rich sequence binding (SH3) proteins, A-kinase-anchoring proteins, receptor for activated c kinase, β arrestins 1 and 2, and phosphatidic acid-rich regions of cellular membranes. These specific interactions are considered to play an important role in the regulation of cAMP signalling by these enzymes [2]. Many long forms are selectively activated by protein kinase A-dependent phosphorylation, whereas both long and short forms are differentially regulated (either inhibited or activated) by extracellular signal-regulated kinase-1-mediated phosphorylation of a conserved catalytic domain serine residue [2].

PDE4 gene products have a broad tissue distribution and can be found in the brain, gastrointestinal tract, spleen, lung, heart, testis and kidney [4]. In addition PDE4 is expressed in almost all cell types except blood platelets (Table 2.1) [5–25].

In vitro *pharmacology of PDE4 inhibitors – focus on human cells*

A number of PDE4 selective inhibitors, and also short-interfering RNAs (siRNAs) to specific PDE4 subtypes [9], have been used to investigate the effect of inhibition of this enzyme on aspects of cellular function. In general,

Table 2.1 EXPRESSION OF PDE4 ISOFORMS AND SPLICE VARIANTS WITHIN HUMAN CELLS

	4A	4B	4C	4D
Eosinophils	+	+	−	+
Neutrophils	+ (A4, A7, A10)	+ (B1, B2)	−/+	+ (D1, D2, D5)
Monocytes	+ (A4, A7, A10)	+ (B1, B2)	−/+	+ (D1, D2, D3, D5)
Macrophages	+ (A4, A7, A10)	+ (B1, B2)	+	+ (D1, D2, D3)
T cells	+ (A4, A7, A10)	+ (B1, B2)	−	+ (D1, D2, D3, D5)
CD4$^+$	+	+ (B2)	−	+
CD8$^+$	+	+ (B2)	−	+
B cells	+	+ (B2)	−	+
Airway epithelial cells	+ (A5)	−	+ (C1)	+ (D2, D3)
Endothelial cells	+	+ (B2)	nt	+ (D3, D5)
Basophils	PDE4 present; no information on subtypes			
Mast cells	PDE4 present; no information on subtypes			
Platelets	−	−	−	−
Fibroblast	PDE4 present; no information on subtypes			
Airway smooth muscle cells	−	+	−	+ (D5)
Pulmonary artery smooth muscle cells	+ (A10, A11)	+ (B2)	+	+ (D5)
Vascular smooth muscle cells[a]	−	−	−	+ (D3, D5)

Note: −, absent; +, present; −/+, where conflicting reports have been published; nt, not tested.
[a]Rat cells.

PDE4 inhibition potently prevents the release of pro-inflammatory mediators from a range of cell types. PDE4 inhibitors have been shown to inhibit adhesion molecule expression, chemotaxis, proliferation, migration and differentiation, and the release of re-modelling factors. They also inhibit or promote apoptosis, the latter appearing to be a selective effect on lymphocytic leukaemia cells. In addition, PDE4 inhibitors relax inherent airway smooth muscle tone and can activate cystic fibrosis transmembrane conductance regulator (CFTR)-mediated Cl$^-$ secretion. This diverse spectrum of biological effects has thus implicated PDE4 as a potential therapeutic target for a range of disease indications. The consequence of PDE4 inhibition in individual cell types is described below.

Eosinophils. Eosinophils are thought to play an important role in asthma and also in exacerbations of chronic obstructive pulmonary disease (COPD) through their ability to release a plethora of pro-inflammatory mediators, which cause tissue injury, remodelling and contraction of smooth muscle [26].

PDE4A, B and D have been detected in human eosinophils, with evidence to suggest that PDE4A is exclusively located in all eosinophil granules [8]. A range of PDE4 inhibitors has been shown to inhibit N-formyl-methionyl-leucyl-phenylalanine (fMLP)-stimulated release of reactive oxygen species from human eosinophils [27] as well as inhibiting complement 5a (C5a) and platelet activating factor (PAF)-stimulated leukotriene (LT)C_4 synthesis [28]. In addition, PDE4 inhibitors can suppress immunoglobulin (Ig)-induced degranulation [29]. PDE4 inhibitors can also inhibit PAF-induced CD11b expression and L-selectin shedding by ~50% [30], and both C5a- and PAF-stimulated eosinophil chemotaxis [28]. In addition, they have also been shown to delay spontaneous human eosinophil apoptosis [31].

Neutrophils. Neutrophils are thought to play a pivotal role in the chronic lung inflammation and tissue destruction present in COPD, severe asthma and cystic fibrosis, through their ability to release many toxic substances such as proteases and oxygen radicals which cause tissue injury and remodelling [26, 32].

PDE4A, B and D are expressed in human neutrophils [6, 8, 25], with evidence that PDE4B2 is the predominant PDE4 isoform in human neutrophils [6]. PDE4A is exclusively located within a subset of myeloperoxidase (MPO) containing neutrophil granules [8]. PDE4 inhibitors have been shown to inhibit the release of a range of pro-inflammatory mediators from human neutrophils. For example, they inhibit fMLP-stimulated release of LTB_4 and reactive oxygen species [27, 33, 34], together with the fMLP/ tumor necrosis factor-α (TNF-α)-stimulated release of matrix metalloproteinase-9 (MMP-9) and neutrophil degranulation products such as neutrophil elastase and MPO [35]. PDE4 inhibitors can also inhibit PAF-induced CD11b expression and L-selectin shedding by ~50% [30] and both TNF-α- and fMLP-mediated neutrophil adhesion to human umbilical vein endothelial cells [35, 36]. PDE4 inhibitors have also been shown to delay spontaneous human neutrophil apoptosis [31].

Monocytes and macrophages. Tissue macrophages differentiate from a common precursor, the circulating monocyte, and play a major role in defence. Alveolar macrophages have, however, been implicated in tissue injury associated with inflammatory diseases of the lung including asthma, acute respiratory distress syndrome and COPD [37].

PDE4A, B, C and D have been detected in human lung macrophages and in peripheral blood monocytes [25, 38]. Interestingly, PDE4A4 was upregulated in lung macrophages from smokers with COPD compared with control smokers [25] and PDE4A4 as well as PDE4B2 were detected in higher amounts in the peripheral blood monocytes of smokers compared with non-smokers [25]. Indeed, PDE4B2 appears to be the predominant PDE isoform in human monocytes [6], and is selectively induced by lipopolysaccharide (LPS). This induction is inhibited by interleukin-4 (IL-4) and IL-10. PDE4 inhibitors are capable of completely abolishing LPS-stimulated TNF-α release from peripheral blood monocytes [39, 40]. Interestingly, in PDE4B deficient mice (but not PDE4D deficient mice), there is a marked reduction in the ability of LPS to stimulate TNF-α release from peripheral blood leukocytes [41], suggesting a key role for PDE4B in this response. Further support for this role can be derived from a separate study demonstrating that mean IC_{50} values for inhibition of LPS-stimulated TNF-α release are significantly correlated with compound potency against the catalytic activity of recombinant human PDE4B (and PDE4A), but not the catalytic activity of recombinant human PDE4D [39]. In contrast to studies with monocytes, only a partial inhibitory effect of PDE4 inhibitors is observed on LPS-stimulated TNF-α release from human alveolar macrophages [42]. This appears to be due to the presence of PDE3 in macrophages, as a dual PDE3/4 inhibitor can completely suppress this effect [42].

Lymphocytes. Hyperactive Th_1-mediated immune responses are thought to be involved in the pathogenesis of many autoimmune diseases, including multiple sclerosis, rheumatoid arthritis [43] and Crohn's disease [44], whereas agents targeting Th_2 cells are sought for diseases such as bronchial asthma [45, 46] and ulcerative colitis [44].

PDE4A, B and D (but not PDE4C) have been detected in human T and B lymphocytes [17, 19, 20, 24, 25]. The PDE4 inhibitor, rolipram, has been shown to partially inhibit (by 40–60%) mitogen-stimulated IL-2 release from $CD4^+$ and $CD8^+$ human T cells [19, 27]. In a recent study, PDE4 subtype-specific siRNAs were used to investigate the functional impact of subtype-specific knockdown on anti-CD3/CD28-stimulated cytokine release from $CD4^+$ T cells. Knockdown of PDE4B or PDE4D (but not PDE4A) inhibited IL-2 release, whereas knockdown of PDE4D showed the most predominant inhibitory effect on interferon-γ (IFN-γ) and IL-5 release [9]. PDE4 inhibitors have also been shown to partially inhibit IL-4 and IL-5 gene expression in Th_2 cells [47], together with IL-4 and IL-5 release from human $CD4^+$ T cells [27]. In contrast, a separate study demonstrated that specific inhibition of PDE4 had no significant effect on Th_2 cell mediated IL-4 or IL-13 generation, but preferentially inhibited Th_1 cell cytokine generation (IFN-γ) [48]. PDE4 inhibitors have also been shown to partially inhibit phytohaemagglutinin and anti-CD3/anti-CD28-stimulated proliferation of $CD4^+$ and $CD8^+$ T cells [19, 27]. In a separate study, dual PDE4A/B and PDE4D selective inhibitors inhibited antigen-stimulated human T cell proliferation. Mean IC_{50} values significantly correlated with compound potency against the catalytic activity of recombinant PDE4A or B, but not with catalytic activity of recombinant PDE4D [39]. In contrast, a PDE4D siRNA (but not PDE4A or B siRNAs) significantly inhibited anti-CD3/CD28-stimulated $CD4^+$ proliferation [9]. The reason for the apparent difference in PDE4 subtype involvement in this proliferative response is unclear, but could be related to either the fact that different T cell populations were used, or that different stimuli were used to elicit proliferation.

Lastly, inhibition of PDE4 results in a potent induction of apoptosis in chronic lymphocytic leukaemia cells, suggesting that PDE4 inhibitors may be of benefit in lymphoid malignancies. Importantly, this effect appears to be relatively specific, as comparable treatment of human peripheral blood T cells does not induce apoptosis [49].

Airway epithelial cells. Airway epithelial cells are not only important barrier cells, but also play an integral role in the pathophysiology of airway diseases through their ability to release multiple pro-inflammatory and pro-remodelling mediators [5]. In addition, CFTR is the primary cAMP-activated chloride channel on the apical membrane of airway epithelia, thereby playing an integral role in controlling the electrolyte/fluid balance and mucociliary clearance [50]. CFTR activation has the potential to enhance mucociliary clearance, which may be of benefit in diseases such as COPD, asthma and cystic fibrosis.

PDE4A, C and D are expressed in human airway epithelial cells [5, 21]. The PDE4 inhibitor, rolipram, partially inhibited granulocyte-macrophage colony-stimulating factor release from IL-1β-stimulated human airway epithelial cells; an effect which was maximally inhibited by dual inhibition of PDE3 and PDE4 [5]. In addition, rolipram inhibited LPS-stimulated IL-6 release from human airway epithelial cells, although relatively high concentrations were required to see an inhibitory effect, with an IC_{50} value of 24 µM, suggesting the effect could have been mediated through other PDE enzymes [51]. Inhibition of PDE4 (in particular the D isoform) has been shown to activate CFTR-mediated chloride secretion in an epithelial cell line [50]. More recently, it has been demonstrated that this effect is likely to be due to an interaction of PDE4 with AMP-activated kinase [52].

Endothelial cells. The endothelium acts a major permeability barrier of the blood vessel wall, and facilitates the transmigration of blood cells to tissue, through expression of adhesion molecules. During inflammation, however, leukocytes may damage endothelial cells, resulting in increased vascular permeability [53]. Endothelial cells also play an important physiological role in angiogenesis. Excessive angiogenesis, however, allows vascularisation of solid tumours and provides routes through which cancer cells may metastasise. Thus agents, which can limit processes involved in angiogenesis (e.g. cell migration), may represent novel therapies for pathologies such as cancer [12].

Human aortic, umbilical vein and microvascular endothelial cells express PDE4A, B and D [11, 12, 22]. Inhibition of PDE4 in combination with appropriate activation of adenylate cyclase inhibits TNF-α-induced E selectin expression on human lung microvascular endothelial cells [54]. In addition, rolipram potently blocked H_2O_2-induced endothelial permeability when combined with prostaglandin E_1 [53]. Inhibition of PDE4 decreased vascular endothelial growth factor-induced migration of endothelial cells [12].

Mast cells and Basophils. Anti-IgE stimulation of mast cells and basophils is a central event in acute allergic disorders such as anaphylactic shock, asthma, allergic rhinitis, and some skin conditions including urticaria and atopic dermatitis [55].

PDE4 has been detected in human basophils [10], although no data is published on which isoforms are present. There does not appear to be any published data describing the nature of PDE4 subtypes present in human mast cells, but the PDE4 inhibitor rolipram inhibits cAMP hydrolysis in mast cells by $\sim 50\%$ [56]. PDE4 inhibitors have been shown to suppress IgE and IL-3-dependent generation of IL-4, IL-13 and histamine [57] and also anti-IgE-induced histamine and LTC_4 release from basophils [10, 56]. PDE4 inhibitors, however, have no effect on anti-IgE-stimulated histamine or LTC_4 release from human lung mast cells [56, 58].

Smooth muscle cells

Airway smooth muscle. Airway smooth muscle cells may contribute to airway remodelling observed in lung diseases such as asthma and COPD, through the release of growth factors, cytokines and extracellular matrix proteins [59].

PDE4B and D are expressed in human airway smooth muscle cells [16, 23]. Roflumilast has been shown to be capable of inhibiting transforming growth factor-β (TGF-β)-induced fibronectin deposition in human airway smooth muscle cells and also TGF-β-induced connective tissue growth factor, collagen I and fibronectin expression in human bronchial tissue rings [59]. PDE4 inhibitors have also been shown to relax inherent tone in isolated human bronchial muscle [60, 61]; but a combination of PDE3 and PDE4 inhibitors is necessary to inhibit allergen or LTC_4-induced contraction [61]. Interestingly, in a study using siRNA targeted to PDE4D5, this PDE4 splice variant was shown to be the key physiological regulator of $β_2$-adrenoceptor-induced cAMP turnover within human airway smooth muscle [16, 23].

Pulmonary artery smooth muscle cells. Aberrant regulation of smooth muscle cell proliferation and apoptosis in distal pulmonary arteries is a characteristic feature of pulmonary artery hypertension [62].

Pulmonary artery smooth muscle cells have been shown to express all PDE4 isoforms. Interestingly several PDE4 splice variants (PDE4A10, A11, B2 and D5) have been shown to be upregulated following hypoxia [13]. PDE4 inhibitors have been shown to inhibit the proliferation of TNF-α/phorbol 12-myristate-13-acetate (PMA)-stimulated increase in MMP-2 and MMP-9 activity of distal human pulmonary artery smooth muscle cells [63].

Vascular smooth muscle cells. Vascular smooth muscle cells (VSMCs) alter their phenotype in response to vascular injury displaying a reduced contractile capacity and increased proliferative, migratory and synthetic capabilities [64].

PDE4D is the dominant PDE4 isoform in VSMCs [7, 15] and PDE4 inhibitors have been shown to inhibit VSCM proliferation [65]. Given the effects of PDE4 inhibitors on VSMC and the selective expression of the PDE4D isoform in this cell type, the potential utility of selective PDE4D inhibitors in adjunctive pharmacotherapy after percutaneous coronary interventions has been suggested, and is reviewed by Houslay [66].

Fibroblasts. Therapies that mitigate the fibrotic process may be able to slow progressive loss of airways function in many lung diseases. Pulmonary fibroblast to myofibroblast conversion is a pathophysiological feature of idiopathic pulmonary fibrosis and COPD [67].

PDE4 is expressed in human fibroblasts; although the subtype(s) have not yet been defined [67]. Lung fibroblast to myofibroblast differentiation (induced by TGF-β) is inhibited by the PDE4 inhibitor piclamilast [67]. PDE4 inhibitors can promote inhibition of TNF-α-stimulated pro-MMP1 and pro-MMP2 release from human lung fibroblasts [68]. Cilomilast and rolipram inhibit chemotaxis of foetal lung fibroblasts towards fibronectin and inhibit contraction of three-dimensional collagen gels [69].

In vivo *pharmacology of PDE4 inhibitors*

Asthma. Exposure of animals to bronchoconstrictor agents (e.g. histamine and methacholine) or exposure of previously sensitised animals to an allergic stimulus, e.g. ovalbumin (OVA) are widely used models to mimic characteristic features of asthma. Several studies have demonstrated that both oral and inhaled administration of PDE4 inhibitors can inhibit allergen-induced pulmonary eosinophilia in guinea pig [70–72], mouse [73, 74] and rat [75, 76]. Local administration of PDE4 inhibitors such as roflumilast (5), AWD 12-281 (10) or UK-500,001 inhibits allergen [72, 74], histamine [33, 72], methacholine and LTD$_4$-induced bronchoconstriction in guinea pigs [72], allergen-induced bronchial hyper-reactivity in mice [74] and antigen-induced reductions in forced vital capacity [75]. PDE4 inhibitors have also been shown to inhibit OVA-induced mucus production and goblet cell hyperplasia in allergic rat and mouse models [73, 76] and subepithelial collagenisation and airway wall thickening in a murine chronic asthma model [77]. PDE4D appears to be the subtype in the airways mediating the response to cholinergic and antigenic responses, at least in the mouse, as airways of PDE4D-deficient mouse are no longer responsive to cholinergic stimulation or antigen-induced airway hyper-reactivity [78].

COPD. Various pre-clinical species (most commonly rats and mice) have been exposed to LPS or cigarette smoke in order to try and induce some of the pathological features observed in COPD, specifically inflammatory cell infiltration and emphysematous changes. In acute cigarette smoke exposure studies in mice, oral treatment with cilomilast inhibited recruitment of neutrophils to the lung together with increases in BAL macrophage inflammatory protein-1α [79], and roflumilast partially inhibited neutrophil influx to the lung [80]. In more chronic smoke exposure studies, roflumilast (oral treatment for 7 months) has been shown to fully prevent emphysema in mice [80], and the PDE4 inhibitor, GPD-1116 (22) (oral treatment for 8 weeks) has also been shown to markedly attenuate the development of cigarette smoke-induced emphysema in senescence-accelerated P1 mice [81]. Local administration of PDE4 inhibitors directly to the lung has also been shown to be effective in inhibiting LPS-induced neutrophil recruitment to the lung in a range of species: rats [33, 74], ferrets [74] and pigs [74].

In a lung injury model utilising PDE4A, B and D knockout mice, PDE4B and PDE4D (but not PDE4A) appear to be important in mediating LPS-induced neutrophil transepithelial migration; an effect which is mediated in part by upregulation of neutrophil CD18 expression [74, 82].

Cough. The allergen-induced increase of cough response to inhaled capsaicin in sensitised animals, and normal cough response in unsensitised animals, have both been shown to be reduced by a PDE4 inhibitor [74, 83, 84].

Multiple sclerosis. In an animal model of multiple sclerosis, experimental autoimmune encephalomyelitis (EAE), a non-brain penetrant PDE4 inhibitor (L-826,141) (1) reduced the severity of EAE and also delayed disease onset [85]. Furthermore, in a relapsing-remitting EAE model of the SJL mouse, rolipram reduced the clinical signs of EAE during both the initial episode of the disease and also during subsequent relapses. Rolipram also markedly reduced demyelination, central nervous system (CNS) inflammation, and secretion of Th$_1$ cytokines [86].

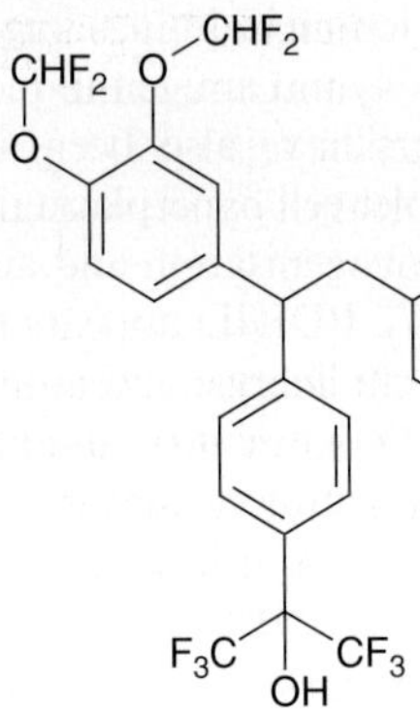

(1) L-826,141

Inflammatory bowel disease. A number of studies have demonstrated that PDE4 inhibitors can prevent colitis and reverse established colitis in a range of pre-clinical disease models. Dextran sodium sulphate, trinitrobenzene sulphonic acid (TNBS) and indomethacin-induced colitis mimic aspects of Crohn's disease and ulcerative colitis, the two major forms of human inflammatory bowel disease (IBD) (as reviewed by Banner and Trevethick [87]). In addition to the well described anti-inflammatory effects of PDE4 inhibitors in these disease models, one rat TNBS-induced colitis study demonstrated that treatment with rolipram could prevent intestinal collagen deposition [88], suggesting that PDE4 inhibitors may also be able to inhibit tissue remodelling in this setting.

Rheumatoid arthritis. Administration of PDE4 inhibitors suppressed the pannus-like inflammation by inhibition of cytokine production from macrophages and synovial fibroblast proliferation in a mouse model of rheumatoid arthritis [89].

Memory. Rolipram is effective in animal models of memory [90]. Specifically, it facilitates the establishment of long-lasting long-term potentiation and improves memory [91] and ameliorates experimentally induced impairments of learning and memory in rodents [92].

Depression. PDE4 inhibitors, including rolipram, produce anti-depressant-like effects in several pre-clinical models. Specifically, they have been shown to: reduce the time of immobility in the forced-swim test; decrease response rate and increase reinforcement rate under a differential-reinforcement-of-low-rate schedule; reverse the effects of chronic, mild stress; normalise the behavioural deficits observed in Flinders sensitive-line and olfactory-bulbectomized rats; antagonise the effects of reserpine and potentiate yohimbine-induced toxicity [93–98]. Utilising PDE4 subtype deficient mice, it appears that PDE4D is an essential mediator of the antidepressant effects of PDE4 inhibitors [99].

Pulmonary artery hypertension. Rolipram prevented hypoxia-induced PDE4 and PDE1 gene upregulation and interfered with the development of pulmonary artery hypertension in transgenic sickle cell mice, most likely through modulation of vascular tone and inflammatory factors [100].

Renal diseases. The therapeutic potential of PDE4 inhibitors in a range of renal indications, such as glomerulonenephritis, renal transplantation and renal failure is reviewed in detail by Dousa [101] and Cheng and Grande [102]. Of particular interest, the PDE4 inhibitor RO-20-1724 (2) alone has been shown to be effective in both the preventative [103] and therapeutic [104] setting in endotoxin-induced acute renal failure in rats.

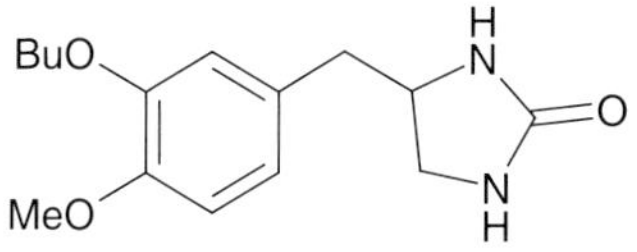

(2) RO-20-1724

Additional potential indications for PDE4 inhibitors

There is also interest in PDE4 inhibitors as potential therapeutic agents for premature birth [105], psychiatric illness [106], hypertension [107], idiopathic pulmonary fibrosis [67], lymphoid malignancies [49], allergic skin diseases and psoriasis [108].

Adverse effects of PDE4 inhibitors

The therapeutic window of orally administered selective PDE4 inhibitors in clinical trials is limited by gastrointestinal side effects of nausea, vomiting, diarrhoea, abdominal pain and dyspepsia, although some of these appear to resolve with continued treatment. Regulatory agencies are, however, particularly concerned by the development of mesenteric vasculitis in laboratory animals.

Mesenteric vasculitis has, however, never been seen in man, and generally has not been seen in non-human primates. Indeed mesenteric vasculitis has never been seen in human patients treated for many years with bronchodilator doses of theophylline, a regime which produces medial necrosis of mesenteric vessels in rats [109, 110]. Rats and dogs may have an increased susceptibility to drug-induced vascular legions as arteriopathies commonly occur in these species [111, 112], and species differences have been shown to exist for both PDE4 expression and functional effects of PDE4 inhibitors. For example, a recent study demonstrated that levels of PDE4 enzyme activity are much higher in rats than humans in multiple tissues, making rats more susceptible to PDE4 inhibitor-induced toxicities [113]. In addition, the PDE4 inhibitor IC542 (structure not available) markedly enhanced LPS-induced IL-6 release from rat whole blood [114], but not from human or non-human primate blood. Nevertheless vasculitis requires careful monitoring in man, and indeed current research is focused on identifying potential predictive biomarkers. To this end, tissue inhibitor of metalloproteinase-1 appears to be an early and sensitive predictive biomarker of PDE4 inhibitor-induced vascular injury in rats [115].

PUBLISHED PDE4 INHIBITORS

MARKETED COMPOUNDS

No PDE4 inhibitors as such have progressed to the market so far; however, some launched drugs do have some PDE4 inhibitory activity. The obvious example is theophylline (3), which has been in use since 1937 in the treatment of asthma [116]. The mechanism of action of the drug is not entirely understood, but as well as antagonising the adenosine A_1 receptor, the compound is an unspecific and weak PDE inhibitor. The drug is not without side effects, possibly due to its rich pharmacology, and is not so widely used as more specific and efficacious drugs with other mechanisms of action. A compound which matched or bettered the efficacy of theophylline without the liabilities would surely find a market. An interesting recent development is a clinical trial, which Argenta announced in February 2008, in which inhaled theophylline (ADC-4022) is to be co-administered with budesonide [117] to COPD patients. The belief is presumably that some synergistic effects of PDE4 inhibition and the steroid will be observed.

A similarly promiscuous compound is ibudilast (4), launched for asthma in 1989. As well as inhibiting PDE4 isoforms A, B and D with IC_{50} value at around 3 µM [118], the drug is also an LTD4 antagonist [119].

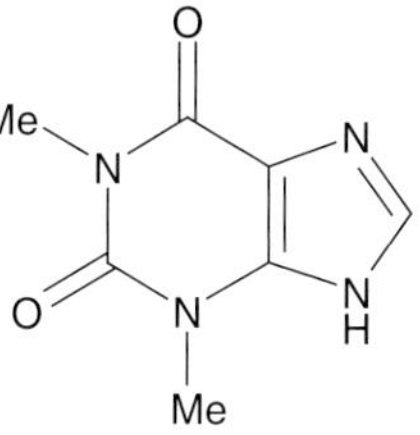

(3) Theophylline

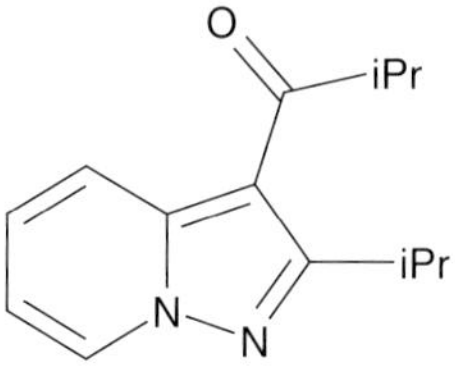

(4) Ibudilast

It is interesting to note that after decades of research on PDE4 as a target, the only successful drugs to date are those which are not specific for PDE4, but which show additional mechanisms of action. It may well be that in order to reduce the side effects of PDE4 inhibitors and make them clinically useful, it is necessary to incorporate into the molecule some other mode of action, either synergistic with, or additive to PDE4 inhibition, which allows efficacy with an improved therapeutic index.

Compounds in phase III clinical trials

The most advanced compounds to date are roflumilast (5) and cilomilast (6).

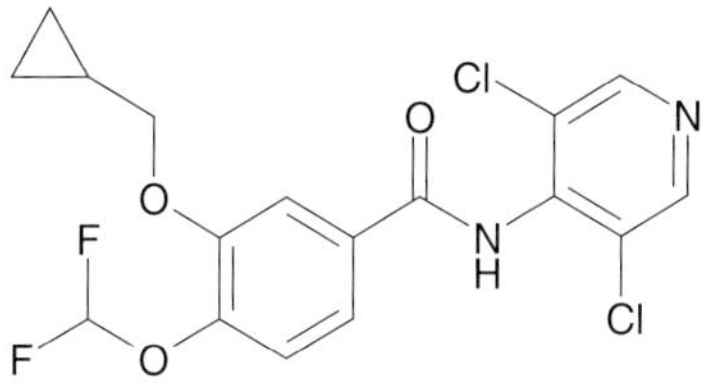

(5) Roflumilast

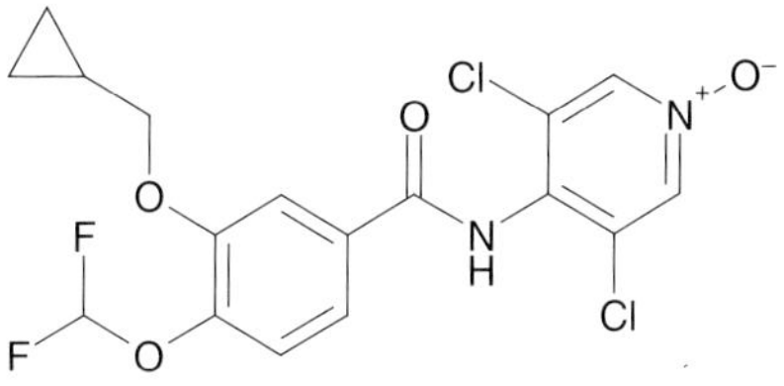

(5a) Roflumilast N-oxide

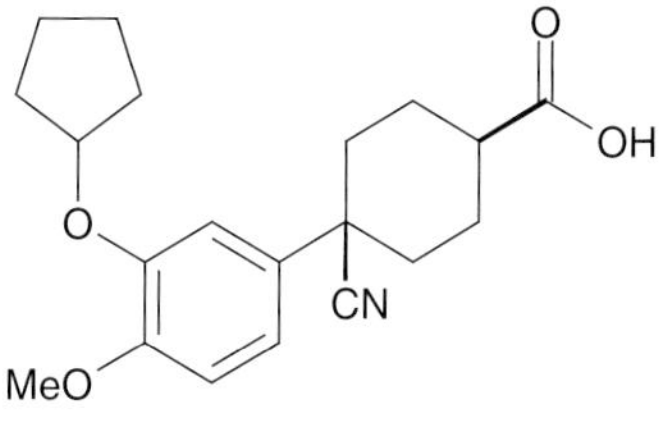

(6) Cilomilast

Both compounds have been published on quite extensively. Cilomilast [120] was originally published as a 'second generation' PDE4 inhibitor, i.e. one which would have efficacy along with a reduced propensity towards a nausea/emesis side effect profile. The hypothesis behind this 'second generation' profile was originally based around selectivity against the rolipram-binding site of PDE4 [121]. It was suggested that the nausea centre of the brain, and the stomach and GI tract were rich in PDE4 enzyme in the high-affinity state, and that inhibition of this 'high-affinity rolipram-binding site' (HARBS) was concomitant with emesis [122–125]. This effect could be measured *in vitro* by an increase in acid production from rabbit gastric glands [126]. Thus, a series of compounds was produced, culminating in cilomilast, with selectivity for low-affinity PDE4 over HARBS. Cilomilast has been in numerous clinical trials, for both COPD and asthma [118] and has been shown to demonstrate some efficacy in reducing inflammatory cells in tissues [127] and to improve lung function [128–132]. Cilomilast is reported to be well tolerated, however, the side effects of nausea and emesis do occur at higher doses (the clinically effective dose is 15 mg b.i.d.). Thus, efficacy of this drug appears to be marginal and limited by a narrow therapeutic index [133]. In 2003, GlaxoSmithKline (GSK) received an approval letter from the FDA for cilomilast, but is believed to have discontinued development of the compound in 2007.

Roflumilast, in contrast, is still reported to be in development, being in Phase III trials for asthma in Japan, and is probably the closest PDE4 inhibitor to the market [134]. In 2002, Tanabe Seiyaku obtained rights to develop roflumilast in Japan for COPD and asthma. Like cilomilast, roflumilast is from the class of inhibitors that appear structurally to have been derived from rolipram (7) [135]. There are many published results from clinical trials with roflumilast; two of the biggest (each having over 1,000 patients treated for 24 weeks or 1 year) have been for COPD, and both have shown good tolerability, improved quality of life, reduced exacerbations and an improvement in lung function compared to placebo [136–139].

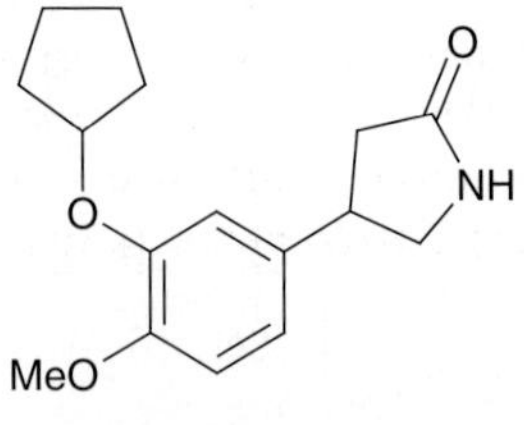

(7) Rolipram

Despite the positive results reported from such trials, roflumilast has clearly had some problems reaching the market. In late November 2005, Altana (acquired by Nycomed in 2007) withdrew a marketing authorisation application filed in the E.U. seeking approval of roflumilast for the treatment of COPD and asthma following consultation with the European Medicines Agency (EMEA). The maximum tolerated dose of the drug appears to be 500 μg q.d. given orally, which is a very small dose. While the compound is reported to be very potent (IC_{50} = 800 pm at PDE4 [140]), it is also known to cause emesis in animal models (ED_{50} = 0.3 mg/kg i.v. in the pig [74]). It seems likely that the low maximum tolerated dose (MTD) in humans is again indicative of a small therapeutic index, which perhaps makes it difficult to demonstrate a clear efficacious effect.

It is interesting to note that roflumilast has a primary metabolite in man – the N-oxide (5a) [141] – which is almost as potent an inhibitor of the PDE4 enzyme as the parent is (IC_{50} at PDE4 = 2 nM [27]). In this case it is highly likely that the primary metabolite is also contributing to any pharmacological effects observed in man. Indeed, it could be speculated that the nature of production of the metabolite (effectively slow delivery i.v., avoiding direct exposure to the GI tract) could result in a reduced side effect profile for that active component. It remains to be seen if roflumilast will make it to market, but in any case it appears that a competitor with a wider therapeutic index would certainly have an advantage.

The third PDE4 inhibitor of interest to have reached Phase III clinical trials is tetomilast (8) (OPC-6535) from Otsuka. This drug is reported to be in Phase III for the treatment of Crohn's disease and ulcerative colitis [142–144]. An earlier Phase II trial of 186 patients with ulcerative colitis receiving 25 mg or 50 mg of oral tetomilast did not succeed in achieving statistical significance in remission rates, although secondary (disease activity index) readouts were positive [145]. The main adverse events were emesis and nausea. The two Phase III clinical trials (FACT I and FACT II) began in 2003, with a patient population of >750 based on the readouts from Phase II. The results of these trials have recently been published [146], and unfortunately a statistically significant improvement was neither demonstrated for the primary efficacy end point nor for most secondary end points. It is claimed that one possible reason for this lack of drug superiority was a very high placebo response rate. It is unclear at this time where this leaves the further development potential of tetomilast.

(8) Tetomilast

Compounds in phase II clinical trials

One of the more advanced compounds in Phase II clinical trials is oglemilast (9), also known as GRC-3886, which is reported to have an IC_{50} value of around 1 nM at PDE4, and to be effective in animal models of arthritis [147]. Glenmark Pharmaceuticals are pursuing this compound for the oral treatment of COPD and asthma in Europe, however, little seems to have been published following trials in asthma and COPD which began in 2006.

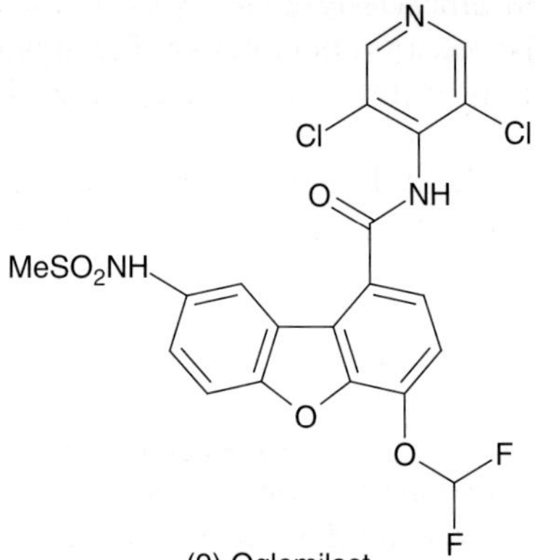

(9) Oglemilast

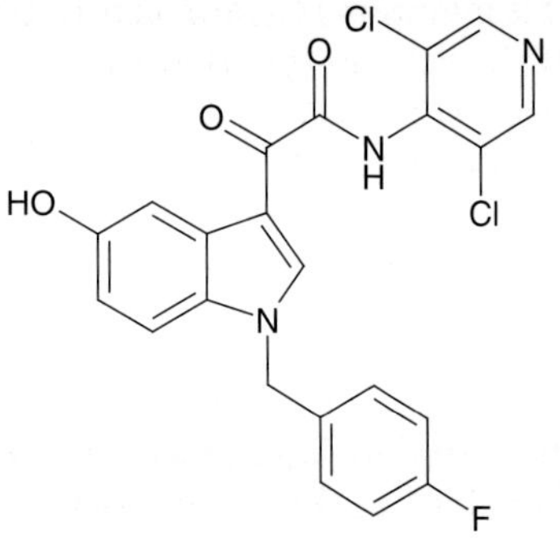

(10) AWD-12-281

Of similar topology to oglemilast is AWD-12-281 also known as GW842470 (10), a PDE4 inhibitor on which rather more pre-clinical work has been published [74]. Both compounds have the familiar dichloropyridine-amide motif which is also present in roflumilast, and both have an acidic moiety on a heterocyclic system (indole in the case of AWD-12-281). The phenol in (10) is indicative of a design towards a topical application, the phenol presumably providing a point for glucuronidation and rapid systemic clearance. This compound was originally from Elbion, and is a potent and selective PDE4 inhibitor designed to act locally on inflammatory cells of the respiratory tract or skin. Pre-clinically, it was demonstrated that AWD-12-281 suppressed antigen-induced late-phase eosinophilia in Brown-Norway rats following intratracheal administration, with an ID_{50} of 7 mg/kg compared to 0.1 mg/kg for beclomethasone. The compound also inhibited LPS-induced acute lung neutrophilia in rats ($ID_{50} = 0.02$ μg/kg i.t.), ferrets ($ID_{50} = 10$ μg/kg i.t.) and pigs ($ID_{50} \sim 100–250$ μg/kg i.t.) [74]. Use of the ferret and pig models allowed the estimation of a therapeutic index, and AWD-12-281 displayed a much lower emetic potential compared to cilomilast in ferrets and roflumilast in pigs following oral, i.v. or i.t. administration, and no emesis was seen in dogs treated at the highest dose achievable by inhalation (15 mg/kg). Thus, it was anticipated that the low therapeutic index problem generally associated with oral PDE4 inhibitors may be obviated by an i.t. delivery, while maintaining efficacy against inflammation readouts [148].

A collaboration agreement between Elbion and GSK was signed in July 2002, and the partners had been conducting Phase II clinical trials for the topical treatment of atopic dermatitis in 2006. It is believed that they were also pursing an inhaled delivery for the treatment of asthma, allergic rhinitis and COPD. However, no recent development has been reported for these indications and it must be suspected that development has been discontinued. The GSK pipeline does however still contain an inhaled PDE4 inhibitor in more recent Phase II trials for the treatment of asthma, COPD and allergic rhinitis, although identified by compound number '256066', the structure is currently unclear.

Continuing with the theme of PDE4 inhibitors designed to be dosed by inhalation, is a compound from Pfizer, UK-500,001. UK-500,001 potently inhibited PDE4A, B and D activities with IC_{50} values of 1.9, 1.01 and 0.38 nM, respectively [149]. In cellular assays, UK-500,001 blocked the release of TNF-α from human peripheral blood mononuclear cells ($IC_{50} = 0.13$ nM) and leukotriene B4 from neutrophils ($IC_{50} = 0.24$ nM). This compound was dosed by inhalation in Phase II studies for asthma, but results have recently been published which suggest that the compound will not progress further. After dosing i.t. at 0.4 mg b.i.d. and 2 mg b.i.d., UK-500,001 did not attenuate airway responses to allergen and histamine challenges in asthmatics [150]. A second study in COPD patients, in which the compound was dosed i.t. at 0.1, 0.4 or 1 mg b.i.d. for 6 weeks showed no significant difference in the change from baseline in trough FEV1 (the forced expiratory volume of air blown out in 1 s) at week 6. Secondary endpoints were also not significantly affected. UK-500,001 was safe and well tolerated, although the incidence of treatment-related adverse events was increased in the higher doses compared to placebo [151]. Although the structure of UK-500,001 has not yet been disclosed, several recent patent applications from Pfizer, which were published in 2005, describe large *cis*-dicarboxamide cyclohexanes such as the structures (11) PDE4 $IC_{50} = 0.6$ nM [152] and (12) PDE4 $IC_{50} = 0.07$ nM [153]. Patent applications exemplifying individual compounds may indicate compounds of particular interest. Compounds (13) and (14) were described in a 2004 application [154], in which a controlled release formulation was claimed to potentially reduce the side effect profile. The preparation of polymorphic crystalline forms of nicotinamide (15) was described in a 2006 application [155]; the emphasis on crystalline form here could imply a connection with an inhaled drug, since such forms are easier to mill or micronise to a respirable fraction size, compared to amorphous compounds. The presence of metabolisable or 'soft' sites in the drug, such as the phenol and the sulfide, further imply a drug designed not to be systemically available.

(11)

(12)

(13)

(14)

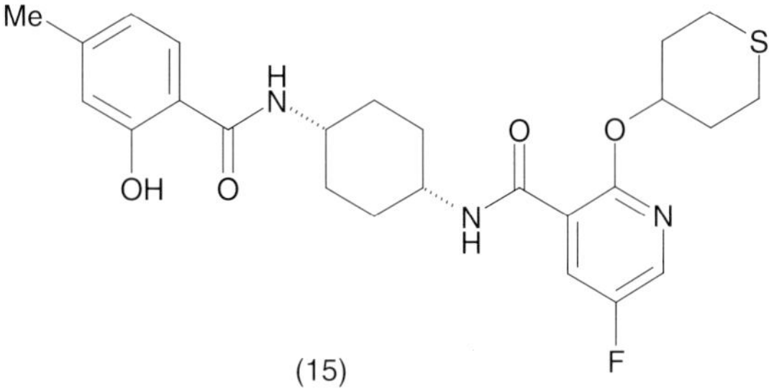

(15)

Memory Pharmaceuticals are looking at different indications for their PDE4 inhibitors, primarily Alzheimer's disease and dementia [156]. The company's PDE4 inhibitor programme includes MEM 1414, along with MEM 1917 and several backup compounds. The project appears to be in current development and there has been little published data on these compounds outside of the patent literature; however, a company press release in March 2008 states that MEM 1414 has demonstrated efficacy in a broad range of pre-clinical cognition and anti-inflammatory models [157]. In addition, Phase I studies have demonstrated a favourable safety profile for the compound overall, and particularly with respect to nausea and vomiting, which has limited the development of other PDE4 compounds. The company plans to progress MEM 1414 into a Phase IIa trial by the end of 2008.

Memory enhancement is also the indication being looked at by Helicon Therapeutics with the compound HT-0712, whose structure is suggested by patent application activity to be (16) [158]. The compound, originally known as IPL-455903, was developed at Inflazyme and licensed by Helicon in January 2003. The compound has shown efficacy in animal models of memory whether dosed before or after training [159, 160]. In March 2006, Inflazyme released a statement announcing the initiation of a Phase IIa study to assess the safety, tolerability and efficacy of IPL-455903, in patients with age-associated memory impairment. It is believed that increasing and prolonging the levels of cAMP in the brain may assist in memory consolidation. The drug was dosed over a 28-day period from 15 to 90 mg/day/subject and has been reported to be safe and well tolerated, although preliminary indications are that the effects on memory and learning failed to show a positive statistical effect.

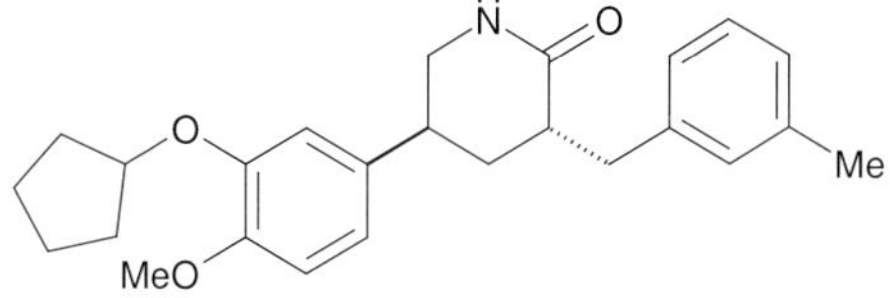

(16) Possible structure of HT-0712 aka IPL-455903

Merck & Co. did appear to have a compound designed for the treatment of COPD in Phase II trials – MK-0873 – a compound whose structure has not yet been disclosed, but which has been recently described as having completed an oral pharmacokinetic study in man [161]. The drug was given as a 2.5 mg dose over a 6-day period to a limited number of subjects with theophylline, and was described as being well tolerated. This compound was in the Merck pipeline as of December 2005, but is no longer included (see web site www.merck.com, 2008).

Other drugs which have proceeded to Phase II include piclamilast (RPR-73401) (17) [162–165], CDP-840 (18) [166–169], tofimilast (CP-325366) (19) [170–173], CDC-801 and CC-1088 [174], arofylline (LAS-31025) [175, 176] and lirimilast (BAY-19-8004) [177, 178]. However, these drugs have been slow to move forward and there are no obvious compounds from this set looking likely to progress to Phase III.

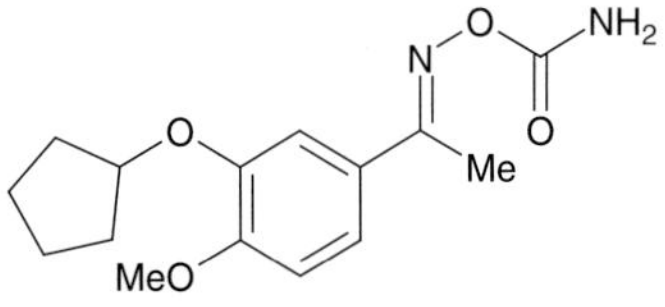

(17) Piclamilast

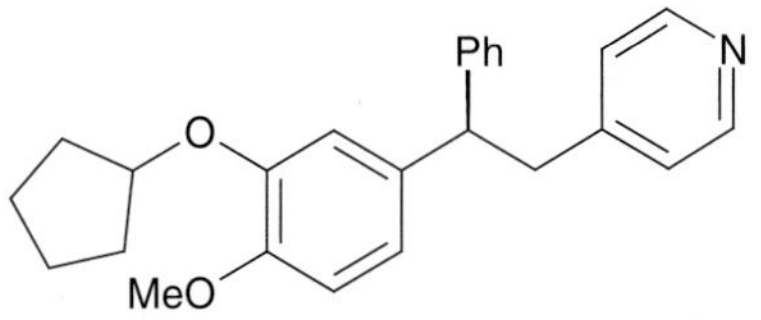

(18) CDP-840

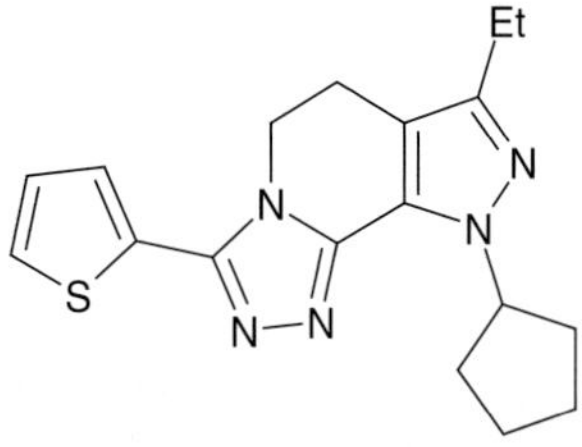

(19) Tofimilast

Compounds in phase 1 clinical trials

Information on compounds currently undergoing Phase I clinical trials is unsurprisingly sketchy. However, one compound which has been published on fairly extensively is YM-976 (20), the PDE4 inhibitor from Yamanouchi (Astellas) [179–182]. This potent, selective and competitive inhibitor of PDE4 gives an IC_{50} value of 2.2 nM (in human peripheral leukocytes) and was found to inhibit TNF-α production in LPS-stimulated human peripheral blood mononuclear cells (IC_{50} = 9.4 nM). In the carrageenan-induced pleurisy model in rats, it inhibited cell infiltration with an oral ED_{30} value of 9.1 mg/kg and in models of antigen-induced eosinophil accumulation in the lungs of sensitised mice, rats and ferrets, the compound dose-dependently inhibited eosinophilia with respective ED_{50} values of 3.6, 1.7 and 1.2 mg/kg p.o. In mice, the compound also inhibited IL-5 production with an ED_{50} value of 5.8 mg/kg p.o. In a model of eosinophilia induced by repeated exposure to antigen in rats, chronic administration of (20) was more effective than single doses, with ED_{50} values of 0.32 and 1.4 mg/kg p.o., respectively. Importantly, one of the main touted advantages of YM-976 was the pre-clinical evidence of an improvement in side effect profile – unlike rolipram and RP-73401, it was devoid of emetic effects in ferrets at up to 10 mg/kg p.o. [183]. The actual reason for the lowered emetogenic effect is not clearly understood, but reduced brain penetration is suggested as the most likely explanation.

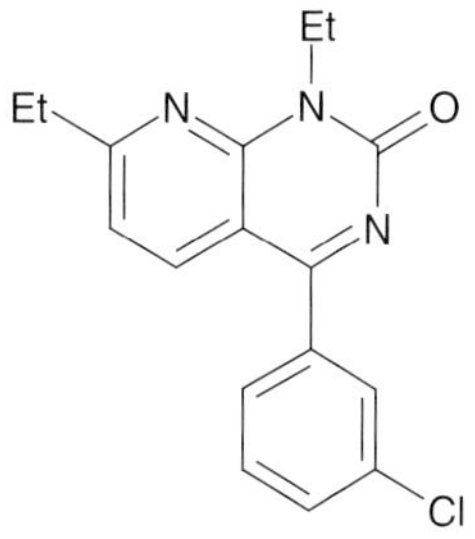

(20) YM-976

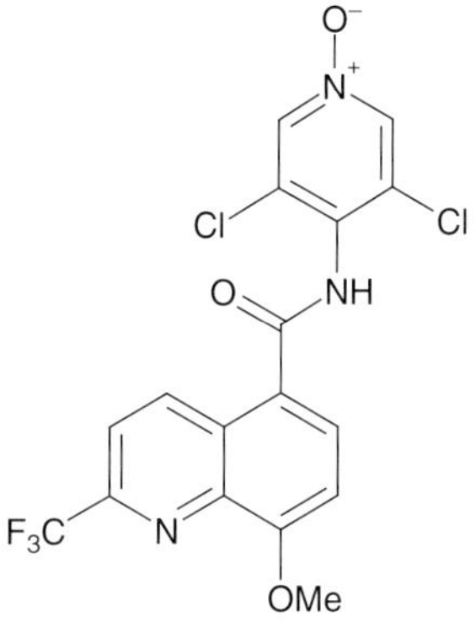

(21) SCH-351591

YM-976 was reported to be in Phase I trials for asthma in 1998, but no further development has been disclosed, and it seems likely that the compound has been discontinued. It is a similar story for SCH-351591 (21) [184–186]; another compound with the familiar dichloropyridine motif, in this case already incorporating the N-oxide seen in the roflumilast metabolite. Schering-Plough, under license from Celltech Group was developing SCH-351591 for the potential treatment of asthma and COPD. Phase I trials for these indications were ongoing in July 2002, however by October 2002, Schering-Plough had discontinued development of the compound and all rights had been returned to Celltech.

GRC-4039, a PDE4 inhibitor from Glenmark was reported to have begun Phase I trials in February 2008, with rheumatoid arthritis as the primary indication. The compound is claimed to have an IC_{50} value of 2.7 nM at PDE4, with >3,700-fold selectivity for this subtype. The molecule is reported to have good bioavailability, and to show 'favourable' results in early toxicology studies, good efficacy in rheumatoid arthritis models and no emesis in pre-clinical models [187].

The Aska Pharmaceuticals compound GPD-1116 (22), whose efficacy in a pre-clinical mouse model of emphysema is described above, is believed to be in Phase I clinical trials for the treatment of asthma and COPD.

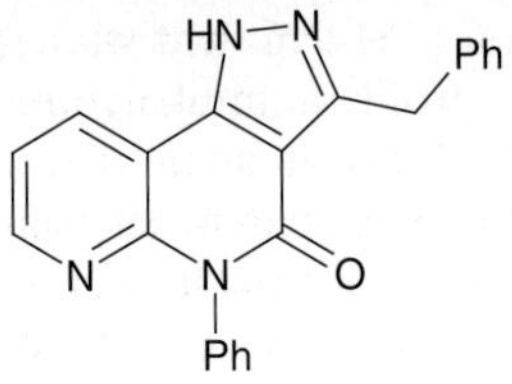

(22) GPD-1116

Other drugs which are believed to be currently undergoing, or to have recently completed Phase I clinical trials, but for which there is little further information include ELB-353 (AWD-12-353, Elbion) [188] and AVE-8112 for Alzheimer's disease (Sanofi-Aventis) [189].

Compounds in pre-clinical evaluation

While PDE4 has been pursued as a target for the potential treatment of several indications for many years now, the level of challenge has been high enough to warrant continued research, and the number of publications in the field remains high. Researchers at the Merck Frosst Center for Therapeutic Research have published in the field for the past 9 years, and a recent paper [190] confirms their continued interest. This group [191] and others [6, 82, 192, 193] had previously speculated on the advantages of a PDE4B selective compound in achieving a good therapeutic window; in this recent publication however, the inhibitors described have little or no isozyme specificity, with IC_{50} values < 10 nM. Instead their strategy was to identify a novel scaffold of PDE4 inhibitors by combining known pharmacophores in the public domain which display low incidence of emesis, with highly potent ones. In a human whole blood *in vitro* assay, they inhibit the LPS-induced release of the cytokine TNF-α (IC_{50} < 0.5 µM). Optimised inhibitors were evaluated *in vivo* for efficacy in an ovalbumin-induced bronchoconstriction model in conscious guinea pigs. Their propensity to produce an emetic response was evaluated by performing pharmacokinetic studies in squirrel monkeys. Compounds (23) and (24) are particularly highlighted and are follow-up compounds to L-454560 (25), Merck Frosst's first development candidate [194, 195], which had issues with CYP2C9 inhibition, and isomerisation of the olefin core in humans through the addition and elimination of glutathione. Cyclopropane carboxylic acid (24) proved to be a superior PDE4 inhibitor in regard to its therapeutic index since it showed the greatest ratio of plasma concentration at which emesis was observed (65 μM C_{max} after oral dosing) compared to the whole blood IC_{50} in squirrel monkeys (34 nM vs. TNF-α release). Thus, plasma concentration causing emesis:plasma concentration giving efficacy at $IC_{50} = 65000:34 = 1911$ [190]. Recent patent application filing activity by Merck Frosst confirms the interest in structures of this class. The cyclopropane carboxylic acid motif is seen again in 1,8-naphthyridine (26), which is the subject of two 2007 process patent applications [196, 197]. A further compound, the 1,8-naphthyridine N-oxide (27), is the individual subject of a 2007 application relating to its dry powder formulation with lactose for use in an inhaler [198]. Polymorphic forms of this compound had been described in a previous 2005 application [199].

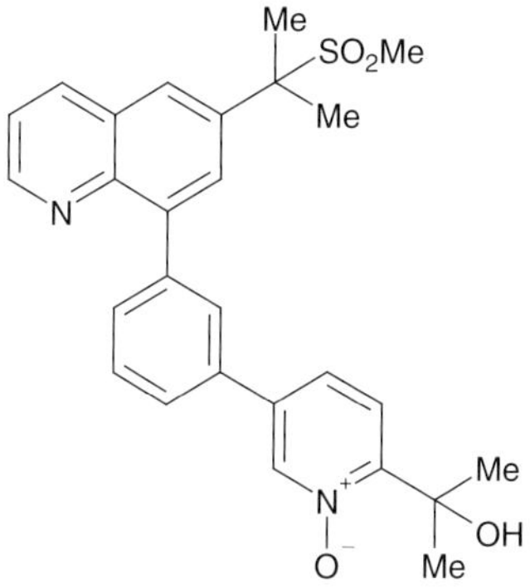

(23)

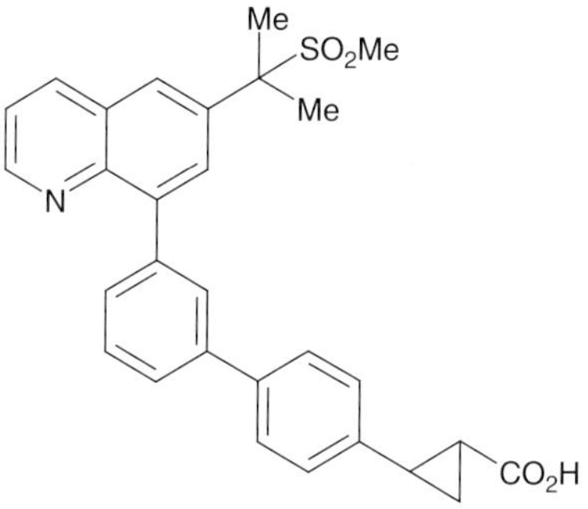

(24)

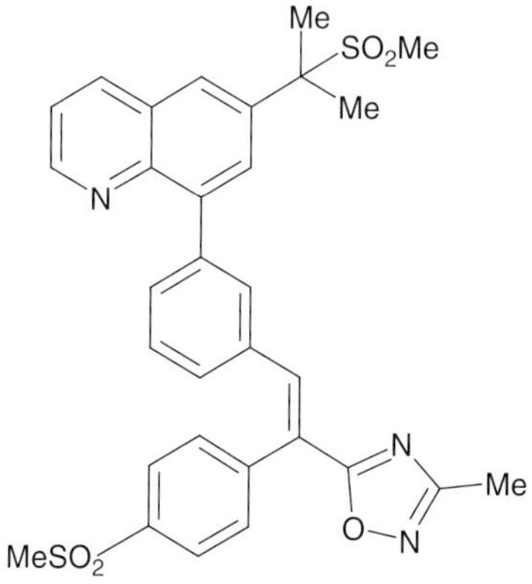

(25) L-454560

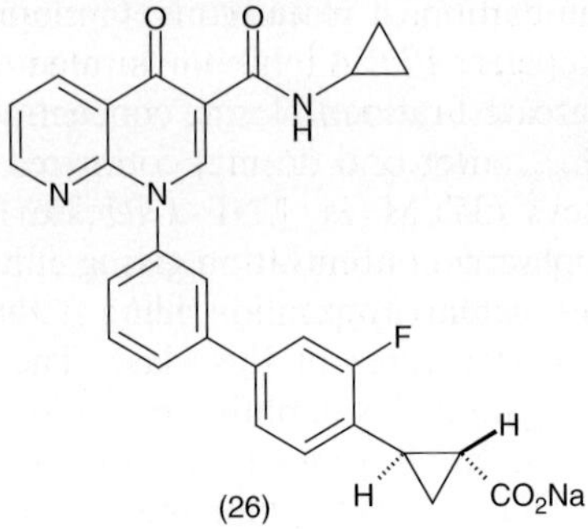

Schering-Plough have published a number of papers describing the chemistry and pharmacology of SCH35191 (21) [184–186], which was discontinued after Phase I clinical trials (*vide supra*). Issues with this compound included the results from a rising dose study in monkeys, in which it was shown to cause vasculopathy [200], and (in a curious reversal of the metabolism seen in roflumilast) the metabolic conversion of the N-oxide to the more potent pyridine compound (28). In order to address the pharmacokinetics issue, bioisosteres for the amido-pyridine group were sought. A recent publication [201] describes how this search led to the discovery of oxazole-based compounds as a new class of PDE4 inhibitors. This paper describes the medicinal chemistry around these compounds, in particular the use of structure-based design to optimise the interactions in the highly polar region of the PDE4 active site, where the metal ions Zn^+ and Mg^{2+}, conserved water molecules and polar residues are located. The optimised compound thus obtained is oxazole (29), $IC_{50} =$ 19 nM (PDE4B). Subtype specificity over PDE4D was not found in this series, which also tends to show some activity at PDE10 and PDE11. Due to favourable PK ($C_{max} = 9.5\,\mu M$ at 10 mg/kg p.o.) compound (29) was taken forward into a rat LPS-induced pulmonary inflammation model, in which it showed an 64% inhibition at 3 mg/kg. Encouragingly no emetic effect was observed in the cynomolgous monkey after oral dosing at 30 mg/kg ($C_{max} = 13\,\mu M$).

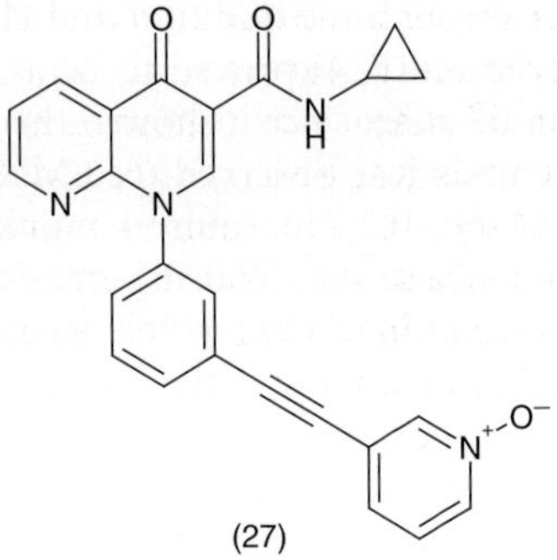

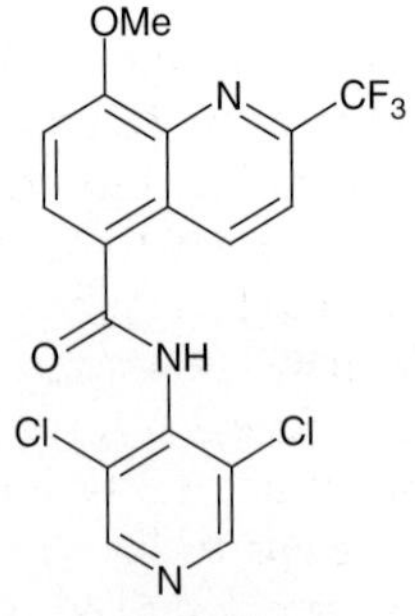

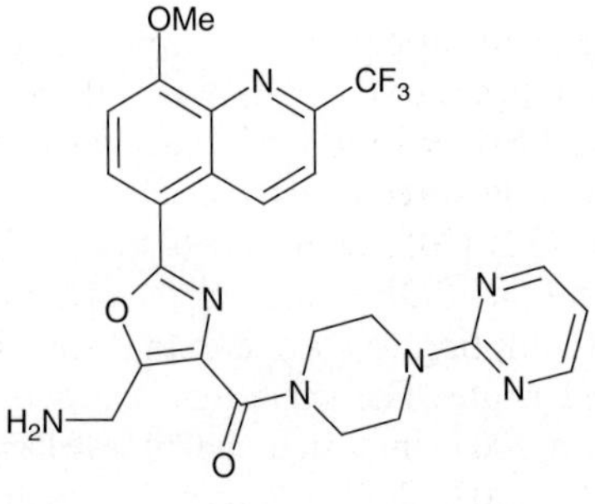

Other recent patenting activity. The continued interest of pharmaceutical and academic researchers into this target is readily demonstrated by the number of patent applications submitted in recent years. Applications of interest include three from AstraZeneca which contain pyridopyrimidines such as (30) [202, 203] and diazapinones such as (31) [204] (PDE4B2 IC_{50} = 0.33 nM). These are the first patent application filings from this group and it will be interesting to follow the progress of the compounds, and any rationale for their design.

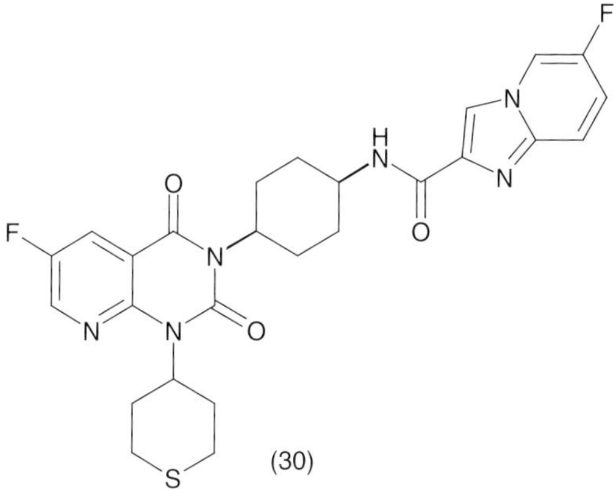

(30)

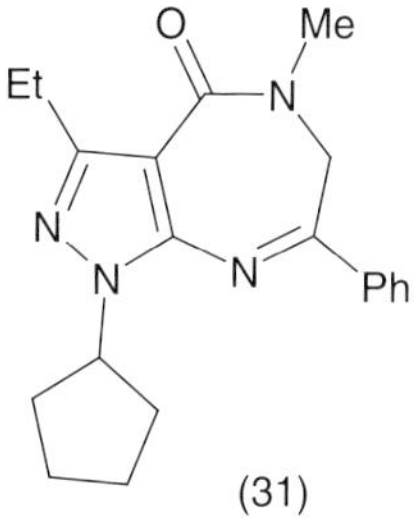

(31)

Despite its apparent ongoing interest in cilomilast and GSK256066, GSK still appears to have an active research programme in PDE4 inhibition. GSK have filed 19 PDE4 inhibitor applications worldwide since the start of 2006, including three to date in 2008. Of these most recent three, the drugs described, (32) [205, 206] and (33) are claimed to have very high potency (picomolar) at PDE4 and have comparatively high molecular weights, making it tempting to speculate that they have been designed for delivery by the inhaled route. The same pyrazolopyridine core structural type is seen in filings from 2007, in which the claimed emphasis seems to be on selectivity for PDE4B [207, 208]; compounds such as (34) are described as having

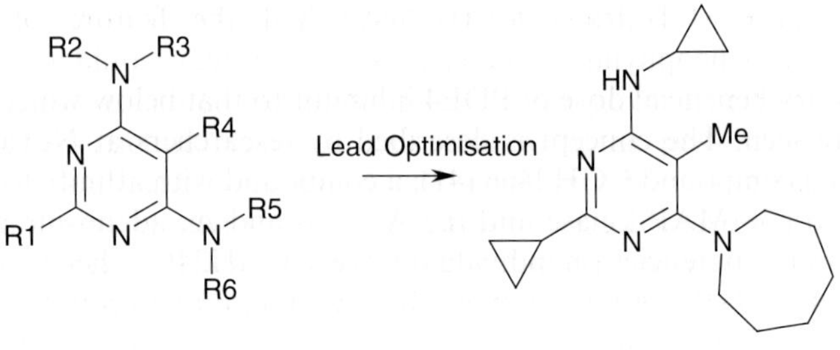

(42)

Lead Optimisation →

(43) UCB-101333-3

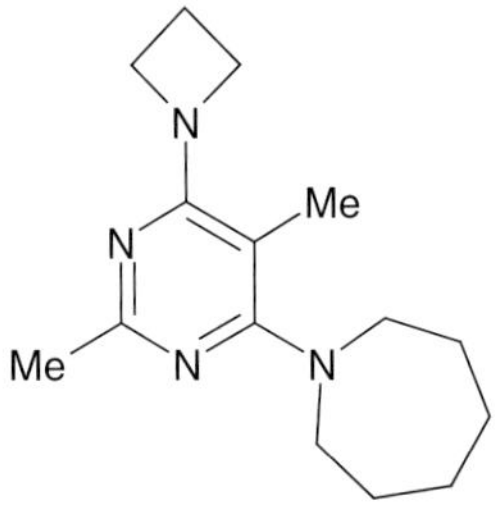

(44)

Thus, while PDE4 inhibition has been a target for many years now, researchers continue to present more hypotheses to be tested by the rational drug design that will bring us to our goal of a therapeutically beneficial drug. We are confident that pharmacological intervention at this pleiotropic moderator will be key to the positive treatment of many different disease processes, and we look forward to further developments in the field ultimately leading to the first safe, efficacious PDE4 inhibitor to be available to patients.

REFERENCES

[1] Beavo, J.A. and Brunton, L.L. (2002) *Nat. Rev. Mol. Cell Biol.* **3**, 710–718.
[2] Conti, M., Richter, W., Mehats, C., Livera, G., Park, J.-Y. and Jin, C. (2003) *J. Biol. Chem.* **278**, 5493.
[3] Bingham, J., Sudarsanam, S. and Srinivasan, S. (2006) *Biochem. Biophys. Res. Commun.* **350**, 25–32.
[4] Zhang, K.Y.J., Ibrahim, P.N., Gillette, S. and Bollag, G. (2005) *Expert Opin. Ther. Targets* **9**, 1283–1305.
[5] Wright, L.C., Seybold, J., Robichaud, A., Adcock, I. and Barnes, P. (1998) *Am. J. Physiol.* **275**, L694–L700.

[6] Wang, P., Wu, P., Ohleth, K.M., Egan, R.W. and Billah, M.M. (1999) *Mol. Pharmacol.* **56**, 170–174.
[7] Tilley, D.G. and Maurice, D.H. (2005) *Mol. Pharmacol.* **68**, 596–605.
[8] Pryzwansky, K.B. and Madden, V.J. (2003) *Cell Tissue Res.* **312**, 301–311.
[9] Peter, D., Jin, S.L., Conti, M, Hatzelmann, A. and Zitt, C. (2007) *J. Immunol.* **178**, 4820–4831.
[10] Peachell, P.T., Undem, B.J., Schleimer, R.P, MacGlashan, D., Lichtenstein, L., Cieslinski, L.B. and Torphy, T.J. (1992) *J. Immunol.* **148**, 2503–2510.
[11] Netherton, S.J., Sutton, J.A., Wilson, L.S, Carter, R.L. and Maurice, D.H. (2007) *Circ. Res.* **101**, 768–776.
[12] Netherton, S.J. and Maurice, D.H. (2005) *Mol. Pharmacol.* **67**, 263–272.
[13] Millen, J., MacLean, M.R. and Houslay, M. (2006) *Eur. J. Cell Biol.* **85**, 679–691.
[14] Maurice, D.H., Palmer, D., Tilley, D.G., Dunkerley, H.A., Netherton, S.J., Raymond, D.R., Elbatarny, H.S. and Jimmo, S.L. (2003) *Mol. Pharmacol.* **64**, 533–546.
[15] Liu, H. and Maurice, D.H. (1999) *J. Biol. Chem.* **274**, 10557–10565.
[16] LeJeune, I.R., Shepherd, M., Van Heeke, G., Houslay, M.D. and Hall, I.P. (2002) *J. Biol. Chem.* **277**, 35980–53989.
[17] Landells, L.J., Szilagy, C.M., Jones, N.A., Banner, K.H., Allen, J.M., Doherty, A., O'Connor, B.J., Spina, D. and Page, C.P. (2001) *Br. J. Pharmacol.* **133**, 722–729.
[18] Ito, M., Nishikawa, M., Fujioka, M., Miyahara, M., Isaka, N., Shiku, H. and Nakano, T. (1996) *Cell. Signal.* **8**, 575–581.
[19] Giembycz, M.A., Corrigan, C.J., Seybold, J., Newton, R. and Barnes, P.J. (1996) *Br. J. Pharmacol.* **118**, 1945–1958.
[20] Gantner, F., Gotz, C., Gekeler, V, Schudt, C., Wendel, A. and Hatzelmann, A. (1998) *Br. J. Pharmacol.* **123**, 1031–1038.
[21] Fuhrmann, M., Jahn, H.U., Seybold, J., Neurohr, C., Barnes, P.J., Hippenstiel, S., Kraemer, H.J. and Suttorp, N. (1999) *Am. J. Respir. Cell Mol. Biol.* **20**, 292–302.
[22] Campos-Toimil, M., Keravis, T., Orallo, F, Takeda, K. and Lugnier, C. (2008) *Br. J. Pharmacol.* **154**, 82–92.
[23] Billington, C.K., LeJeune, I.R., Young, K.W and Hall, I.P. (2008) *Am. J. Respir. Cell Mol. Biol.* **38**, 1–7.
[24] Baroja, M.L., Cieslinski, L.B., Torphy, T.J., Wange, R.L., Madrenas, J., Gantner, F., et al. (1999) *J. Immunol.* **162**, 2016–2023.
[25] Barber, R., Baillie, G.S., Bergmann, R, Shepherd, M.C., Sepper, R., Houslay, M.D. and Heeke, G.V. (2004) *Am. J. Physiol. Lung Cell Mol. Physiol.* **287**, L332–L343.
[26] Watt, A.P., Schock, B.C. and Ennis, M. (2005) *Curr. Drug Targets Inflamm. Allergy* **4**, 415–423.
[27] Hatzelmann, A. and Schudt, C. (2001) *J. Pharmacol. Exp. Ther.* **297**, 267–279.
[28] Tenor, H., Hatzelmann, A., Church, M.K, Schudt, C. and Shute, J.K. (1996) *Br. J. Pharmacol.* **118**, 1727–1735.
[29] Kita, H., Abu-Ghazleh, R.I., Gleich, G.J. and Abraham, R.T. (1991) *J. Immunol.* **146**, 2712–2718.
[30] Berends, C., Dijkhuizen, B., de Monchy, J.G, Dubois, A.E., Gerritsen, J. and Kauffman, H.F. (1997) *Eur. Respir. J.* **10**, 1000–1007.
[31] Parkkonen, J., Hasala, H., Moilanen, E., Giembycz, M.A. and Kankaanranta, H. (2008) *Pulm. Pharmacol. Ther.* **21**, 499–506.
[32] Tirouvanziam, R. (2006) *Drug News Perspect.* **19**, 609–614.

[33] Trevethick, M., Philip, J., Chaffe, P, Sladen, L., Cooper, C., Nagendra, R., Douglas, G., Clarke, N., Perros-Huguet, C. and Yeadon, M. (2007) *ATS Abstr.* **A927**,

[34] Trevethick, M., Banner, K.H., Ballard, S, Barnard, A., Lewis, A., Browne, J., et al. (2007) *ATS Abstr.* **A116**,

[35] Jones, N.A., Boswell-Smith, V., Lever, R and Page, C.P. (2005) *Pulm. Pharmacol. Ther.* **18**, 93–101.

[36] Derian, C.K., Santulli, R.J., Rao, P.E, Solomon, H.F. and Barrett, J.A. (1995) *J. Immunol.* **154**, 308–317.

[37] Lee, T.H. and Lane, S.J. (1992) *Am. Rev. Respir. Dis.* **145**, S27–S30.

[38] Tenor, H., Hatzelmann, A., Kupferschmidt, R, Stanciu, L., Djukanoviv, R., Schudt, C., Wendel, A., Church, M.K. and Shute, J.K. (1995) *Clin. Exp. Allergy* **25**, 625–663.

[39] Manning, C.D., Burman, M., Christensen, S.B., Cieslinski, L.B., Essayan, D.M., Grous, M., Torphy, T.J. and Barnette, M.S. (1999) *Br. J. Pharmacol.* **128**, 1393–1398.

[40] Molnar-Kimber, K., Yonno, L., Heaslip, R and Weichman, B. (1993) *Agents Actions* **39**, C77–C79.

[41] Jin, S.L. and Conti, M. (2002) *Proc. Natl. Acad. Sci. USA* **99**, 7628–7633.

[42] Schudt, C., Tenor, H., Loos, U., Mallmann, P., Szamel, M. and Resch, K. (1993) *Eur. Respir. J.* **6**, 367S.

[43] Dolhain, R.J., van der Heiden, A.N., ter Haar, N.T., Breedveld, F.C. and Miltenburg, A.M. (1996) *Arthritis Rheum.* **39**, 1961–1969.

[44] Neurath, M.F., Finotto, S. and Glimcher, L.H. (2002) *Nat. Med.* **8**, 567–573.

[45] Bielekova, B., Lincoln, A., McFarland, H. and Martin, R. (2000) *J. Immunol.* **164**, 1117–1124.

[46] Caramori, G., Gronberg, D., Ito, K, Casolari, P., Adcock, I.M. and Papi, A. (2008) *J. Occup. Med. Toxicol.* **3**, S6.

[47] Essayan, D.M., Kagey-Sobotka, A., Lichtenstein, L.M. and Huang, S.R. (1997) *J. Pharmacol. Exp. Ther.* **282**, 505–512.

[48] Claveau, D., Chen, S.L., O'Keefe, S., Styhler, A., Liu, S., Huang, Z., Nicholson, D.W. and Mancini, J.A. (2004) *J. Pharmacol. Exp. Ther.* **310**, 752–760.

[49] Lerner, A., Kim, D.H. and Lee, R. (2000) *Leuk. Lymphoma* **37**, 39–51.

[50] Liu, S., Veilleux, A., Zhang, L, Young, A., Kwok, E., Laliberté, F., et al. (2005) *J. Pharmacol. Exp. Ther.* **314**, 846–854.

[51] Haddad, J.J., Land, S.C., Tarnow-Mordi, W.O., Zembala, M., Kowalczyk, D. and Lauterbach, R. (2002) *J. Pharmacol. Exp. Ther.* **300**, 559–566.

[52] Kongsuphol, P., Hieke, B., Mehta, A., Treharne, K.J., Viollet, B., Jiraporn, O., Schrieber, R. and Kunzelmann, K. (2008) European cystic fibrosis society meeting, April.

[53] Suttorp, N., Weber, U., Welsch, T and Schudt, C. (1993) *J. Clin. Invest.* **91**, 1421–1428.

[54] Blease, K., Burke-Gaffney, A. and Hellewell, P.G. (1998) *Br. J. Pharmacol.* **124**, 229–237.

[55] Foreman, J.C. (1989) *In* "Textbook of Immunopharmacology". Dale, M. and Foreman, J.C. (eds), pp. 19–37. Blackwell, Oxford.

[56] Weston, M.C., Anderson, N. and Peachell, P.T. (1997) *Br. J. Pharmacol.* **121**, 287–295.

[57] Eskandari, N., Wickramasinghe, T. and Peachell, P.T. (2004) *Br. J. Pharmacol.* **142**, 1265–1272.

[58] Banner, K.H., Moriggi, E., Da Ros, B, Schioppacassi, G., Semeraro, C. and Page, C.P. (1996) *Br. J. Pharmacol.* **119**, 1255–1261.

[59] Burgess, J.K., Oliver, B.G., Poniris, M.H, Ge, Q., Boustany, S., Cox, N., Moir, L.M., Johnson, P.R. and Black, J.L. (2006) *J. Allergy Clin. Immunol.* **118**, 649–657.

[60] Naline, E., Qian, Y., Advenier, C., Raeburn, D. and Karlsson, J.A. (1996) *Br. J. Pharmacol.* **118**, 1939–1944.
[61] Schimdt, D.T., Watson, N., Dent, G, Rühlmann, E., Branscheid, D., Magnussen, H. and Rabe, K.F. (2000) *Br. J. Pharmacol.* **131**, 1607–1618.
[62] Runo, J.R. and Loyd, J.E. (2003) *Lancet* **361**, 1533–1544.
[63] Growcott, E.J., Spink, K.G., Ren, X, Afzal, S., Banner, K.H. and Wharton, J. (2006) *Respir. Res.* **7**, 9.
[64] Owens, G.K., Kumar, M.S. and Wamhoff, B.R. (2004) *Physiol. Rev.* **84**, 767–801.
[65] Phillips, P.G., Long, L., Wilkins, M.R and Morrell, N.W. (2005) *Am. J. Physiol.* **288**, L103–L115.
[66] Houslay, M.D. (2005) *Mol. Pharmacol.* **68**, 563–567.
[67] Dunkern, T.R., Feurstein, D., Rossi, G.A., Sabatini, F. and Hatzelmann, A. (2007) *Eur. J. Pharmacol.* **572**, 12–22.
[68] Martin-Chouly, C.A., Astier, A., Jacob, C, Pruniaux, M.P., Bertrand, C. and Lagente, V (2004) *Life Sci.* **75**, 823–840.
[69] Kohyama, T., Liu, X., Wen, F.Q., Wang, H., Kim, H.J., Takizawa, H., Cieslinski, L.B., Barnette, M.S. and Rennard, S.I. (2002) *Am. J. Respir. Cell Mol. Biol.* **26**, 694–701.
[70] Banner, K.H., Marchini, F., Buschi, A, Moriggi, E., Semararo, C. and Page, C.P. (1995) *Pulm. Pharmacol.* **8**, 37–42.
[71] Banner, K.H. and Page, C.P. (1995) *Br. J. Pharmacol.* **114**, 93–98.
[72] Raeburn, D., Underwood, S.L., Lewis, S.A, Battram, C.H., Tomkinson, A., Sharma, S., Jordan, R., Souness, J.E. and Webber, S.E. (1994) *Br. J. Pharmacol.* **113**, 1423–1431.
[73] Deng, Y.M., Xie, Q.M., Tang, H.F, Sun, J.G., Deng, J.F., Chen, J.Q. and Yang, S.Y. (2006) *Eur. J. Pharmacol.* **547**, 125–135.
[74] Kuss, H., Hoefgen, N., Johanssen, S., Kronbach, T. and Rundfeldt, C. (2003) *J. Pharmacol. Exp. Ther.* **307**(1), 373–385.
[75] Chapman, R.W., House, A., Richard, J, Celly, C., Prelusky, D., Ting, P., Hunter, J.C., Lamca, J. and Phillips, J.E. (2007) *Eur. J. Pharmacol.* **571**, 215–221.
[76] Tang, H.F., Chen, J.Q., Xie, Q.M, Zheng, X.Y., Zhu, Y.L., Adcock, I. and Wang, X. (2006) *Biochim. Biophys. Acta* **1762**, 525–532.
[77] Kumar, R.K., Herbert, C., Thomas, P.S, Beume, R., Yang, M., Webb, D.C. and Foster, P.S. (2003) *J. Pharmacol. Exp. Ther.* **307**, 349–355.
[78] Hansen, G., Jin, S., Umetsu, D.T and Conti, M. (2000) *Proc. Natl. Acad. Sci. USA* **97**, 6751–6756.
[79] Leclerc, O., Lagente, V., Planquois, J.-M., Berthelier, C., Artola, M., Eichholtz, T., Bertrand, C.P. and Schmidlin, F. (2006) *Eur. Respir. J.* **27**, 1102–1109.
[80] Martorana, P.A., Beume, R., Lucattelli, M, Wollin, L. and Lungarella, G. (2005) *Am. J. Respir. Crit. Care Med.* **172**, 848–853.
[81] Mori, H., Nose, T., Ishitani, K, Kasagi, S., Souma, S., Akiyoshi, T., et al. (2008) *Am. J. Physiol. Lung Cell. Mol. Physiol.* **294**, L196–L204.
[82] Ariga, M., Neitzert, B., Nakae, S., Mottin, G., Bertrand, C., Pruniaux, M.P., Jin, S.-L.C. and Conti, M. (2004) *J. Immunol.* **173**, 7531–7538.
[83] Fujimura, M. and Liu, Q. (2007) *Pulm. Pharmacol. Ther.* **20**, 543–548.
[84] Hanjing, L., Qui, Z., Wei, W, Yu, L., Liu, R. and Zhang, M. (2004) *Chin. Med. J.* **117**, 1620–1624.
[85] Moore, C.S., Earl, N., Frenette, R, Styhler, A., Mancini, J.A., Nicholson, D.W., Hebb, A.L., Owens, T. and Robertson, G.S. (2006) *J. Pharmacol. Exp. Ther.* **319**, 63–72.

[86] Sommer, N., Martin, R., McFarland, H.F., Cannella, B., Raine, C.S., Scott, D.E., Löschmann, P.A. and Racke, M.K. (1997) *J. Neuroimmunol.* **79**, 54–61.
[87] Banner, K.H. and Trevethick, M.A. (2004) *Trends Pharmacol. Sci.* **25**, 430–436.
[88] Videla, S., Vilaseca, J., Medina, C, Guarner, F., Salas, A. and Malagelada, J.R. (2006) *J. Pharmacol. Exp. Ther.* **316**, 940–945.
[89] Kobayashi, K., Suda, T., Manabe, H. and Miki, I. (2007) *Mediators Inflamm.* **2007**, Article ID: 58901 (http://www.hindawi.com/journals/mi/contents.html).
[90] Blokland, A., Schreiber, R. and Prickaerts, J. (2006) *Curr. Pharm. Des.* **12**, 2511–2523.
[91] Barad, M., Bourtchouladze, R., Winder, D.G, Golan, H. and Kandel, E. (1998) *Proc. Natl. Acad. Sci. USA* **95**, 15020–15025.
[92] Imanishi, T., Sawa, A., Ichimaru, Y, Kato, S., Yamamoto, T. and Ueki, S. (1997) *Eur. J. Pharmacol.* **321**, 273–278.
[93] Mizokawa, T., Kimura, K., Ikoma, Y, Hara, K., Oshino, N., Yamamoto, T. and Ueki, S. (1988) *Jpn. J. Pharmacol.* **48**, 357–364.
[94] O'Donnell, J.M. (1993) *J. Pharmacol. Exp. Ther.* **264**, 1168–1178.
[95] O'Donnell, J.M. and Zhang, H.T. (2004) *Trends Pharmacol. Sci.* **25**, 158–163.
[96] Overstreet, D., Double, K. and Schiller, G.D. (1989) *Pharmacol. Biochem. Behav.* **34**, 691–696.
[97] Saccomano, N.A., Vinick, F.J., Koe, B, Nielsen, J.A., Whalen, W.M., Meltz, M., et al. (1991) *J. Med. Chem.* **34**, 291–298.
[98] Wachtel, H. (1983) *Neuropharmacology* **22**, 267–272.
[99] Zhang, H.T., Huang, Y., Jin, S.L, Frith, S.A., Suvarna, N., Conti, M. and O'Donnell, J.M. (2002) *Neuropsychopharmacology* **27**, 587–595.
[100] De Franceschi, L., Platt, O.S., Malpeli, G, Janin, A., Scarpa, A., Leboeuf, C., Beuzard, Y., Payen, E. and Brugnara, C. (2008) *FASEB J.* **22**, 1849–1860.
[101] Dousa, T.P. (1999) *Kidney Int.* **55**, 29–62.
[102] Cheng, J. and Grande, J.P. (2007) *Exp. Biol. Med.* **232**, 38–51.
[103] Begany, D.P., Carcillo, J.A., Herzer, W.A., Mi, Z. and Jackson, E.K. (1996) *J. Pharmacol. Exp. Ther.* **278**, 37–41.
[104] Carcillo, J.A., Herzer, W.A., Mi, Z, Thomas, N.J. and Jackson, E.K. (1996) *Pharmacol. Exp. Ther.* **279**, 1197–1204.
[105] Mehats, C., Oger, S. and Leroy, M.J. (2004) *Eur. J. Obstet. Gynecol. Reprod. Biol.* **117**, S15–S17.
[106] Millar, J.K., Mackie, S., Clapcote, S.J, Murdoch, H., Pickard, B.S., Christie, S., et al. (2007) *J. Physiol.* **584**, 401–405.
[107] Wang, D. and Wang, T. (2005) *Curr. Opin. Investig. Drugs* **6**, 283–288.
[108] Baumer, W., Hoppmann, J., Runfeldt, C and Kietzmann, M. (2007) *Inflamm. Allergy Drug Targets* **6**, 17–26.
[109] Collins, J.J., Elwell, M.R., Lamb, J.C., Manus, A.G., Heath, J.E. and Makovec, G.T. (1988) *Fundam. Appl. Toxicol.* **11**, 472–484.
[110] Nyska, A., Herbert, R.A., Chan, P.C, Haseman, J.K. and Hailey, J.R. (1998) *Arch. Toxicol. Sci.* **72**, 731–737.
[111] Bishop, S.P. (1989) *Toxicol. Pathol.* **17**, 109–117.
[112] Ruben, Z., Deslex, P., Nash, G, Redmond, N.I., Poncet, M. and Dodd, D.C. (1989) *Toxicol. Pathol.* **17**, 145–152.
[113] Bian, H., Zhang, J., Wu, P., Varty, L.A., Jia, Y., Mayhood, T., Hey, J.A. and Wang, P. (2004) *Biochem. Pharmacol.* **68**, 2229–2236.

[114] Dietsch, G.N., Dipalma, C.R., Eyre, R.J, Eyre, R.J., Pham, T.Q., Poole, K.M., et al. (2006) *Toxicol. Pathol.* **34**, 39–51.
[115] Dagues, N., Pawlowski, V., Sobry, C., Hanton, G., Borde, F., Soler, S., Freslon, J.-L. and Chevalier, S. (2007) *Toxicol. Sci.* **100**, 238–247.
[116] Boswell-Smith, V., Cazzola, M. and Page, C.P. (2006) *J. Allergy Clin. Immunol.* **117**, 1237–1243.
[117] Effect of ADC4022 Co-Administered with Budesonide on Pulmonary Inflammation in Subjects with Moderate to Severe COPD. (2008) ClinicalTrials.gov identifier: NCT00634413. http://clinicaltrials.gov/ct2/show/NCT00634413
[118] Gibson, L.C., Hastings, S.F., McPhee, I., Clayton, R.A., Darroch, C.E., Mackenzie, A., Mackenzie, F.L., Nagasawa, M., Stevens, P.A. and Mackenzie, S.J. (2006) *Eur. J. Pharmacol.* **538**(1–3), 39–42.
[119] Souness, J.E., Villamil, M.E., Scott, L.C., Tomkinson, A., Giembycz, M.A. and Raeburn, D. (1994) *Br. J. Pharmacol.* **111**(4), 1081–1088.
[120] Kroegel, C. and Foerster, M. (2007) *Expert Opin. Investig. Drugs* **16**(1), 109–124.
[121] Souness, J.E. and Rao, S. (1997) *Cell. Signal.* **9**, 227.
[122] Zhao, Y., Zhang, H.T. and O'Donnell, J.M. (2003) *J. Pharmacol. Exp. Ther.* **305**(2), 565–572.
[123] Christensen, S.B., Guider, A., Forster, C.J., Gleason, J.G., Bender, P.E., Karpinski, J.M., et al. (1998) *J. Med. Chem.* **41**, 821–835.
[124] Giembycz, M.A. (2001) *Expert Opin. Investig. Drugs* **10**, 1361–1379.
[125] Barnette, M.S., Christensen, S.B., Underwood, D.C. and Torphy, T.J. (1996) *Pharmacol. Rev. Commun.* **8**, 65–73.
[126] Barnette, M.S., Grous, M., Cieslinski, L.B., Burman, M., Christensen, S.B. and Torphy, T.J. (1995) *J. Pharmacol. Exp. Ther.* **273**, 1396–1402.
[127] Down, G., Siederer, S., Lim, S. and Daley-Yates, P. (2006) *Clin. Pharmacokinet.* **45**, 217–233.
[128] Giembycz, M.A. (2006) *Br. J. Clin. Pharmacol.* **62**, 138–152.
[129] Rennard, S.I., Schachter, N., Strek, M., Rickard, K. and Amit, O. (2006) *Chest* **129**, 56–66.
[130] Baeumer, W., Szelenyi, I. and Kietzmann, M. (2005) *Expert Rev. Clin. Immunol.* **1**, 27–36.
[131] Grootendorst, D.C., Gauw, S.A., Baan, R., Kelly, J., Murdoch, R.D., Sterk, P.J. and Rabe, K.F. (2003) *Pulm. Pharmacol. Ther.* **16**, 115–120.
[132] Gamble, E., Grootendorst, D.C., Brightling, C.E., Troy, S., Qiu, Y.g., Zhu, J., et al. (2003) *Am. J. Respir. Crit. Care Med.* **168**, 976–982.
[133] Michel, O., Dentener, M., Cataldo, D., Cantinieaux, B., Vertongen, F., Delvaux, C. and Murdoch, R.D. (2007) *Pulm. Pharmacol. Ther.* **20**, 676–683.
[134] Bardin, P.G. (2007) *Expert Rev. Clin. Immunol.* **3**, 469–476.
[135] McKenna, J.M. and Muller, G.W. (2007) *In* "Cyclic Nucleotide Phosphodiesterases in Health and Disease". pp. 667–699. CRC Press LLC, Boca Raton, FL.
[136] Boswell-Smith, V. and Spina, D. (2007) *Int. J. Chron. Obstruct. Pulm. Dis.* **2**, 121–129.
[137] Calverley, P.M., Sanchez-Toril, F., McIvor, R.A., Teichmann, P., Bredenbroeker, D. and Fabbri, L.M. (2006) *Proc. Am. Thorac. Soc.* **3**(Abstr. Issue), A725.
[138] Rabe, F., O'Donnell, D., Muir, F., Jenkins, C., Witte, S., Bredenbroeker, D. and Bethke, D. (2004) *Eur. Respir. J.* **24**(Suppl. 48), Abstr. 267.
[139] O'Donnell, D., Muir, J.S. and Jenkins, C. (2004) *Am. J. Respir. Crit. Care Med.* **169**(Suppl. 7), A602.

[140] Bundschuh, D.S., Barsig, J., Beume, R., Eltze, M., Schudt, C., Wollin, L. and Hatzelmann, A. (2001) *Am. J. Respir. Crit. Care Med.* **163**(Suppl. 5), A431.

[141] Bethke, T.D., Boehmer, G.M., Hermann, R., Hauns, B., Fux, R., Moerike, K., David, M., Knoerzer, D., Wurst, W. and Gleiter, C.H. (2007) *J. Clin. Pharmacol.* **47**, 26–36.

[142] Feagan, B.G. (2007) *Am. J. Gastroenterol.* **102**(S1), S7–S13.

[143] O'Mahony, S. (2005) *IDrugs* **8**, 502–507.

[144] McIntyre, J.A., Castaner, J. and Castaner, R.M. (2004) *Drugs Future* **29**(10), 1003–1006.

[145] Schreiber, S., Keshavarzian, A., Isaacs, K.L., Schollenberger, J., Guzman, J.P., Orlandi, C. and Hanauer, S.B. (2007) *Gastroenterology* **132**, 76–86.

[146] Keshavarzian, A., Mutlu, E., Guzman, J.P., Forsyth, C. and Banan, A. (2007) *Expert Opin. Investig. Drugs* **16**, 1489–1506.

[147] Gullapalli, S., Karande, V., Amrutkar, D., Offner, H. and Narayanan, S. (2005) GRC3886 – A selective PDE4 inhibitor with potential effect in rheumatoid arthritis, 7th World Congr Inflamm, Abstr. 5022, August 20–24, Melbourne.

[148] Gutke, H.-J., Guse, J.-H., Khobzaoui, M., Renukappa-Gutke, T. and Burnet, M. (2005) *Curr. Opin. Investig. Drugs* **6**, 1149–1158.

[149] Trevethick, M., Banner, K., Ballard, S., Barnard, A., Lewis, A., Browne, J., et al. (2007) *Am. J. Respir. Crit. Care Med.* **175**, A116.

[150] Phillips, P., Bennetts, M., Banner, K., Ward, J., Wessels, D. and Fuhr, R. (2007) *Eur. Respir. J.* **30**, 491s Abstr. 2964.

[151] Vestbo, J., Tan, L. and Atkinson, G. (2007) *Eur. Respir. J.* **30**, Abstr. P3598.

[152] Bunnage, M.E. and Mathias, J.P. (2004) *PCT Int. Appl.*, WO 2005009965.

[153] Mathias, J.P. (2005) *PCT Int. Appl.*, US 2005026952.

[154] Berchielli, A., Daugherity, P.D., Shamblin, S.L., Thombre, A.G. and Waterman, K.C. (2004) *PCT Int. Appl.*, WO 2004004684.

[155] Murtagh, L.M., Taylor, S.C.J. and Willis, N.J. (2006) *PCT Int. Appl.*, WO 2006077497.

[156] Rose, G.M., Hopper, A., De Vivo, M. and Tehim, A. (2005) *Curr. Pharm. Des.* **11**, 3329–3334.

[157] Memory Pharmaceuticals Corp, Memory to focus on key R&D programs and alliances. (2008) Web site: http://www.memorypharma.com/

[158] Hallam, T.M. (2007) *PCT Int. Appl.*, WO 2007137181.

[159] MacDonald, E., Van der Lee, H., Pocock, D., Cole, C., Thomas, N., VandenBerg, P.M., Bourtchouladze, R. and Kleim, J.A. (2007) *Neurorehabil. Neural Repair.* **21**, 486–496.

[160] Bourtchouladze, R., Lidge, R., Catapano, R., Stanley, J., Gossweiler, S., Romashko, D., Scott, R. and Tully, T. (2003) *Proc. Natl. Acad. Sci. USA* **100**, 10518–10522.

[161] Boot, J.D., De Haas, S.L., Van Gerven, J.M.A., De Smet, M., Leathem, T., Wagner, J., et al. (2008) *Pulm. Pharmacol. Ther.* **21**, 573–577.

[162] Cook, D.C., Jones, R.H., Kabir, H., Lythgoe, D.J., McFarlane, I.M., Pemberton, C., Thatcher, A.A., Thompson, D.M. and Walton, J.B. (1998) *Org. Process Res. Dev.* **2**, 157–168.

[163] Boichot, E., Germain, N., Lugnier, C., Lagente, V. and Bourguignon, J.J. (1998) *Am. J. Respir. Crit. Care Med.* **157**, A141.

[164] Beeh, K.M., Beier, J., Schulz, A.K., Lerch, C. and Buhl, R. (2003) *Eur. Respir. J.* **22**, Abstr. P740.

[165] Raeburn, D., Underwood, S.L., Lewis, S.A., Woodman, V.R., Battram, C.H., Tomkinson, A., Sharma, S., Jordan, R., Souness, J.E., Webber, S.E. and Karlsson, J.-A. (1994) *Br. J. Pharmacol.* **113**, 1423–1431.

[166] Aggarwal, V.K., Bae, I., Lee, H.-Y., Richardson, J. and Williams, D.T. (2003) *Angew. Chem. Int. Ed.* **42**, 3274–3278.
[167] Lynch, J.E., Choi, W.-B., Churchill, H.R.O., Volante, R.P., Reamer, R.A. and Ball, R.G. (1997) *J. Org. Chem.* **62**, 9223–9228.
[168] Alexander, R.P., Warrellow, G.J., Eaton, M.A.W., Boyd, E.C., Head, J.C., Porter, J.R., et al. (2002) *Bioorg. Med. Chem. Lett.* **12**, 1451–1456.
[169] Nishikata, T., Yamamoto, Y. and Miyaura, N. (2007) *Tetrahedron Lett.* **48**, 4007–4010.
[170] Fregonese, L., Grootendorst, D.C., Gauw, S.A., Wei, C.G., Bennetts, M., Ward, J.K., Phillips, P.G. and Rabe, K.F. (2007) *Am. J. Respir. Crit. Care Med.* **175**, A486.
[171] Danto, S., Wei, G.C. and Gill, J. (2007) *Am. J. Respir. Crit. Care Med.* **175**(Abstr. Issue), A131.
[172] Watson, J.W. (2007) *J. Med. Chem.* **50**, 344–349.
[173] Urban, F.J. (2001) *Org. Process Res. Dev.* **5**, 575–580.
[174] Molostvov, G., Morris, A., Rose, P., Basu, S. and Muller, G. (2004) *Br. J. Haematol.* **124**, 366–375.
[175] Gupta, R., Kumar, G. and Kumar, R.S. (2005) *Methods Find. Exp. Clin. Pharmacol.* **27**, 101–118.
[176] Ferrer, P., Xuan, T.D., Chanal, J., Lockhart, A., Bousquer, J. and Luria, X. (1997) *Am. J. Respir. Crit. Care Med.* **155**, A660.
[177] Sturton, R.G. (2000) *Am. J. Respir. Crit. Care Med.* **161**, A200.
[178] Grootendorst, D.C., Gauw, S.A., Verhoosel, R., van der Veen, H., van der Linden, A., Moesker, H., Hiemstra, P.S. and Rabe, K.F. (2002) *98th International Conference of the Am. Thorac. Soc.* (May 17–22, Atlanta), Abstr. 308.
[179] Moriuchi, H., Nakahara, T., Maruko, T., Sakamoto, K. and Ishii, K. (2003) *Eur. J. Pharmacol.* **470**, 57–64.
[180] Aoki, M., Yamamoto, S., Kobayashi, M., Ohga, K., Kanoh, H., Miyata, K., Honda, K. and Yamada, T. (2001) *J. Pharmacol. Exp. Ther.* **297**, 165–173.
[181] Aoki, M., Fukunaga, M., Kitagawa, M., Hayashi, K., Morokata, T., Ishikawa, G., Kubo, S. and Yamada, T. (2000) *J. Pharmacol. Exp. Ther.* **295**, 1149–1155.
[182] Aoki, M., Kobayashi, M., Ishikawa, J., Saita, Y., Terai, Y., Takayama, K., Miyata, K. and Yamada, T. (2000) *J. Pharmacol. Exp. Ther.* **295**, 255–260.
[183] Aoki, M., Fukunaga, M., Sugimoto, T., Hirano, Y., Kobayashi, M., Honda, K. and Yamada, T. (2001) *J. Pharmacol. Exp. Ther.* **298**, 1142–1149.
[184] Billah, M., Cooper, N., Cuss, F., Davenport, R.J., Dyke, H.J., Egan, R., et al. (2002) *Bioorg. Med. Chem. Lett.* **12**, 1621–1623.
[185] Billah, M., Buckley, G.M., Cooper, N., Dyke, H.J., Egan, R., Ganguly, A., et al. (2002) *Bioorg. Med. Chem. Lett.* **12**, 1617–1619.
[186] Billah, M.M., Cooper, N., Minnicozzi, M., Warneck, J., Wang, P., Hey, J.A., et al. (2002) *J. Pharmacol. Exp. Ther.* **302**, 127–137.
[187] Glenmark Pharmaceuticals Ltd. (2008) Glenmark starts phase I trial of GRC-4039. http://www.glenmarkpharma.com/media/pdf/releases/4039_in_Phase_I_Feb08_3.pdf
[188] Elbion Pipeline. (2008) http://www.elbion.com/pipeline.htm
[189] Sanofi-Aventis pipeline. (2008) http://en.sanofi-aventis.com/rd/portfolio/p_rd_portfolio_snc.asp
[190] Gallant, M., Chauret, N., Claveau, D., Day, S., Deschenes, D., Dube, D., et al. (2008) *Bioorg. Med. Chem. Lett.* **18**, 1407–1412.
[191] Robichaud, A., Stamatiou, P.B., Jin, S.-L., Catherine, L., Nicholas, M., Dwight, L., France, L.S., Huang, Z., Conti, M. and Chan, C.C. (2002) *J. Clin. Invest.* **110**, 1045–1052.

[192] Ma, D., Wu, P., Egan, R.W., Billah, M.M. and Wang, P. (1999) *Mol. Pharmacol.* **55**, 50–57.
[193] Jin, S.-L.C., Richter, W. and Conti, M. (2007). *In* "Cyclic Nucleotide Phosphodiesterases in Health and Disease". pp. 323–346. CRC Press LLC, Boca Raton, FL.
[194] Macdonald, D., Mastracchio, A., Perrier, H., Dube, D., Gallant, M., Lacombe, P., et al. (2005) *Bioorg. Med. Chem. Lett.* **15**, 5241–5246.
[195] Huang, Z., Dias, R., Jones, T., Liu, S., Styhler, A., Claveau, D., et al. (2007) *Biochem. Pharmacol.* **73**, 1971–1981.
[196] Dube, D., Gallant, M. and Lacombe, P. (2007) *PCT Int. Appl.*, WO 2007048225.
[197] Cameron, M. (2007) *PCT Int. Appl.*, WO 2007050576.
[198] Thibert, R., Meisner, D., Rossi, J. and Tanfara, H. (2007) *PCT Int. Appl.*, US 2007071692.
[199] Clas, S.-D., Naccache, R., Yu, H., Murry, J. and Variankaval, N. (2005) *PCT Int. Appl.*, US 2005105084.
[200] Losco, P.E., Evans, E.W., Barat, S.A., Blackshear, P.E., Reyderman, L., Fine, J.S., Bober, L.A., Anthes, J.C., Mirro, E.J. and Cuss, F.M. (2004) *Toxicol. Pathol.* **32**, 295–308.
[201] Kuang, R., Shue, H.J., Blythin, D.J., Shih, N.Y., Gu, D., Chen, X., et al. (2007) *Bioorg. Med. Chem. Lett.* **17**, 5150–5154.
[202] Austin, R., Bonnert, R., Hunt, F., Nikitidis, G., Sanganee, H., Sjoe, P. and Warner, D. (2007) *PCT Int. Appl.*, WO 2007108750.
[203] Lisius, A., Nikitidis, G. and Sjoe, P. (2007) *PCT Int. Appl.*, WO 2007004958.
[204] Henriksson, K., Lisius, A., Sjoe, P. and Storm, P. (2007) *PCT Int. Appl.*, WO 2007040435.
[205] Allen, D.G., Aston, N.M., Barnett, R.P., Chudasama, R.M., Day, C.J., Edlin, C.D., Kindon, L.J. and Trivedi, N. (2008) *PCT Int. Appl.*, WO 2008015416.
[206] Allen, D.G., Aston, N.M., Edlin, C.D. and Trivedi, N. (2008) *PCT Int. Appl.*, WO 2008015437.
[207] Edlin, C.D., Holman, S., Jones, P.S., Keeling, S., Lindvall, M., Kristian, M., Charlotte J. and Trivedi, N. (2007) *PCT Int. Appl.*, WO 2007036733.
[208] Christensen, S.B.IV, Holman, S., Keeling, S.E. and Sayani, A. P. (2007) *PCT Int. Appl.*, WO 2007036734.
[209] Hamblin, J.N., Angella, T.D.R., Ballantinea, S.P., Cooka, C.M., Coopera, A.W.J., Dawsona, J., et al. (2008) *Bioorg. Med. Chem. Lett.* **18**, 4237–4241.
[210] Aston, N.M., Robinson, J.E. and Trivedi, N. (2007) *PCT Int. Appl.*, WO 2007045861.
[211] Edlin, C.D. and Holman, S. (2007) *PCT Int. Appl.*, WO 2007107499.
[212] Eldred, C.D. and Robinson, J.E. (2006) *PCT Int. Appl.*, WO 2006097340.
[213] Wang, H., Peng, M.S., Chen, Y., Geng, J., Robinson, H., Houslay, M.D., Cai, J. and Ke, H. (2007) *Biochem. J.* **408**(2), 193–201.
[214] Zhang, K.Y.J. (2007) *In* "Cyclic Nucleotide Phosphodiesterases in Health and Disease". pp. 583–605. CRC Press LLC, Boca Raton, FL.
[215] Griffiths, C.E.M., Van Leent, E.J.M., Gilbert, M. and Traulsen, J. (2002) *Br. J. Dermatol.* **147**, 299–307.
[216] Hanifin, J.M., Chan, S.C., Cheng, J.B., Tofte, S.J., Henderson, W.R., Jr.., Kirby, D.S. and Weiner, E.S. (1996) *J. Invest. Dermatol.* **107**, 51–56.
[217] Trifilieff, A., Keller, T.H., Press, N.J., Howe, T., Gedeck, P., Beer, D. and Walker, C. (2005) *Br. J. Pharmacol.* **144**, 1002–1010.
[218] Provins, L., Christophe, B., Danhaive, P., Dulieu, J., Durieu, V., Gillard, M., Lebon, F., Lengele, S., Quere, L. and van Keulen, B. (2006) *Bioorg. Med. Chem. Lett.* **16**, 1834–1839.
[219] Provins, L., Christophe, B., Danhaive, P., Dulieu, J., Gillard, M., Quere, L. and Stebbins, K. (2007) *Bioorg. Med. Chem. Lett.* **17**, 3077–3080.

3 H^+/K^+ ATPase Inhibitors in the Treatment of Acid-Related Disorders

MARK BAMFORD

GlaxoSmithKline, Medicines Research Centre, Gunnels Wood Road, Stevenage, Herts SG1 2NY, UK

INTRODUCTION	76
THE IRREVERSIBLE PROTON PUMP INHIBITORS (PPIs)	78
Omeprazole (Losec)	82
Lansoprazole (AG-1479, Prevacid)	83
Pantoprazole (BY1023/SK&F96022)	84
Rabeprazole (Aciphex)	84
MODIFICATIONS TO THE PPI THEME	85
ACID SUPPRESSION – GENERAL CONSIDERATIONS	89
REVERSIBLE INHIBITION OF H^+/K^+ ATPASE	91
Imidazopyridines	91
Tricyclic Imidazopyridine Analogues	103
Alternatives to the Imidazopyridine Template	110
Benzimidazoles	113
Tricyclic Benzimidazole Analogues	117
Triazolopyridines	119
Pyrrolo[2,3-d]pyridazines	120
Pyrrolopyridines	121
4-Amino-Quinolines	126
Monocyclic Derivatives	135
Alternative Types of H^+/K^+ ATPase Inhibitors	142
THE BINDING SITE FOR POTASSIUM-COMPETITIVE INHIBITORS	148
CONCLUDING REMARKS	150
REFERENCES	151

INTRODUCTION

Acid-related diseases (ARD) are highly prevalent in the developed world [1]. They have a significant impact on patient quality of life and are a major burden on health care systems. One of these diseases, gastroesophageal reflux disease (GERD), is a chronic condition affecting approximately 20% of people on a weekly basis. Symptoms include heartburn, acid regurgitation, chest pain, epigastric pain, and respiratory conditions including chronic cough. Approximately 30% of GERD patients have erosion or ulcer formation, but symptoms can be just as severe in non-erosive (NERD) patients. Control of pH of the gastric contents has been shown to correlate with healing and symptom relief in both peptic ulcer and GERD [2]. Early treatments included antacids, which act to neutralise gastric acid, alginates (thought to provide a mechanical barrier to the effects of acid) and (non-selective) acetylcholine antagonists (e.g., dicyclomine). These were then superceded in the early 1980s by H_2 receptor antagonists (H_2RAs) such as cimetidine and ranitidine. By the late 1980s, the proton pump inhibitors (PPIs) began to emerge. PPIs have been widely reported [3] to afford more effective long-term symptom resolution, healing of erosion and ulcer, and prevention of relapse than H_2RAs. PPIs have also been shown to inhibit the urease activity, which protects *Helicobacter pylori* from acid and can contribute to the eradication of this pathogen in combination therapy with antibacterial agents.

Although PPIs are the current treatment of choice [3], many patients (up to 64%) [4] receiving GERD therapy in the primary care setting do not have fully controlled symptoms, and PPIs are often co-prescribed with other acid suppressive agents such as H_2RAs [3]. It is estimated that as many as 30% of GERD patients remain symptomatic on standard once daily PPI dosing, rising to as high as 55% in the functional heartburn category of NERD patients [5].

Gastric acid secretion is a complex process (Figure 3.1) [6]. Acid secretion is effected through the parietal cell, found in the stomach wall. There are more than a billion parietal cells in the human stomach which, when activated, are capable of achieving pH ~ 0.8 for significant periods of time. Activation of the parietal cell results in generation of the canalicular cavity, with trafficking of the H^+/K^+ ATPase proton pump from its resting site within cytoplasmic tubules to microvilli of the canalicular membrane. Acid secretion by the H^+/K^+ ATPase pump is induced synergistically by three key mediators, gastrin, acetylcholine and histamine, through interaction with their respective receptors in the parietal cell. Approaches to inhibition of each of these have been actively investigated with the aim of inhibiting the activation of the parietal cell acid secretion. In addition, induction of acid

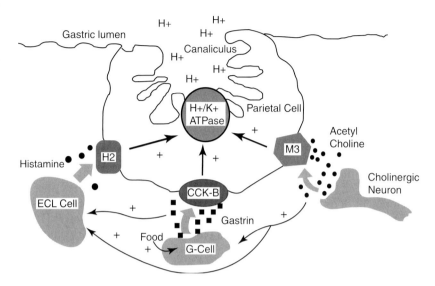

Fig. 3.1 Gastric acid secretion by the H^+/K^+ ATPase in the parietal cell is driven by acetylcholine, gastrin and histamine.

suppressive mechanisms, mainly somatostatin and prostaglandins, have been investigated [7]. It is the gastric H^+/K^+ ATPase proton pump which is the common downstream effector of acid secretion. Its inhibition is therefore likely to give the most effective suppression of gastric acid. This is the protein which is the target of the PPIs such as omeprazole and its analogues.

The gastric H^+/K^+ ATPase is a member of the P-type ATPase family. Other members of this family include the Na^+/K^+ ATPase and the Ca^{2+} ATPase. It is a 100 kDa protein, composed of a large 1,034 aa α-subunit containing the catalytic and ion transport functions, and a 1TM 290 aa β-subunit believed to be important in regulation and trafficking of the protein complex [8]. The catalytic cycle involves ion translocation driven by phosphorylation and dephosphorylation of a conserved aspartate residue [9]. In its phosphorylated state, the protein adopts a conformation [E_2] with a binding site facing outward into the parietal cell canaliculus, which is of high affinity for potassium ion and low affinity for proton. On dephosphorylation, this reverts to the [E_1] conformation with an inward facing, low-affinity potassium ion, high-affinity proton-binding site [10, 11]. The adenosine triphosphatase activity of the protein drives this process, resulting in an electro-neutral exchange of protons and potassium ions across the canalicular membrane [12]. This generates a million-fold proton gradient, so acidifying the gastric lumen [13].

Another shortcoming of PPIs is the heterogeneity of response between patients due to variability in first-pass metabolism and therefore exposure [22–24]. PPIs undergo hepatic metabolism primarily by CyP450s 2C19 and 3A4 [25]. This also means an enhanced risk of drug–drug interactions, with a number of such interactions reported for omeprazole [26]. A considerable number of pyridine-2-ylmethylsulphinyl-1H-benzimidazole PPIs (1–15) are commercially available or in current or discontinued clinical development. Currently licensed PPIs are omeprazole, esomeprazole, lansoprazole, pantoprazole and rabeprazole. The clinical pharmacology of the PPIs has been expertly reviewed [26]. While transforming the treatment of acid-related disorders and providing effective, safe acid suppression and ulcer healing, the primary issues with the PPIs centre around rate of onset of acid suppression, nocturnal acid breakthrough and potential for inter-individual pharmacokinetic variation and drug interactions, particularly those mediated by CyP450 2C19. Relative clinical efficacy in healing and symptom control in GERD and gastric ulcer treatment, and differences in metabolism and drug–drug interaction potential of the different PPIs have also been reviewed [26–28]. Different studies show different conclusions ranging from no difference in clinical effect to significant preference for specific PPIs. The differences in the rate of acid suppression, with pantoprazole being the slowest [20], may, at least partly, be related to differences in the rate of inhibition of the H^+/K^+ ATPase, which in turn is affected by the rate of acid-catalysed conversion to active tetracyclic sulfenamide. This is reported to be pH dependent, and at pH 5.1 the rate of activation ranges from 0.12 h for rabeprazole to 4.7 h for pantoprazole [26]. Furthermore, the potency and rate of inhibition of the parietal cell is pK_a dependent. Mild basicity allows distribution into tissues at physiological pH with concentration under acid conditions. However, the higher pK_a of rabeprazole (5.0) allows greater concentration, potency and rate of onset than for omeprazole (pK_a 4.0) and this may translate to improved clinical effect [26]. However, only esomeprazole has significant supporting evidence to suggest more rapid onset of symptom relief in GERD and superior healing of oesophagitis than omeprazole and lansoprazole [27].

Omeprazole (1)

Esomeprazole (2)

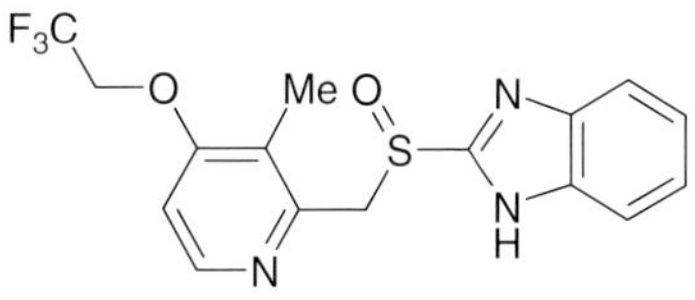

Lansoprazole (3)

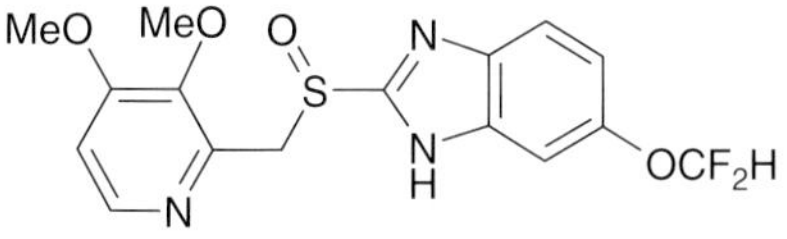

Pantoprazole (4)

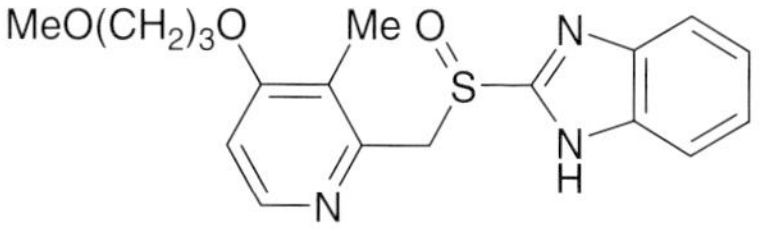

Rabeprazole (5)

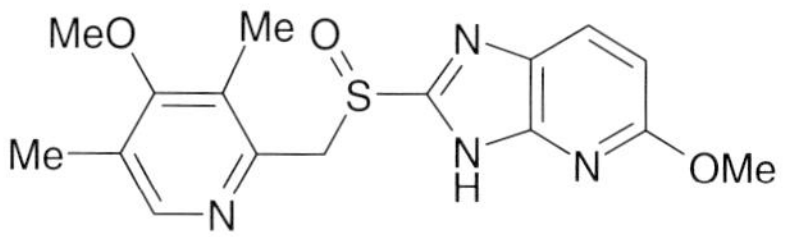

Tenatoprazole (6)

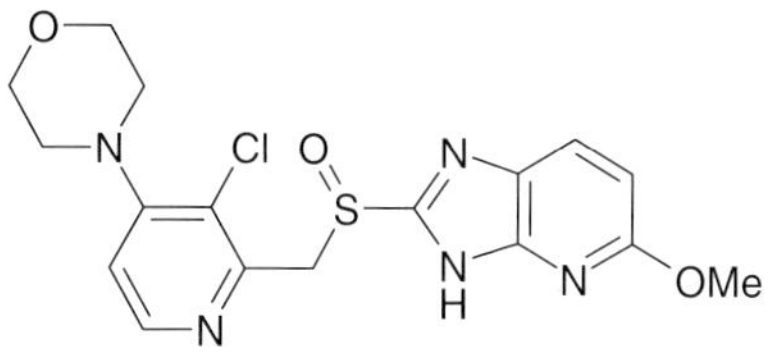

(7)

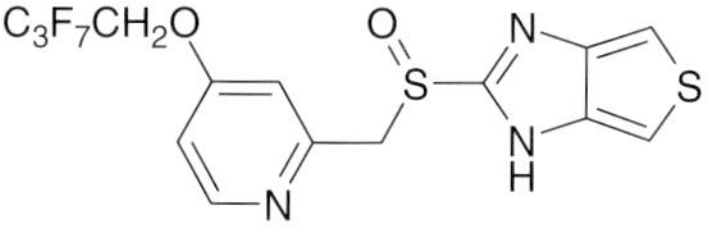

Saviprazole (8)

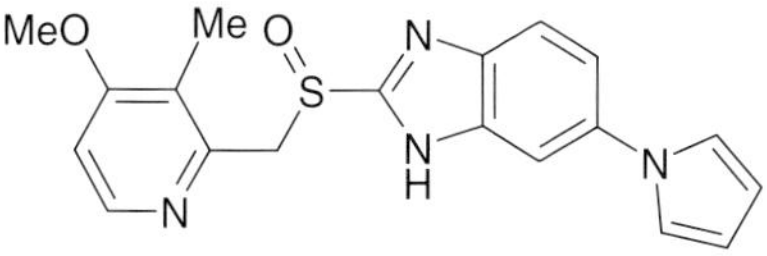

IY81149 (9)

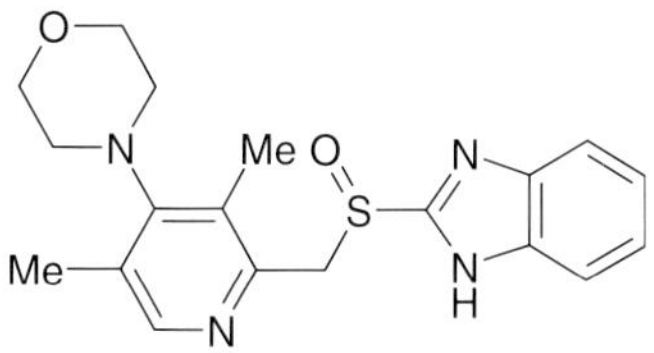

(10)

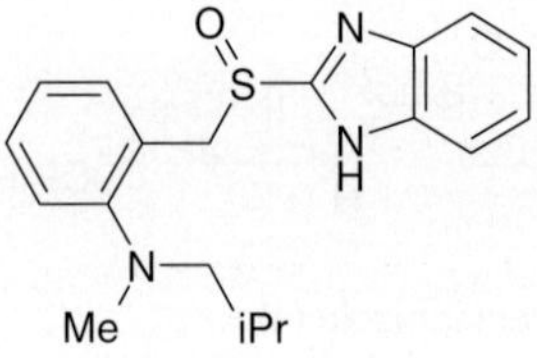

Leminoprazole (11)

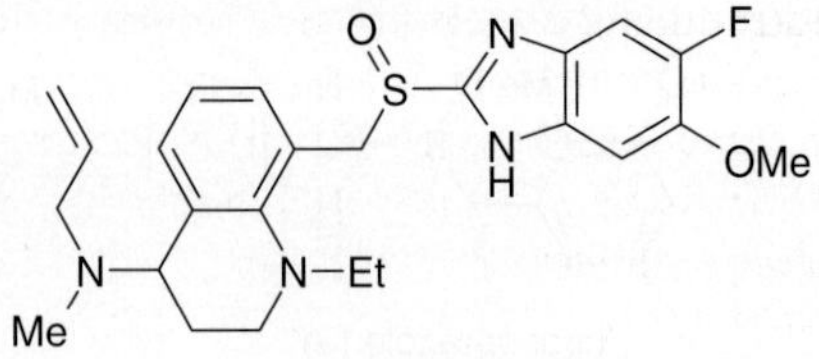

OPC22575 (12)

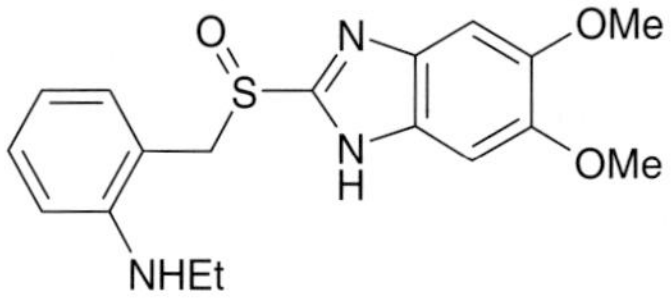

S-3337 (13)

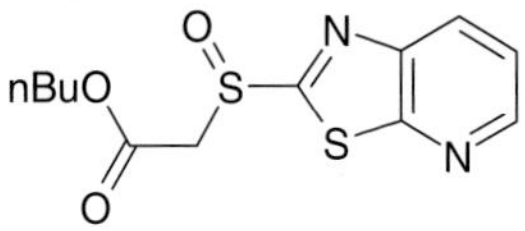

ME3407 (14)

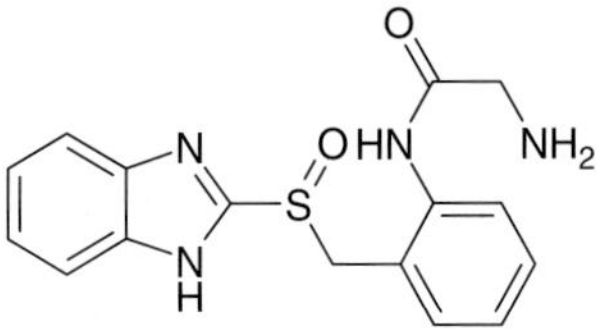

(15)

OMEPRAZOLE (LOSEC)

Omeprazole (1) was the first PPI to reach the market in 1988 and its properties have been extensively documented [15]. Oral bioavailability in man is modest at around 30–40%. It has a duration of action of 4–5 days [15] with an effective half-life of 30–45 h, during which time about 50% of secretory activity is restored through biosynthesis of new proton pumps [16]. (1) has been shown to have clinically important drug–drug interaction liabilities, mainly because it inhibits and is a substrate for CyP450 2C19; the concentrations required for interactions with CyP450 3A4 are however significantly higher than those which are clinically relevant [26, 29]. The selectivity of (1) in the *in vivo* setting for the H^+/K^+ ATPase over the Na^+/K^+ ATPase, which share 60% amino acid identity in the α-subunit [30], derives solely from its requirement for acid activation. Under physiological K^+ conditions *in vitro*, the selectivity of (1) for H^+/K^+ ATPase over Na^+/K^+ ATPase is reported to be only 2.5-fold. As with all acid reducing agents (1) also reduces the absorption of the azole antifungal drugs ketoconazole and itraconazole.

Esomeprazole (2) (Nexium) is the single (S)-enantiomer of omeprazole (1) [31]. It shows significantly higher bioavailability (89%) of the active isomer in man following repeat dosing, although somewhat lower at 64% following a single dose. (2) gives superior efficacy to that of (1) at a similar dose; for example, in symptomatic GERD patients at 40 mg/day, (2) maintained intragastric pH >4 for 12 h in 50 and 88% of patients on days 1 and 5, respectively, while (1), at the same dose, gave the same response in 34 and 77% of patients, respectively [26]. Similar superiority has also been claimed for (2) compared with other PPIs, such as pantoprazole (4); for example, 87% erosive GERD patients taking (2) 40 mg/day were still in remission 6 months into the maintenance phase of treatment compared with 75% of patients using (4) 40 mg/day. This is perhaps not surprising since the amount of active drug in mg/mg equivalent doses of (2) is twice that in (1) or other racemic PPIs. However, (2) does have a lower affinity, and is a poorer substrate, for CyP450 3A4 than the (R)-enantiomer and so shows a diminished first-pass metabolism, leading to greater plasma concentrations than those achieved with omeprazole.

LANSOPRAZOLE (AG-1479, PREVACID)

Lansoprazole (3), developed by Takeda, was first launched in 1991 and licensed to a number of companies for marketing in different countries [32]. Oral bioavailability is somewhat higher than (1) at around 80%. (3) is reported to be twice as potent as (1) in isolated canine parietal cells [33]. It is also suggested that although inhibition is achieved through activation to its cyclic sulfenamide which then reacts with the H^+/K^+ ATPase thiols in the same way as (1) [34], in contrast, with (3) acid inhibition in the parietal cell is reversible. It is believed that *de novo* protein synthesis is not involved in this reversal, instead endogenous glutathione is involved in the recovery process. The performance of i.v. dosed (3) has been compared with that of (1) in rats and dogs [35]. In rat, (3) is an inhibitor of gastric bleeding induced by haemorrhagic shock ($ID_{50} = 0.47$ mg/kg), gastric mucosal lesions induced by aspirin ($ID_{50} = 0.3$ mg/kg) or indomethacin ($ID_{50} = 2.03$ mg/kg), and basal ($ID_{50} = 0.7$ mg/kg), histamine-induced ($ID_{50} = 0.17$ mg/kg) or 2-deoxyglucose-induced ($ID_{50} = 2.23$ mg/kg) acid secretion. In the Heidenhain pouch dog, histamine-stimulated acid secretion was inhibited with an ID_{50} of 0.14 mg/kg, and acid secretion induced by histamine, pentagastrin, bethanechol or peptone meal with ID_{50} values of 0.2–0.7 mg/kg [36]. On this basis, (3) is reported to be between two and ten times more potent in the rat, and equipotent in the dog compared with (1) [36]. Sepracor recently reported that they were developing the single (S)-isomer of lansoprazole (see Sepracor corporate website).

PANTOPRAZOLE (BY1023/SK&F96022)

Pantoprazole (4) was first launched in Germany by Byk Gulden in 1994 and subsequently in the rest of the world [37–39]. The additional 3-methoxy group flanking the 4-methoxy group is reported to result in the appropriate balance of potency and stability, especially at neutral pH [39], although all PPIs have half-lives of minutes at pH 1. At pH 5.1, (4) has an activation half-life of 4.7 h compared with 1.4 h for (1) and 0.12 h for rabeprazole (5) [40]. The compound is reported to be of similar overall potency and efficacy both pre-clinically and clinically, although its oral/i.v. dose ratio is smaller than that for (1) and is therefore suggested to be more stable in the acidic stomach, with higher oral bioavailability than (1) of $\sim 80\%$ [38, 39]. As for (1), it binds irreversibly to Cys813, but also to Cys822 [41]. It has also been suggested [42, 43] that the potential for drug–drug interactions both in rat and in man are less for (4) than for (1) (which inhibits CyP450 2C19) or (3) (which is a metabolic inducer). Thus, in man (4) shows no interaction with a broad spectrum of metabolic probes, nor was its own metabolism influenced [44] and this is likely due to a lower affinity for CYP3A4 and CYP2C19 [41]. In healthy male volunteers, pentagastrin ($0.6\,\mu g/h/kg$)-stimulated acid secretion was completely inhibited by a single dose of 60 mg of (4) [45]. Dose linearity was observed in the range 5–80 mg, and the compound was well tolerated.

RABEPRAZOLE (ACIPHEX)

Rabeprazole (5) is reported to have a more rapid onset of action than (1), with a rate of symptom relief in duodenal ulcer or GERD patients also at least as good at 20 or 40 mg/day as with 20 mg/day of (1). However, a similar rate of healing of duodenal ulcers was observed. It was also comparable to (1) in maintenance of healing and symptom control in erosive GERD patients, and was equivalent to (1) or (3) as a combination (with antibiotic) treatment for eradication of *H. pylori* infection [46]. Its metabolism by CyP450 2C19 is reported to be somewhat less than for (1), (3) or (4). It therefore shows less variability in effect across the 2C19 genotype patient population and so lower potential for drug–drug interactions [5].

MODIFICATIONS TO THE PPI THEME

Modifications of the pyridine-2-ylmethylsulfinyl-1H-benzimidazoles have been explored with the aim of improving their stability in acid or neutral media. While most of the pyridylmethylsulfinylbenzimidazoles carry a 4-alkoxy substituent to modulate the reactivity of the pyridyl nitrogen, as in (1–9), this substituent has also been replaced by an alkylamino group. Thus, compound (10) was reported by the Beecham group to afford 72% inhibition of carbachol or histamine-stimulated acid production in the perfused stomach of the anaesthetised rat at an i.v. bolus dose of 0.5 mg/kg [47].

Tenatoprazole (6) (TU-199) is a pyridoimidazole, which is currently in Phase II clinical trials [48]. It dose-dependently inhibited histamine- (0.1–0.4 mg/kg p.o.), carbachol- and tetragastrin-induced (0.2–0.8 mg/kg p.o.) acid secretion in the gastric fistula dog over 24 h. The effect reached a maximum after 3–4 doses and was claimed to be more pronounced than that of (1) or (3), with longer duration of action at 0.6 mg/kg. It is also reported to bind irreversibly to Cys813, as well as to Cys822 of the H^+/K^+ ATPase α-subunit [49]. The decay of binding to Cys813 was shown to be relatively fast, with a half-life of 3.9 h while that to Cys822 was resistant to reducing agents *in vitro* and afforded inhibition duration dependent only on protein turnover. This, together with the long 8 h plasma half-life which approaches the rate of new proton pump synthesis, led the authors to propose that this may result in prolonged acid inhibition compared with (1) and (2). Indeed, it is suggested from Phase I trials that prolonged acid control is achieved with 20 and 40 mg doses, extending to reduced nocturnal acid breakthrough compared to 40 mg of (2) [50].

More recently, the Altana group have combined both the features in the disclosure of the amino-halogeno pyridinylmethylsulfinyl-imidazo[4,5-b]pyridine (7), which they claim is more stable in neutral media [51]. When dosed at 2.2 μmol/kg intraduodenally, as either the racemate, or (R)- or (S)-isomers, (7) gives >50% inhibition of pentagastrin-induced acid secretion *in vivo* in the perfused rat stomach.

In attempts to provide rapid onset of action and potentially prolonged duration of action, alternative formulations of (1) have been generated. Santarus has developed Zegerid (previously Rapinex), a powder-formulated or chewable tablet combination of (1), either 20 or 40 mg, with sodium bicarbonate, indicated for use in the treatment of heartburn and GERD, treatment and maintenance of healing in erosive oesophagitis and duodenal ulcers, and treatment of gastric ulcers and prevention of gastrointestinal (GI) bleeding. In Phases II and III studies, Zegerid compared very favourably with (4) and the H_2RA, cimetidine, respectively. In post-marketing studies, Zegerid compared favourably with both (3) and (2), affording significantly higher percentage of time >pH 4 over the night-time period [52]. The rapid antacid effect of the sodium bicarbonate also reduces the acid-induced degradation of (1), so affording a more rapid and prolonged absorption of the drug.

Attempts have also been made to generate PPI prodrugs. One approach to generating longer acting PPIs has been to generate more acid stable amides of anilinomethylsulfinylbenzimidazoles, such as compound (15), which are cleaved to the corresponding anilines by plasma aminopeptidases [53]. The anilines are then capable of generating the cyclic sulfenamide active species. However, there have been no reports of the progression of such an approach to clinical studies.

An alternative approach has been the substitution of the benzimidazole NH with acyloxyalkyl, alkoxycarbonyl, aminoethyl and alkoxyalkyl groups [54]. Sachs and co-workers, and then the Allergan group, have identified N-sulphonyl analogues of PPIs, such as (16) and (17) (AGN 201904-Z), with improved stability under acid and neutral conditions [55–59]. The expectation, therefore, is that such compounds will have improved gastric stability and improved duration of action following oral administration [57]. The compounds have been shown to have extended plasma half-lives in rats and dogs, partly due to enhanced chemical stability [58] and partly due to a slower absorption profile following oral dosing, in turn due to poorer membrane permeability [57]. Thus, following an oral dose in rats of 2 mg/kg, compound (16) affords a later C_{max} (ca. 40 min vs. ca. 20 min for (1)) and measurable drug concentrations at 6.5 h, compared with undetectable levels with (1) at 100 min. This translates to a longer duration of inhibition of acid secretion, with (16) at 1 mg/kg p.o. giving 91% inhibition of acid secretion in histamine-stimulated, pylorus-ligated rats at 5 h post-dose, compared with 45% inhibition at 5 h following oral dosing of (1) omeprazole (2 mg/kg) [58].

Compound (17) AGN 201904-Z shows an elimination half-life of 2.4 h following 10 mg/kg dosing in rats compared with 0.7 h for (1) itself, and this may be driven by the significantly reduced CaCo2 cell membrane permeability of 0.12×10^{-6} cm/s vs. 13×10^{-6} cm/s for (1). In a Phase I clinical trial in healthy male volunteers (see ClinicalTrials.gov web site, 31 May 2006), (17) afforded greater, more prolonged acid suppression than esomeprazole [59]. Thus, once-daily 600 mg enteric-coated (17) capsules, delivered 50 mg equivalent omeprazole, affording 87% of the time at pH >4 compared with 57% of the time at pH >4 for once-daily 40 mg delayed-release esomeprazole tablets, during a 24 h period on the fifth day of dosing.

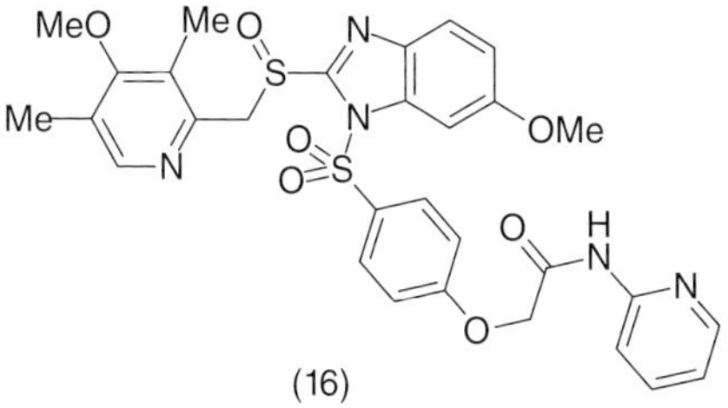

(16)

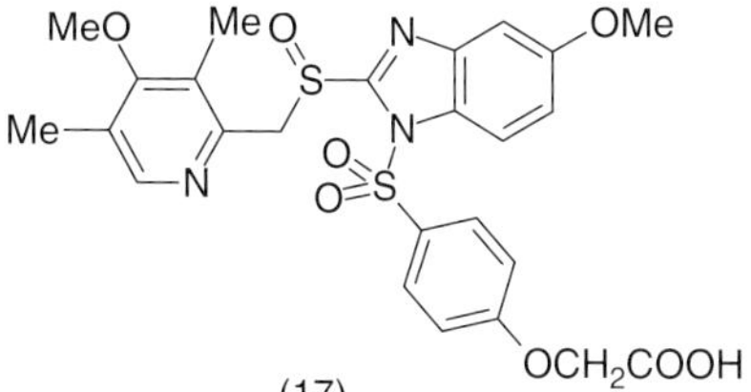

(17)

The Allergan group has claimed that a combination of a PPI (or its prodrug) such as omeprazole (16 mg/kg) with MDR2 inhibitor MK-571 (10 mg/kg) dosed orally in rats affords significantly enhanced C_{max} values [60]. Conversely, however, the systemic half-life was shortened.

Derivatives of PPIs, including lansoprazole (3), have been generated in which functionality has been added to allow release of nitric oxide to enhance antibacterial activity and to facilitate ulcer healing, for example, (18) [61].

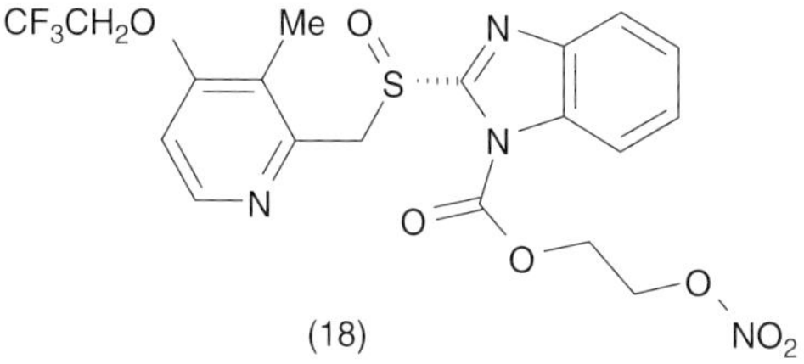

(18)

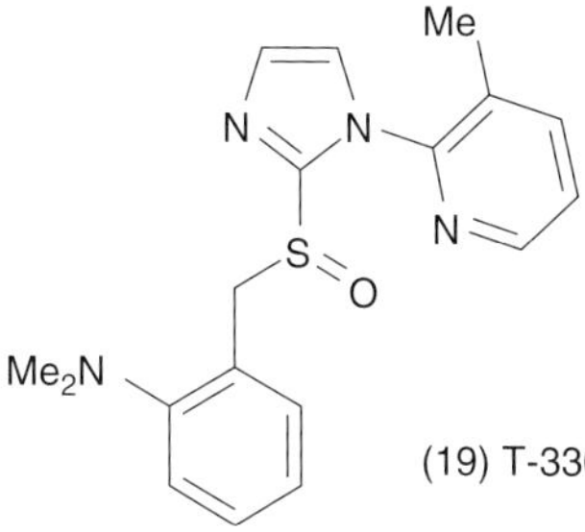

(19) T-330

T330 (19), although having structural features common to the PPIs, is argued to be a reversible H^+/K^+ ATPase inhibitor, but through a mechanism distinct from the K^+-competitive compounds such as SCH-28080 [62]. It is suggested that T330 is converted into a sulfenic acid intermediate under acid conditions rather than the sulfenamide in omeprazole-like compounds, and that the S–S bridge with the sulfenic acid is more easily cleaved, therefore not requiring *de novo* protein synthesis for restoration of H^+/K^+ ATPase activity [63]. (19) was shown to inhibit pentagastrin-stimulated gastric acid secretion in the chronically fistulated rat with an ED_{50} of 0.73 mg/kg p.o. [64]. Although this is more potent than (1), it is also shorter acting, being more similar to that of the H_2RA ranitidine.

Karimian et al. have attempted to generate alternative irreversible H^+/K^+ ATPase inhibitors [65–67]. They used the 1,2,4-thiadiazolo[4,5-a]benzimidazole and imidazo[1,2-d]-1,2,4-thiadizole templates as the species to interact with the H^+/K^+ ATPase thiol, and used the 3-substituent to tune the chemical reactivity of the thiadiazole to thiol nucleophiles at pH 7 (as measured using 2-mercaptoethanol). Thus, piperazine derivative (20) has a half-life of thiol reactivity of ~6.7 h. The reactivity of such species is significantly increased at lower pH, with reactivity half-lives reduced to less than 1 h. Compound (20) has a rather modest acid secretion-inhibitory potency (as measured by ^{14}C-aminopyrine accumulation) against histamine and cAMP-stimulated isolated mouse gastric gland, with an EC_{50} of 50 μM. Compound (21) afforded significant acid inhibition 3–5 h after a 300 μmol/kg oral dose in the peptone-stimulated pylorus-ligated rat. These compounds are designed to undergo nucleophilic attack at the sulfur atom of the thiadiazole, providing similar irreversible inhibition to that afforded by the omeprazole sulfenamide intermediate.

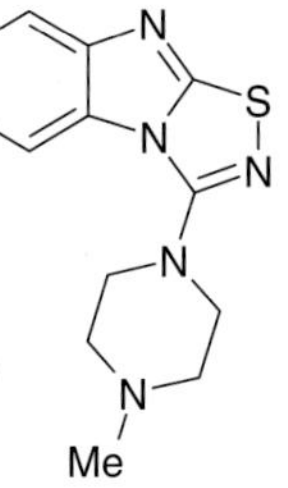

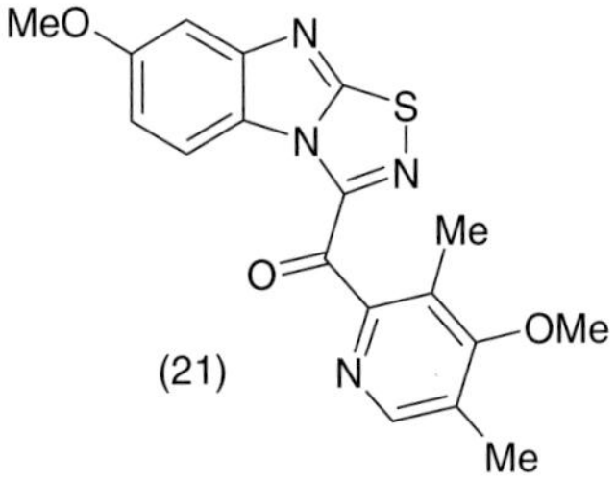

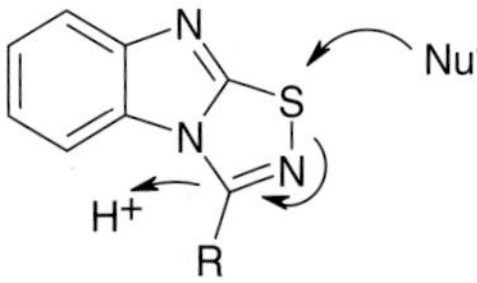

A related imidazo[2,1-b]thiazolo[4,5-g]benzothiazole derivative YJA20379-1 (22) has been reported to inhibit the H^+/K^+ ATPase activity of pig gastric microsomes. This was antagonised by dithiothreitol, but could not be reversed by dilution and washing, suggesting an irreversible mechanism of action as for omeprazole (1) [68]. However, as with (1), (22) is unstable at extremes of pH and, furthermore, the compound has poor oral bioavailability of less than 10%, which was assumed to be due to extensive first pass metabolism [69].

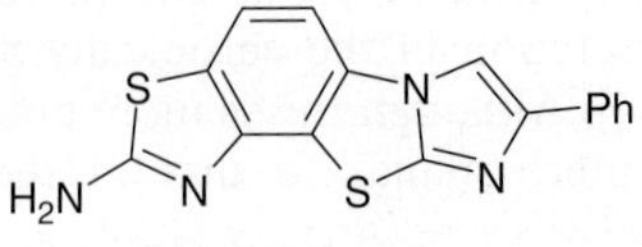

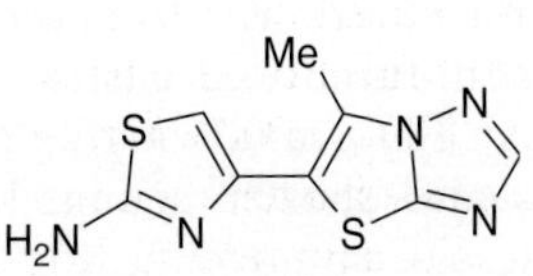

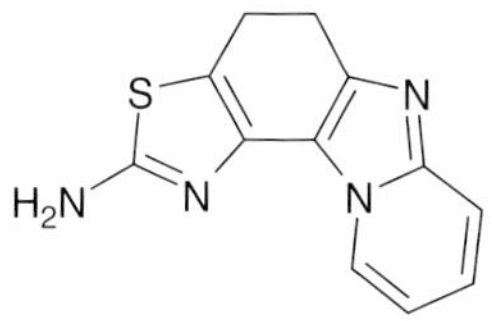

(24) YJA-20379-2

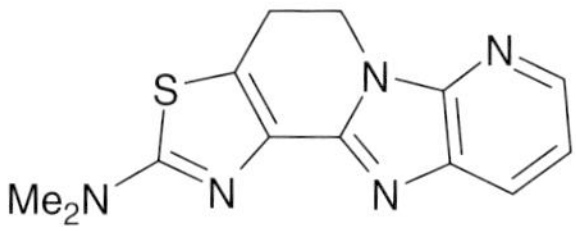

(25)

Other similar compounds include (23), which showed an ED_{50} of 11.1 mg/kg p.o. [70], (24) (YJA-20379-2) with an ED_{50} of 2.9 mg/kg p.o. [71, 72] and (25), which also showed an ED_{50} of 11 mg/kg p.o. [73] in protection of rat stomach from ethanol-induced lesions.

Further compounds with an irreversible mode of action have been identified. RS13232A (26) [74] has an H^+/K^+ ATPase IC_{50} of 0.4 µM. When tested in the rat at 1, 3 and 9 mg/kg orally, it showed significant inhibition of histamine-stimulated acid secretion only at the highest dose. Ebselen (27), a compound known to have glutathione peroxidase-like activity and known to form adducts with thiol-containing compounds such as glutathione, has been shown to inhibit potassium-dependent H^+/K^+ ATPase activity with an IC_{50} of 0.06 µM [75].

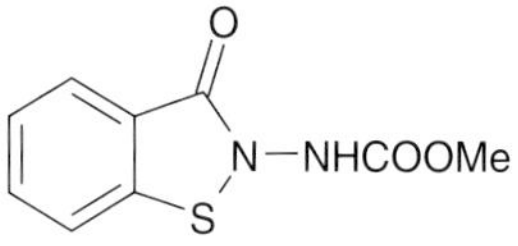

(26) RS13232A

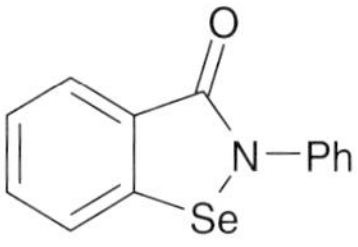

(27) Ebselen

ACID SUPPRESSION – GENERAL CONSIDERATIONS

PPIs have a good safety profile and there is no evidence of direct, clinically relevant toxic effects [76]. The safe clinical use of PPIs for nearly 20 years is testament to this low risk from prolonged acid suppression in man [77–79]. The primary function of acid secretion is the inactivation of ingested microorganisms [80]. Indeed long-term treatment with omeprazole has been shown to cause bacterial overgrowth in the upper gut due to profound long-standing acid suppression [81, 82], but this only rarely leads to clinical disease [83].

Profound acid suppression also results in hypergastrinaemia, and gastrin regulates enterochromaffin (ECL) cells both functionally (Figure 3.1) and trophically. Despite concerns relating to this hypertrophism in rodents, there has been no evidence to date that this translates to the clinic. In preclinical species: (i) tumours originating from the ECL cells have been shown to arise from long-term acid suppression [77]; (ii) carcinoid development resulting from hypergastrinaemia [84] was reported on long-term treatment of rats with omeprazole [15, 85–89]; (iii) H^+/K^+ ATPase knockout (Atp4a$-/-$) mice, studied for up to 20 months, showed a number of changes associated with chronic achlorhydria, including hyperplasia, mucocystic metaplasia and changes in some growth factors, but critical characteristics of gastric neoplasia were absent [90]. However, the effects reported in humans have been considerably less pronounced, with only a slight hyperplasia of ECL cells being observed [91, 92], and no direct evidence of increases in progression to metaplasia or gastric cancer [90]. Indeed, there have even been reports that PPIs such as omeprazole may actually increase the sensitivity of tumours to anticancer drugs due to acidification of the hypoxic tumour resulting from blockade of the vacuolar H^+ ATPase by the PPI [93, 94].

Off-label use of PPIs and investigation of effects other than gastric antisecretory actions are starting to emerge [95, 96]. Thus, there are reports of potential anti-inflammatory, cancer cell apoptosis and pro-regenerative activities of PPIs, and of their clinical use for indications such as non-ulcer dyspepsia and non-specific abdominal pain.

Concerns persist, however, regarding rebound acid activation. While there are studies to suggest the contrary, rebound acid hypersecretion and tachyphylaxis/tolerance associated with long-term dosing of current H_2RAs are widely reported phenomena [97]. On removal of long-term H^+/K^+ ATPase inhibitor-mediated acid suppression, probably due to the hypergastrinaemia and resulting trophic effects on the oxyntic mucosa, there is acid rebound with an overshoot over basal secretion rate [98].

Also, the experience with use of PPIs (as well as H_2RAs) in on-demand or intermittent therapy to date has been mixed. Zacny *et al.* concluded that current H_2RAs are not appropriate in an on-demand setting, and on-demand PPI therapy may have some utility in only a small proportion of non-erosive GERD patients [99]. Also, in studies to evaluate on-demand therapy with PPIs, it was found that high proportions of patients were self-medicating for 7, 14 or even 28 days [26]. When taken on demand for 6 months, only 27% of patients achieved complete relief of heartburn after 1–2 days treatment with rabeprazole, and only 56% of patients were controlled with up to four consecutive days of dosing.

Treatments with improvements over the PPIs are required to afford more complete acid suppression, especially at night. Night-time heartburn is a common cause of sleep disturbance, with recent studies with omeprazole (20 or 40 mg) indicating that only 53% of patients received adequate relief from night-time heartburn [100]. Also, there is frequent off-label use of PPIs at higher doses, and concomitant antacid use among PPI users [101].

REVERSIBLE INHIBITION OF H^+/K^+ ATPASE

Unlike the irreversible PPIs, the reversible H^+/K^+ ATPase inhibitors do not require acid activation and would be expected to be significantly more stable than PPIs under all physiological conditions with implications for their pharmacokinetic half-life. Reversible inhibitors are suggested to offer the potential for full acid suppression from first dose together with long duration of action [102].

IMIDAZOPYRIDINES

In the early 1980s, a group at Schering-Plough discovered that a series of benzyloxy-imidazopyridines afforded potent inhibitory effects on acid secretion, both *in vitro* and *in vivo*, and afforded gastric cytoprotective activity [103, 104]. Although suspected when these molecules were first discovered, the target protein was not known. All optimisation, structure activity relationship (SAR) work and compound development was conducted by *in vivo* assay, primarily in the dog. From this early work SCH-28080 (28) was identified as a clinical candidate molecule and progressed to Phase I studies. Investigations of the mode of action using [^{14}C]-labelled (28) [105] and with photo-affinity ligands [106, 107] subsequently confirmed the H^+/K^+ ATPase as the molecular target. Thus, the radio-ligand was shown to bind saturably to gastric vesicle preparations, to be displaced by K^+ and to bind more avidly in the presence of ATP, but not with a non-hydrolysable analogue of ATP, indicating potassium competitivity and tighter binding to the phosphorylated form of the enzyme. The compound was shown to bind, probably in its protonated form, to the luminal side of the H^+/K^+ ATPase [108–110] at a site which prevented irreversible inhibition by omeprazole (1) [111].

The compound is a weak base with pK_a of 5.5 and this (as for other APAs of this type) may explain the high concentrations observed in gastric tissues [112]. As with the PPIs, cell permeability is facilitated as the unprotonated form, while protonation in the acidic canaliculus may result in the compound no longer being membrane permeable. Compound binding results in inhibition of potassium-stimulated dephosphorylation of the phosphoenzyme [110]. In isolated guinea pig gastric mucosa, (28) abolished acid secretory responses to histamine, methacholine and dibutyryl cyclic AMP plus theophylline, suggesting a direct action on the parietal cells rather than acting by cholinergic or histaminergic stimulatory pathways [113]. *In vivo*, it was demonstrated that (28) afforded an increase in total mucous as well as an increase in bicarbonate secretion, both of which may contribute to its cytoprotective activity [113]. The oral potency for (28) in the Heidenhain pouch dog is reported as 4.4 mg/kg [112, 114].

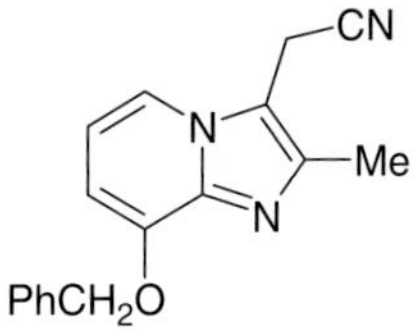

(28) SCH-28080

SCH-28080 was discontinued following Phase I studies due to observations in pre-clinical species of liver toxicity, and elevated liver transaminase enzymes in human volunteers during a rising dose tolerance study [115]. It was also demonstrated that, although the compound is well absorbed, the cyanomethyl and the benzyloxy groups were susceptible to extensive oxidative metabolism, with a significant disparity between the i.v. and p.o. potency in the dog [112]. One hypothesis proposed for the observed toxicological effects was that the C3-cyanomethyl group was oxidised *in vivo* to a reactive metabolite that resulted in generation of cyanide anion which is in turn converted by rhodanase into thiocyanate. Indeed, thiocyanate was measured in plasma from rodents dosed with (28), but this was shown not to be responsible for the observed acid suppressive effect since the parent compound concentration was in direct proportion to, while thiocyanate accumulation was temporally distinct from, acid suppression [115].

The SAR was investigated in the search for an improved molecule. *ortho* F or Cl substitution of the benzyloxy moiety was well tolerated, but substitution of this ring with an electron withdrawing group that could not mesomerically donate electrons afforded reduced activity. Also, although replacing the phenyl ring with an electron-rich thiophene moiety was well tolerated, when replaced with a more electron-deficient ring-like pyridine, the resulting compound was inactive. The C8-O substituent could be replaced with N but not S. The gastric anti-secretory activity resided in the extended conformation of the *trans* vinyl compound (29) rather than the *cis* isomer (30). This was further confirmed by demonstrating equivalent activity in the cyclised imidazo[1,2a]pyrano[2,3-c]pyridine analogue (31) to that of (28) [116].

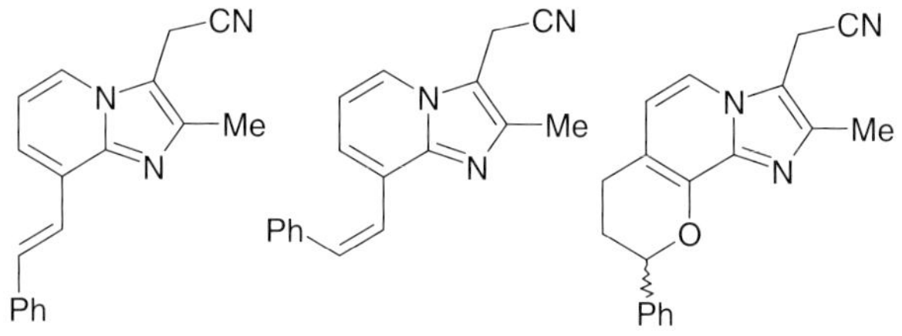

(29)   (30)   (31)

Further studies showed that a two-atom linker between the imidazopyridine ring and the pendant phenyl group was optimal [117]. Small lipophilic substituents are preferred at C2 and a small substituent at C3 is allowed that tolerates substitution with a small heteroatom-containing group. Replacing C7 with N is reported to reduce H^+/K^+ ATPase-inhibitory activity. Pharmacophore modelling suggested an orientation of the pendant phenyl ring, which is orthogonal to the core imidazopyridine ring [117]. Quantitative structure activity relationship (QSAR) approaches best predict *in vivo* activity in the extended series by taking into account the effective concentration of the compound existing in its protonated form at pH 7.4 [117]. In an attempt to overcome the issues with the CH_2CN at C3, this moiety was replaced with an amino group, affording possibly improved oral activity in the Heidenhain pouch dog model ($ED_{50} = 2\,mg/kg$) [112]. This molecule also has a weakly basic pK_a of 5.8. However, it was again shown to generate cyanide and thiocyanate on metabolism *in vivo*, postulated to be due to 5-membered ring opening. Acylation of the 3-amino group or replacement with an aminomethyl group significantly reduced activity.

A number of groups have made attempts to achieve modified imidazopyridines with improved overall profiles. Yamanouchi identified YM-020 (32), but its development was discontinued [118–121]. Although the reason for this is unknown, there was significant discrepancy between *in vivo* potencies obtained by the oral and i.v. routes. Thus, in pylorus-ligated rats (32) had an ED_{50} of 9.5 mg/kg p.o., while in anaesthetised dogs it inhibited pentagastrin-induced acid secretion with an ED_{50} of 0.08 mg/kg i.v. Although approaching the potency of omeprazole, its duration of action was shorter in Heidenhain pouch dogs.

Fujisawa Pharma patented a series of imidazopyridine analogues, and singled out (33) [122]. This compound afforded 96% inhibition of heteroaryl functionality in place of the dimethylbenzyl aryl group, such as those of the general structure (49).

AstraZeneca have progressed compounds AZD-5745 and AZD-9139 into pre-clinical development, and AR-H047108 (48) into Phase I development. The behaviour of the latter compound in the Heidenhain pouch dog model has been characterised. AR-H047108 ($K_d = 11.2$ nM) demonstrates a delay between peak plasma concentration and peak acid suppressive activity, and a delay in decline of acid suppressive effect following decline in plasma concentrations [152]. This has been explained by a combination of factors: distribution to the acidic compartment of the parietal cell of the weakly basic AR-H047108 (pK_a 5.9), a rate-limiting binding interaction between drug and enzyme, and also a contribution by the pre-systemically generated, primary metabolite AR-H047116 (50) (pK_a 5.4, K_d 125 nM). Although the latter is approximately ten times less potent than the parent, and three times less active in parietal cells, it is generated in high concentrations that rapidly exceed those of parent following oral administration and it has a longer plasma half-life (areas under concentration vs. time curves (AUCs) are approximately five to seven times those of the parent following parent dosing at 0.3–2.4 μmol/kg). Although selected for clinical development, AR-H047108 showed liver toxicity after 14 days dosing in the dog [153].

(49) X = O or S
NR^{1}R^2 = eg NHMe, NH(CH$_2$)$_2$OH

(50)

C6-N-hydroxyethyl amide (47) (AZD-0865) [102, 154] (formerly AR-H044277) has been progressed to Phase IIb clinical trials. At 3 μM, AZD-0865 affords potassium-competitive, reversible and dose-dependent inhibition of potassium-induced H$^+$/K$^+$ ATPase activity in membrane vesicles by 88%. It is highly selective, producing only 8% inhibition of the Na$^+$/K$^+$ ATPase at 100 μM [155]. It shows no effect on kidney function, despite the presence of H$^+$/K$^+$ ATPase in the kidney cortical collecting duct [156], suggesting differences between gastric and renal sources of the protein. It shows higher potency in ion-tight vesicle preparations than in ion-leaky preparations, in keeping with the expected mode of action in which the mildly basic compound (pK_a 6.2) concentrates in the acidic compartment [157]. In isolated rat gastric glands (47) afforded rapid onset of complete, reversible inhibition of histamine-stimulated acid secretion at a concentration of 1 µM, whereas omeprazole required 100 µM concentration [158].

In rats, 1 µmol/kg p.o. (47) gave almost complete inhibition of histamine/carbachol-stimulated acid secretion during a period from 2 to 4.5 h post-dose. At 2 µmol/kg p.o. maximal inhibitory effects were maintained for up to 9 h post-dose, and 10 µmol/kg maximally inhibited acid secretion for 24 h [159].

In the Heidenhain pouch dog, (47) (1 µmol/kg p.o.) gave almost 100% inhibition of histamine-stimulated acid output within 3 h, with 40% inhibition still achieved 24 h after both single and repeated doses. This duration of effect was dose-dependent [160]. The maximal effect was correlated with $\log C_{max}$, again showing the dependency of effect on the pharmacokinetics (PK) [161]. Oral bioavailability of 50% and dose-linear PK was observed in the dose range 0.125–16 µmol/kg, and peak plasma concentrations were achieved rapidly within 0.5–1.0 h post-dose [154, 162, 163]. Multiple dosing of 0.5 µmol/kg/day p.o. for 14 days in the same model had no significant effects on the PK profile, although higher gastric juice drug levels were obtained, and resulted in a faster onset of action compared to that following a single dose [164]. In further evidence of the concentrating effect in the acidic parietal cells, the gastric juice levels exceeded the plasma concentration by 2 h after dosing [160]. Indeed, although undetectable in plasma 24 h post-dose, it was still detectable in gastric juice [164].

In a single blind, randomised Phase I study, escalating doses (0.08–4.0 mg/kg) of (47) were rapidly absorbed, with C_{max} achieved within 1 h. AUC and C_{max} of 2.4–107.1 µM/h and 0.4–21.2 µM, respectively, were achieved dose-proportionally [163]. pH >6 was achieved within 1–2 h of first dose. A dose of 0.8 mg/kg afforded full acid suppression over a 15 h period, at which dose the exposure (AUC_{0-t}) was 27 µM/h. Based on this excellent profile, (47) was progressed into randomised, 4-arm parallel group, multi-centre Phase IIb clinical studies in a total of 1,521 GERD patients with reflux oesophagitis [165]. At doses of 25–75 mg/day, equivalent healing to 40 mg esomeprazole (2) was obtained at 2, 4 and 8 weeks, and equivalent intragastric and intra-oesophageal acid suppression to 40 mg (2) was obtained at week 2. The minimally effective dose was not defined. All doses were well tolerated, and no serious adverse events were reported. However, dose-related, reversible increases in alanine transaminase (ALT) were observed in a small number of patients (0.8–2.2%) suggestive of liver toxicity.

In a similarly designed study in 1,459 patients with GERD without reflux oesophagitis, 25–75 mg/day (47) gave similar control of heartburn to 20 mg (2) at weeks 1, 2 and 4, and similar control of intra-gastric and intra-oesophageal acid to 20 mg (2) at week 2 [166]. Again, the drug was well tolerated, but the same sporadic elevations in liver function tests (LFTs) were observed. Although these returned to normal within 1–3 weeks either on continued treatment or on treatment cessation, the extent of the signal increases may have contributed to the decision to terminate development of the compound. Indeed, in 2005 development of AZD-0865 was discontinued, because it 'had not met the required target product profile' (see Astra-Zeneca, company website).

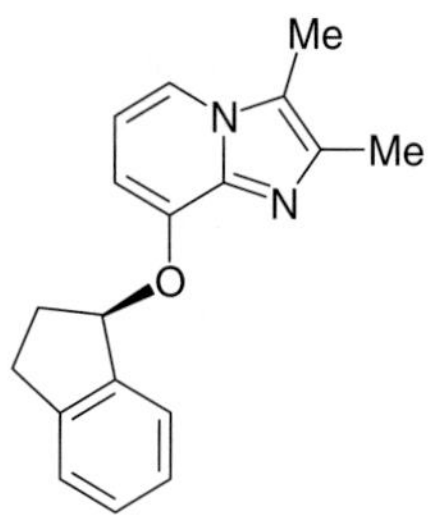

(57)

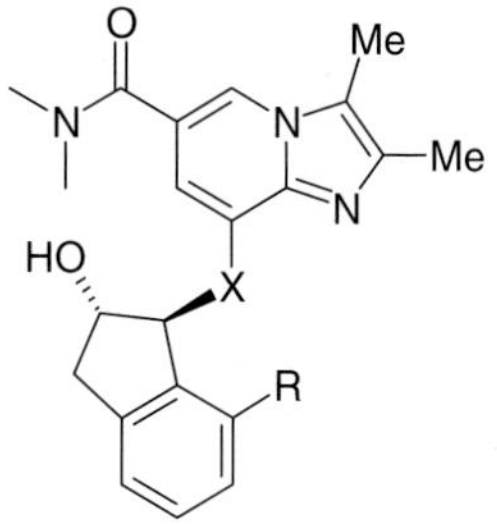

(57a) R = H, X = NH
(57b) R = Me, X = O

The Altana group has exploited further modifications in the C8 substituent through cyclisations to give compounds such as those in Table 3.3. There is a focus on the stereochemistry indicated. These compounds are reported to afford >30% inhibition of pentagastrin-stimulated acid secretion following 1 μmol/kg intraduodenal dose *in vivo* in the perfused rat stomach [167]. The unsubstituted indane also affords *in vivo* activity but data is only reported for dosing by the i.v. route. Thus, (57) gave 62% inhibition of gastric acid secretion and 100% inhibition of aspirin-induced ulcers following 3 μmol/kg i.v. in the rat [168]. Hydroxy-substituted indanes were also targeted [169] with a view to reducing lipophilicity as a means of reducing the metabolism of the imidazopyridines. Some of these, for example, (57a) and (57b), improved the H^+/K^+ ATPase-inhibitory activity compared with AR-H47108 while reducing the $\log P/\log D$ by 2–2.4 log units. The impact of this on the metabolism is not reported.

Table 3.3 EXAMPLES OF IMIDAZOPYRIDINES WITH AN INDANE FUNCTIONALITY

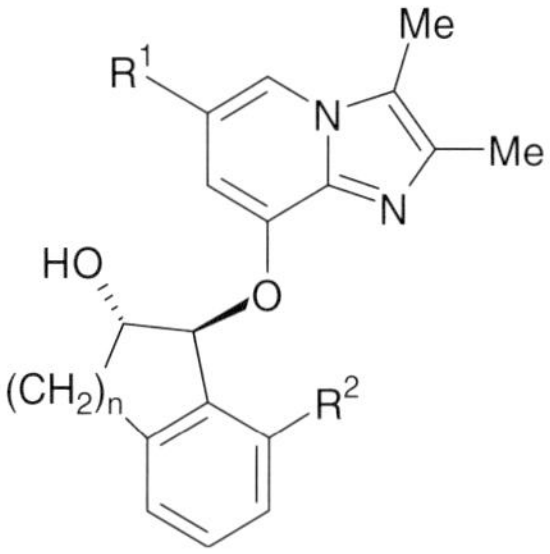

Compound	Stereochemistry	n	R^1	R^2
(51)	trans	1	$CONMe_2$	H
(52)	trans	1	CH_2OMe	H
(53)	1S,2S	1	$CONMe_2$	H
(54)	trans	2	$CONMe_2$	H
(55)	trans	1	$CONMe_2$	OMe
(56)	trans	1	$CONMe_2$	Me

The Altana group has also described deuterated analogues of compounds discussed above, which are potentially more metabolically stable [170]. For example, compound (58) affords H^+/K^+ ATPase inhibition with a pIC_{50} of 6.0 at pH 7.4, although no data regarding metabolic stability or *in vivo* profile are presented.

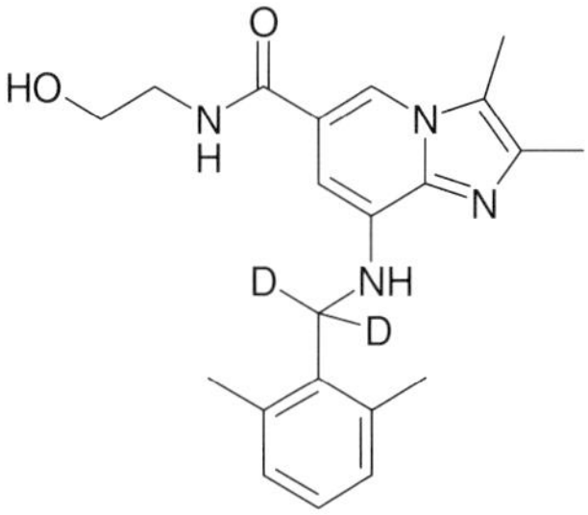

(58)

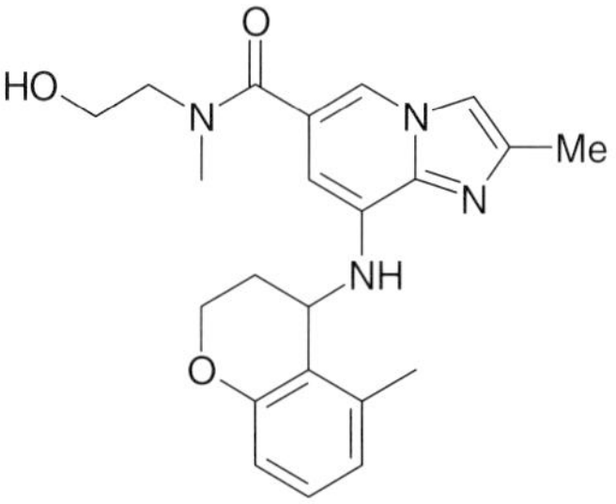

(67)

A Pfizer group has replaced the benzyl group with a chromane moiety (Table 3.4) [171, 172]. Examples of the chromane substituted with H, methyl or halogen are reported. It is evident that similarly high levels of open vesicle pig H^+/K^+ ATPase-inhibitory potencies are obtained for either chromane regioisomer, and that the effect on potency of stereochemistry at the chromane ring junction is not great. The SAR of the chromane substituents appears similar to that observed above, with F and Me substituents being well tolerated, less tolerance of a Cl substituent, and a 7-F substituent being particularly potent. Removal of the C3-Me group results in only a modest loss in potency (67, $IC_{50} = 0.23\,\mu M$) [172], while good potency is retained for the C3-hydroxymethyl analogue (66a) [173]. *In vivo* activity was only referred to for compound (65), which was described as showing good inhibition of acid secretion in the gastric lumen-perfused rat model, but the route of administration is not clear.

Table 3.4   EXAMPLES OF THE PFIZER CHROMANES

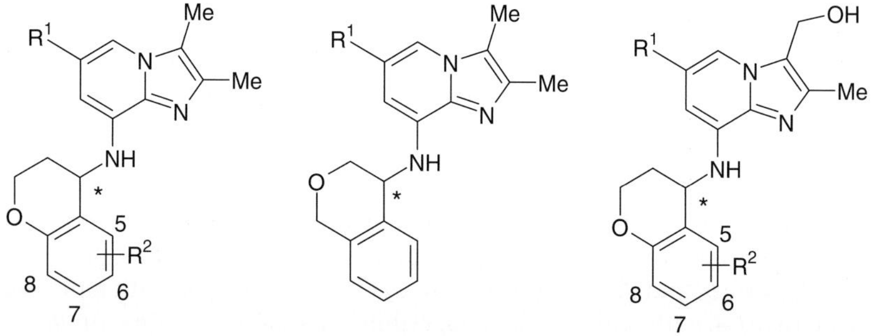

Compound	Template	R^1	R^2	Stereochemistry*	H^+/K^+ ATPase IC_{50} (μM)
(59)	1	CONMe$_2$	H	Racemic	0.085
(60)	1	CONMe$_2$	H	(−)	0.062
(61)	1	CONMe$_2$	H	(+)	0.028
(62)	1	CON(Me)(CH$_2$)$_2$OH	H	Racemic	0.05
(63)	2	CONMe$_2$	H	Racemic	0.103
(64)	2	CON(Me)(CH$_2$)$_2$OH	H	Racemic	0.087
(65)	1	CON(Me)(CH$_2$)$_2$OH	5-Me	(+)	0.093
(66)	1	CONMe$_2$	7-F	Racemic	0.014
(66a)	1	CONMe$_2$	7-F	Racemic	0.029

A GlaxoSmithKline group has explored a variety of aryl, heteroaryl [174] and heterocyclyl [175] substituents at the C6-position of the imidazopyridine template, in combination with various *O*- and *N*-benzyl substituents at C8, although no specific biological data is given.

TRICYCLIC IMIDAZOPYRIDINE ANALOGUES

The early Schering imidazopyridine publications included examples of tricyclic analogues [116, 176] largely with a C3-cyanomethyl group. The Astra group demonstrated that the C3-CH$_2$OH group (e.g., (68) and (69) in Table 3.5), although conferring lower potency in isolated gastric glands, was equally effective *in vivo* as the C3-CH$_2$CN compounds, for example, (70) and (71) [177]. On incorporation into the tricyclic dihydropyrano-imidazopyridine template, enhanced oral bioavailability in the rat and improved potency in both rat and dog was obtained, compared with the bicyclic imidazopyridine analogue.

An Altana group has investigated this tricyclic template. Substitution at either or both of the 7- and 8-positions of the 8,9-dihydro-pyrano[2,3-c]imidazo[1,2-a]pyridine ring as well as the tetrahydro-imidazonaphthyridine ring is well tolerated [178, 179]. Compounds (72) and (73) at 3 μmol/kg i.v. gave 93 and 100% inhibition, respectively, of pentagastrin-induced acid secretion *in vivo* in the perfused rat stomach model [178].

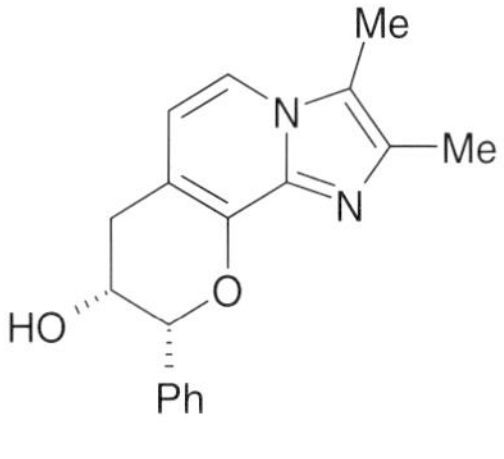

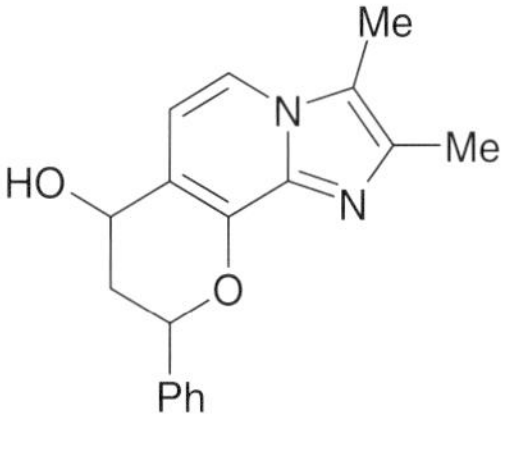

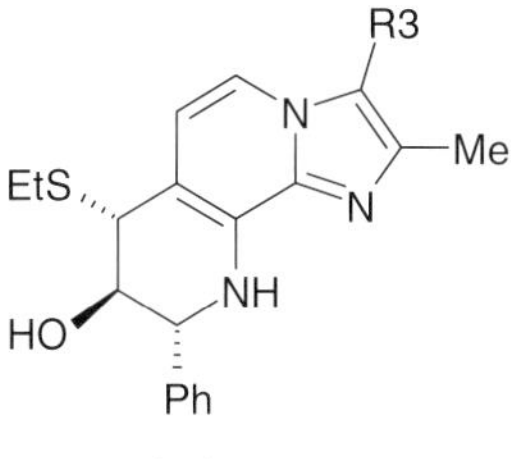

Examples with substitution at both 7- and 8-positions (such as those in Table 3.6) gave 100% inhibition of pentagastrin-induced acid secretion following 3 μmol/kg i.d. or i.v. dose (or 1 μmol/kg i.v. in the case of (74) and (75)) *in vivo* in the perfused rat stomach model [180–184]. A variety of substitution is tolerated at R^1, including a change in stereochemistry, either an ester functionality or a simple hydroxyl group is tolerated at R^2. Methyl, H or halogen at C3 is tolerated. The C3-CH$_2$OH group features among active compounds, such as (76). Also exemplified are fluoroalkoxy

Table 3.5 EXAMPLES OF THE ASTRA IMIDAZOPYRIDINES

Compound	Structure	Gastric glands IC_{50} (μM)	Rat F[a] p.o. (%)	Dog ED_{50} p.o.	Dog F[a] i.d. (%)
(68)	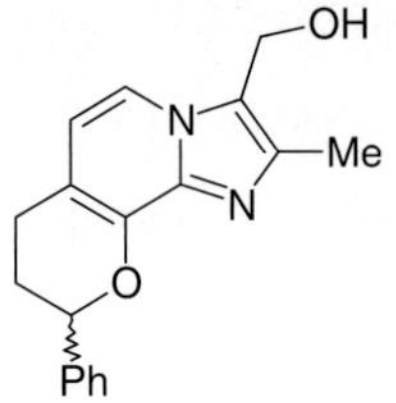	0.25	23	16	18
(69)	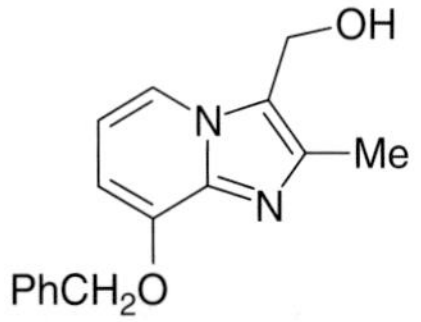	0.43	12	29.8	2.3
(70)	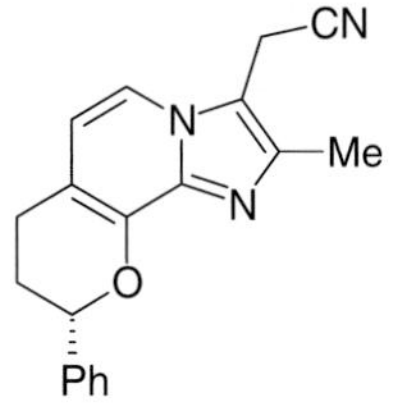	0.064	–	12.2	–
(71)	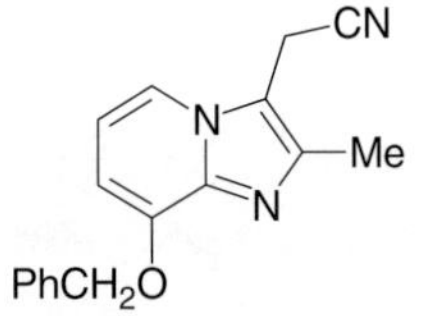	0.065	–	15.9	–

[a]F = Bioavailability

substituents at R^1 and R^2 [185]. A further patent covers sulfur-containing substituents including replacing the ether link at C7 with a thioether (79) [186]. These are again claimed to retain similar *in vivo* potency.

Compound (80) (BYK61359, BY-359, Soraprazan) is a fully reversible inhibitor of gastric H^+/K^+ ATPase with K_i 6.4 nM, K_d 26.4 nM, an IC_{50} in ion-leaky vesicles of 0.1 μM and in isolated gastric glands of 0.19 μM [187]. It shows rapid onset and long duration of action of reversal of pentagastrin-stimulated pH lowering as measured by pH-metry in the gastric-fistulated

Table 3.6 EXAMPLES OF 7-SUBSTITUTION IN TRICYCLIC DERIVATIVES

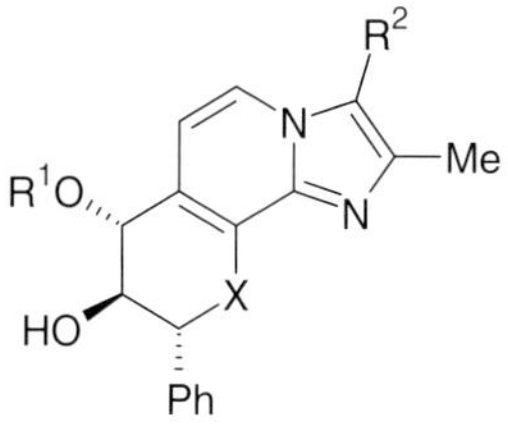

Template A

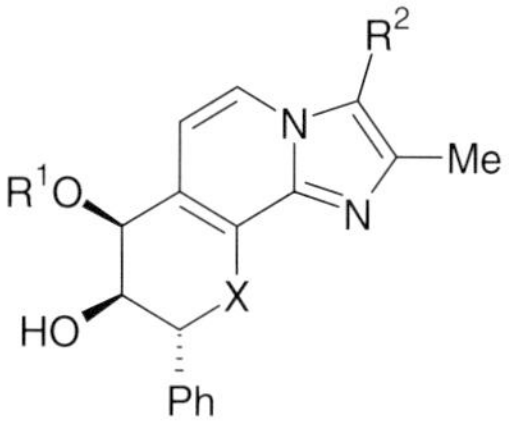

Template B

Compound	Template	X	R^1	R^2
(74)	A	O	MeO(CH$_2$)$_2$	Me
(75)	A	O	Et	Me
(76)	B	NH	MeO(CH$_2$)$_2$	H
(77)	A	NH	MeO(CH$_2$)$_2$	CH$_2$OH
(78)	A	NH	MeS(CH$_2$)$_2$	Me
(80)[a]	A	NH	MeO(CH$_2$)$_2$	H

[a]BY-359, Soraprazan.

dog. Thus, at 27 μmol/kg, a pH greater than 4 was achieved in a mean time of 123 min, and was sustained for a mean of 87% of a subsequent 22 h period, with median pH during that time of 6.7. This compares extremely favourably with an equivalent dose of esomeprazole, with parameters of 290 min, 35.2% and pH 2.2, respectively. Soraprazan was progressed by the Altana group into clinical development for the treatment of acid-induced GI diseases, including GERD. Company communications indicate that in Phase I studies in healthy human volunteers 20 mg/day p.o. for 7 days and 6 mg i.v. were well tolerated, with no serious adverse events or changes in clinical parameters. (80) showed a rapid onset of acid suppression, more rapid than for omeprazole (1), with pH 6 achieved on day 1 of dosing. Acid suppression was dose-related. As with pumaprazole (38), there was no development of tolerance. Oral bioavailability was 50% and was not affected by food or co-administration of antacids. Phase IIa clinical studies provided positive results on proof-of-concept and tolerability, and the compound was in Phase IIb trial to determine best dosage. There has been no recent information on the progression of (80) and Altana recently indicated that they have progressed an additional compound into Phase I, although its structure has not been disclosed. Altana has continued to publish additional investigation of SAR around the tricyclic template. For example, the SAR at C7 and C8 of the Soraprazan-like template has been further expanded. Compounds (81)–(84) were reported to inhibit pentagastrin-stimulated acid secretion by 100% at 1 μmol/kg i.v. in the rat *in vivo*-perfused stomach model [188]. Also, compounds of the general structure (85) afford >85% inhibition of acid secretion in the same model by the intraduodenal route at 1 μmol/kg [189].

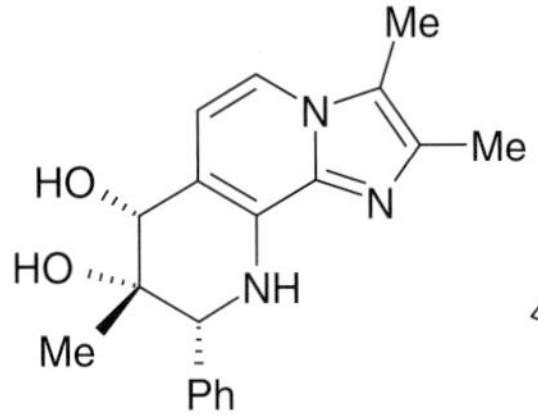

(81)

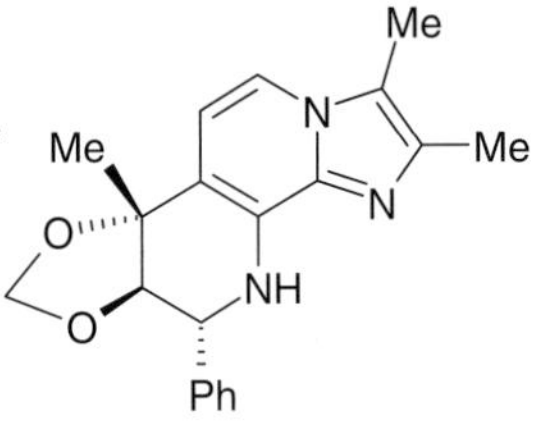

(82)

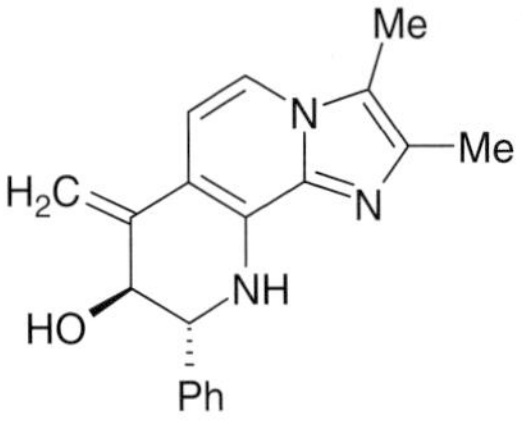

(83)

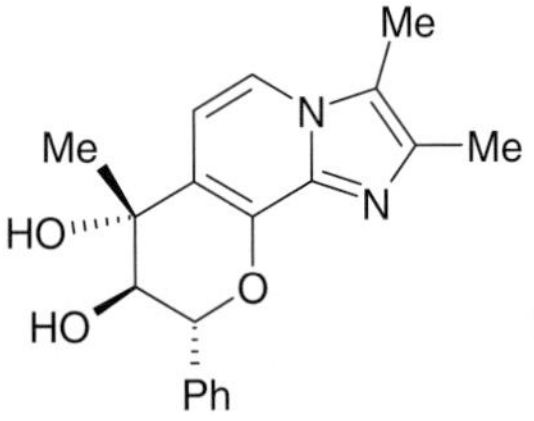

(84)

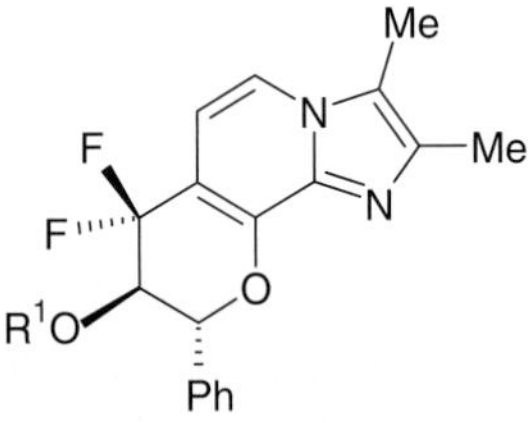

(85) R^1 = H, Ac, MeOCH$_2$CO, Me$_2$NCH$_2$CO, MeOCO

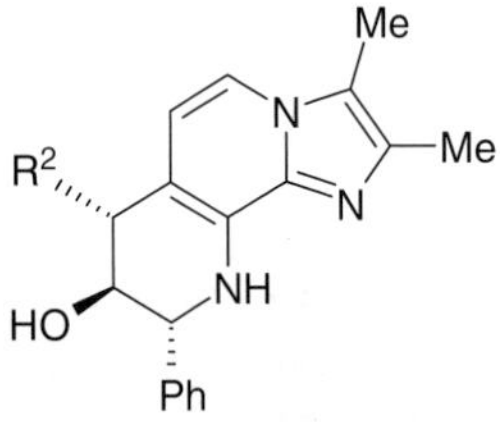

(86) R^2 = (morpholin-4-yl)
(87) R^2 = NMe$_2$
(88) R^2 = Me
(89) R^2 = CH$_2$CN

As well as O-linked functionality pendant to the 7- and 8-positions of the tricyclic core, opening of an intermediate 7,8-epoxide under acid catalysed conditions with amines has furnished a series of *trans* aminoalcohols which are reported to retain *in vivo* activity [190]. Thus, (86) and (87) give 44 and 64% inhibition, respectively, at 3 μmol/kg i.d. of pentagastrin-induced acid secretion *in vivo* in the perfused rat stomach model. Likewise the epoxide can be opened with C-nucleophiles to give molecules such as (88) and (89) which, respectively, afford 93 and 52% inhibition at 1 μmol/kg and 0.3 μmol/kg i.d. in the same *in vivo* model [191].

Derivatisation at positions 7 and 8 of this type of tricyclic template is also compatible with additional substitution at C6. Thus, (90) is reported to inhibit pentagastrin-stimulated acid secretion in the *in vivo* rat-perfused stomach (Ghosh Schild) model following 3 µmol/kg i.v. dose [192].

It is apparent from compounds such as (91)–(95) (Table 3.7) that preferred tricyclic compounds have a small lipophilic group on the imidazopyridine ring, a methyl at R^1 and an *ortho*-methyl-substituted phenyl group on the dihydropyran ring. R^1 as CH_2OH or $COCH_3$ groups appear to reduce *in vitro* and *in vivo* potency. There is a strong preference for the pendant aryl group to be attached with (S)-stereochemistry. It is further evidence that the position of the aryl ring is critical for good H^+/K^+ ATPase activity [193, 194]. Further SAR indicates that the C6-substituent

Table 3.7 6-CARBOXAMIDE SUBSTITUTED O-TRICYCLIC DERIVATIVES

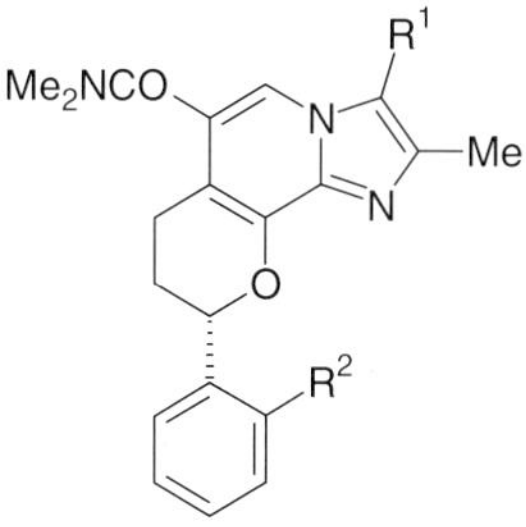

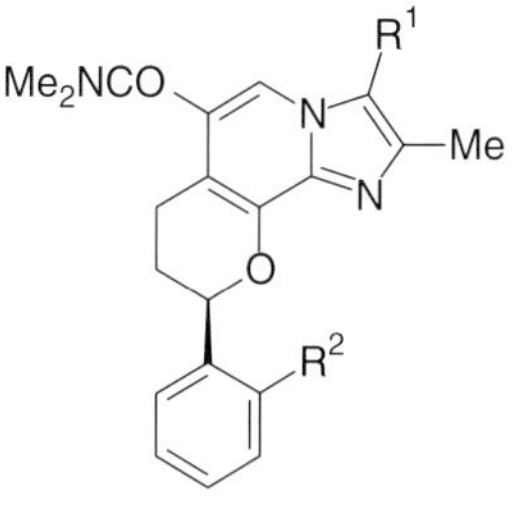

Compound	Template	R^1	R^2	H^+/K^+ ATPase pIC_{50}	% Inhibition (ED_{50} (µmol/kg))
(91)	A	Me	H	6.8	100[a] (0.22)
(92)	B	Me	H	5.4	36[b]
(93)	A	CH_2OH	H	5.8	>50[c] (5.0)
(94)	A	Me	Me	6.3	100[a] (0.4)
(95)	–	–	–	5.3	46[a], 100[b]

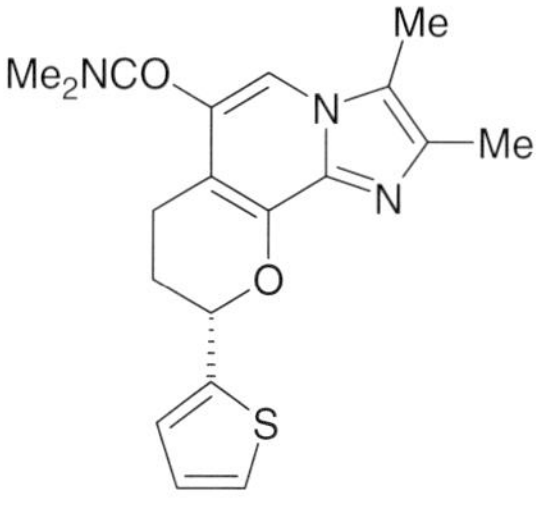

[a] Pentagastrin-stimulated acid secretion in the perfused Ghosh Schild rat stomach following 1 µmol/kg i.d. administration *in vivo*.
[b] Pentagastrin-stimulated acid secretion in the perfused Ghosh Schild rat stomach following 3 µmol/kg i.d. administration *in vivo*.
[c] Pentagastrin-stimulated acid secretion in the perfused Ghosh Schild rat stomach following 6 µmol/kg intraduodenal administration *in vivo*.

has a significant benefit regarding *in vivo* activity, but that the size of the substituent is critical to both *in vitro* and *in vivo* activity [195]. The substituents, especially that at the C3-position, have a significant effect on pK_a and log D; although this is not discussed, it appears that no compound with a pK_a less than 4.6 retains cellular activity.

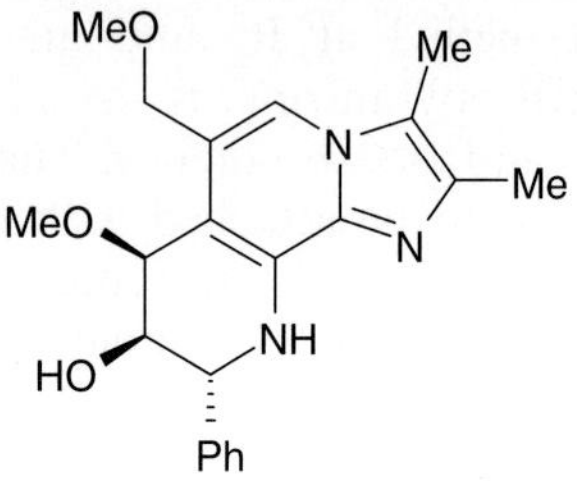

(90)

Similar analogues with the ring oxygen replaced by nitrogen have been generated by the same group; there is again a preference for the (8*S*) stereochemistry (Table 3.8) [196].

Table 3.8 6-CARBOXAMIDE SUBSTITUTED N-TRICYCLIC DERIVATIVES

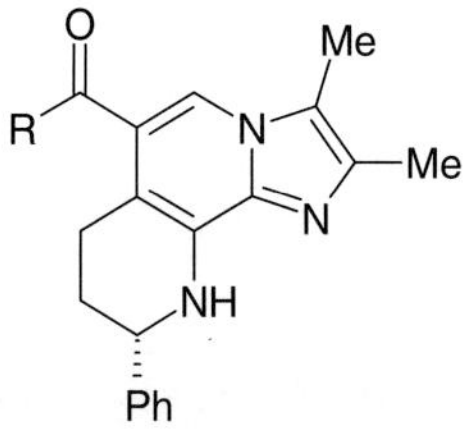

Template A

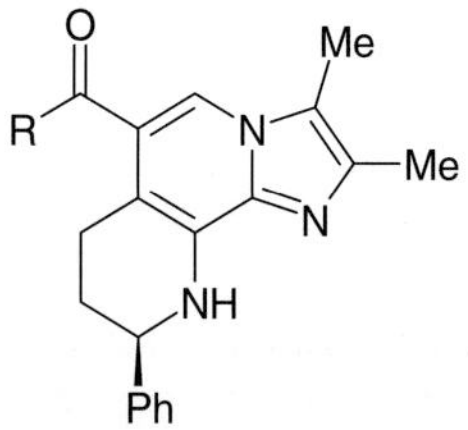

Template B

Compound	Template	R	% Inhibition[a]
(96)	A	Me$_2$N	>50
(97)	A	H$_2$N	>50
(98)	A	*N*-pyrrolidine	>50
(99)	B	Me$_2$N	<50

[a]Pentagastrin-stimulated acid secretion in the perfused rat stomach following 1 µmol/kg intraduodenal administration *in vivo*.

Table 3.9 SUBSTITUTED C-TRICYCLIC DERIVATIVES

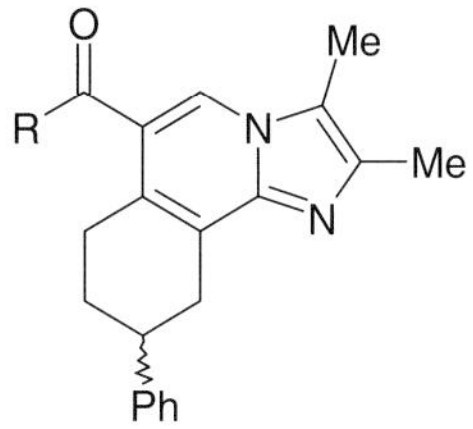

Compound	R
(100)	$MeOCH_2CH_2NH$
(101)	N-morpholine
(102)	MeNH
(103)	$MeOCH_2$

This work has been extended to the all-carbon third ring analogues. The compounds in Table 3.9 (100–103) all show >40% inhibition of pentagastrin-stimulated acid secretion following a 3 μmol/kg intraduodenal *in vivo* dose in the perfused rat stomach model [197]. The stereochemistry at the point of phenyl attachment is not defined in this case.

An interesting variation on the tricyclic imidazopyridine theme has been reported by the Altana group. The series of analogues is exemplified by the spirocyclic derivative (104) which is claimed to afford >40% inhibition of pentagastrin-induced acid secretion of the perfused rat stomach when dosed at 1 μmol/kg intraduodenally [198].

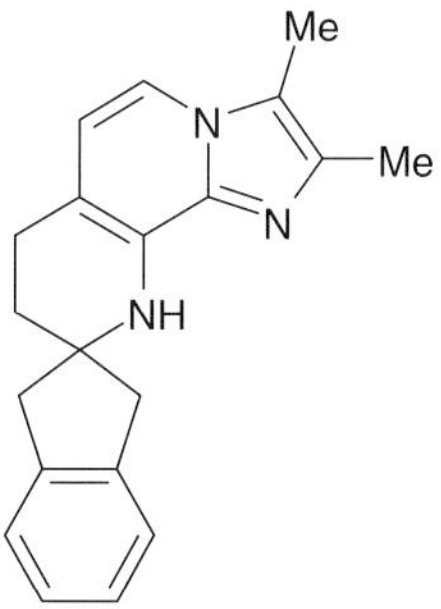

(104)

ALTERNATIVES TO THE IMIDAZOPYRIDINE TEMPLATE

Schering-Plough conducted work aimed at overcoming the apparent metabolic liability of the imidazopyridine template [114]. They inferred that the known metabolism of the structurally related zolimidine (105) at the C7-position could be avoided by insertion of a ring nitrogen at this position. They found that the C3-amino functionality as in SCH-32651 (106) gave the best *in vivo* acid suppressive activity following oral administration to the Heidenhain pouch dog ($ED_{50} = 1.4$ mg/kg, cf. 4.4 mg/kg for SCH-28080 (28)) [114]. Compound (106) is again a mildly basic compound, but with a somewhat reduced pK_a relative to (28) of 4.6. The site of protonation is again likely to be at the imidazole ring nitrogen [112]. This compound was selected as a clinical candidate in replacement for (28), but was not progressed beyond Phase I due to the discovery of ring opening metabolism again resulting in liberation of cyanide. Interestingly, replacement of the C8-O substituent with carbon in this template resulted in a loss of activity. Modification of (106) to incorporate structural features from the imidazopyridine SAR has been recently reported by the Nycomed group [199]. Thus, C6-pyrrolidine amide derivative (106a) with pIC_{50} of 6.9 gave a 10-fold increase in H^+/K^+ ATPase-inhibitory potency relative to (106). There were suggestions, however, of different SAR in this series compared with the imidazopyridines since the primary amide analogue, analogous to that of the potent AR-H47108 (48), was inactive at 100 μM.

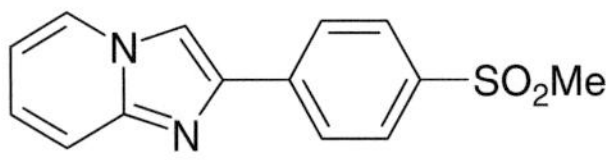

(105) Zolmidine

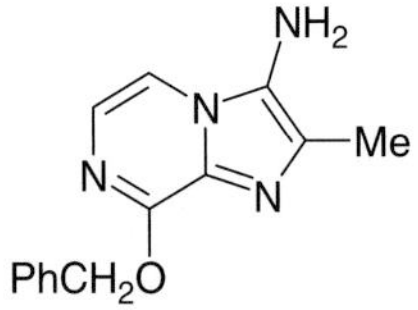

(106)

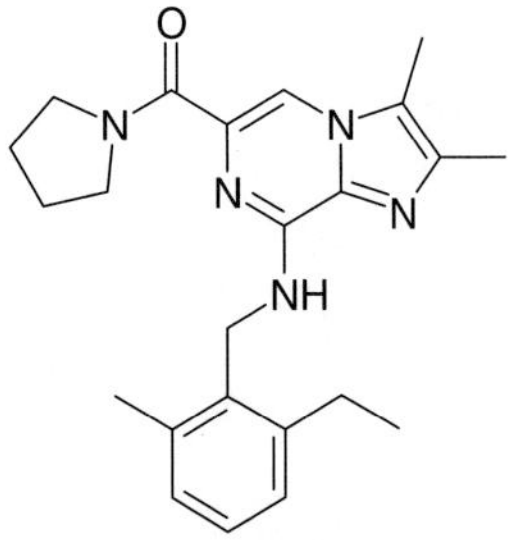

(106a)

Further elaboration of the imidazopyridine template to give alternative heterobicyclic aromatic systems has also been investigated by a number of groups. Early work by the Astra group evaluated a number of templates, including those shown, (107)–(112) [200]. General structure (112), with $R^6 = H$, was also investigated by the Schering group [201]. The Astra group also investigated the imidazo[1,2-a]pyrazine template akin to (106) and, of the examples disclosed, the 8-NH-substituted analogue (113) (H^+/K^+ ATPase $IC_{50} = 0.16\,\mu M$) is more potent than the corresponding 8-O-substituted analogue (114) (H^+/K^+ ATPase $IC_{50} = 2.78\,\mu M$). The position of the most basic centre may now be at the pseudo aminopyridyl centre in (113).

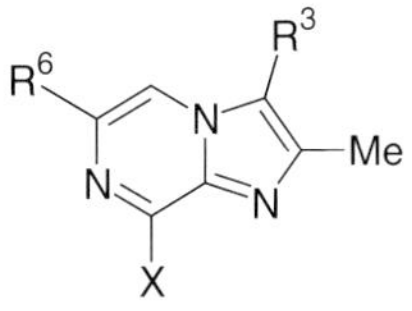

(107) 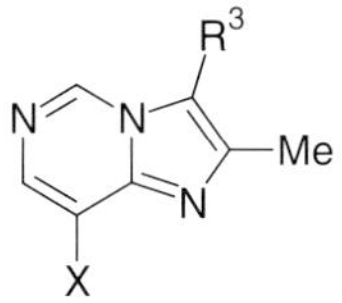 (108) 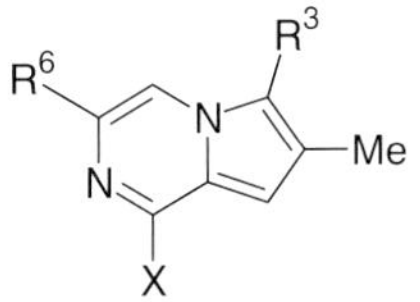 (109)

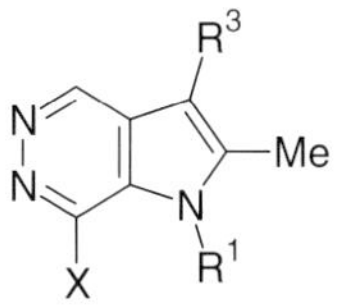

(110) 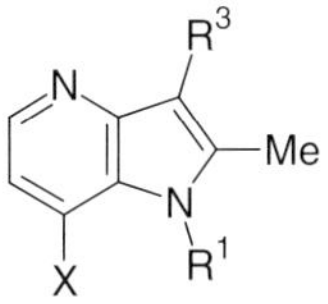 (111)

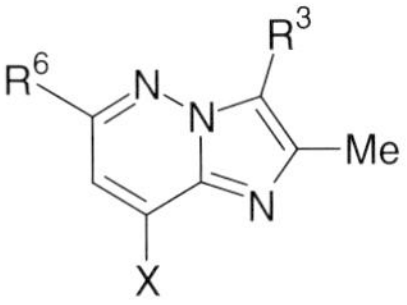

(112)

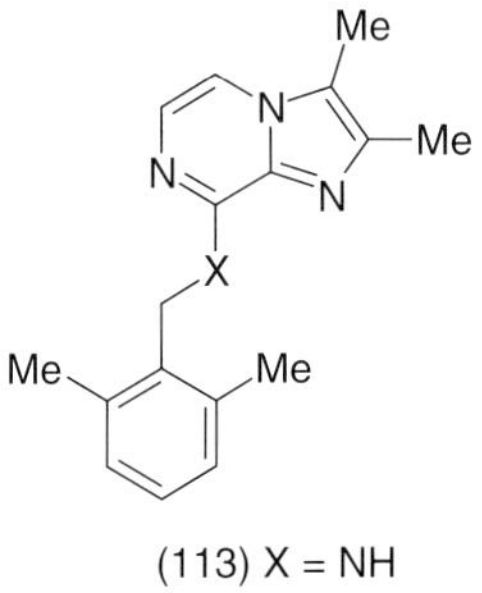

(113) X = NH
(114) X = O

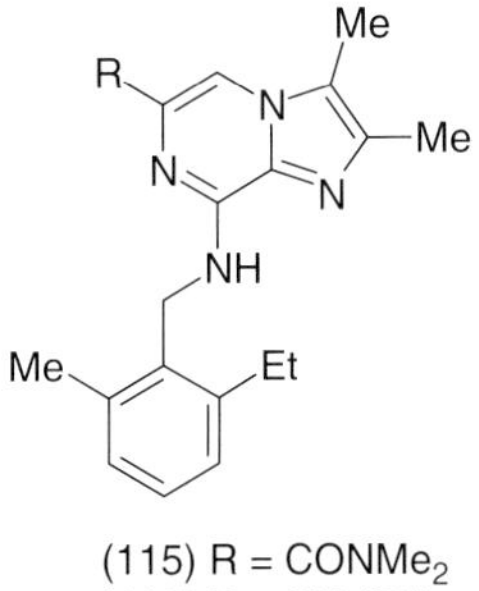

(115) R = $CONMe_2$
(116) R = CH_2OMe

The Altana group has also reported data for compounds with the imidazo[1,2-a]pyrazine template [202]. Thus, 1 μmol/kg i.d. doses of compounds (115) and (116) are claimed to inhibit by 49 and 46%, respectively, the pentagastrin-stimulated acid secretion *in vivo* in the rat-perfused stomach model.

The pyrrolo[1,2-a]pyrazine template (e.g., compound (117)) was investigated by the Syntex group [203], although no specific data was presented. Other related templates attracting attention, but for which there is little data provided, are those of structures (118) [204] and (119) [205].

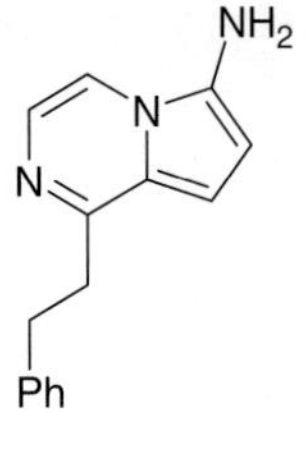

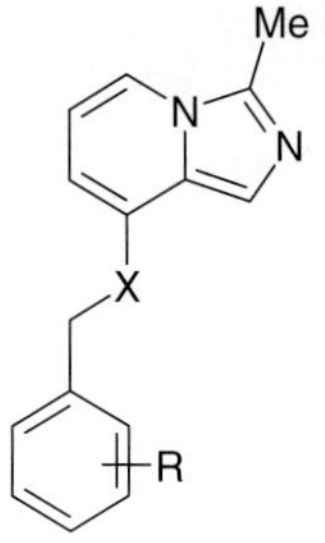

The quaternary ammonium derivative (120) affords further evidence that the active species in all these related systems is the protonated form [206]. (120) afforded an ED_{50} of 1.3 μmol/kg by the i.v. route for suppression of pentagastrin/carbachol-induced acid in the female Sprague–Dawley rat.

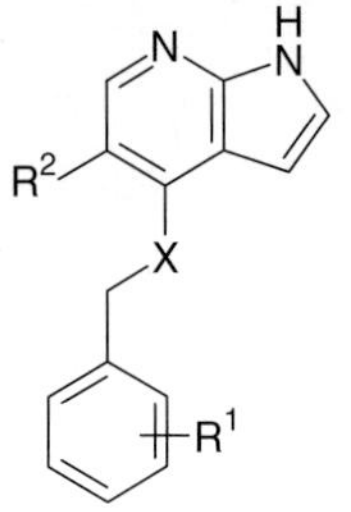

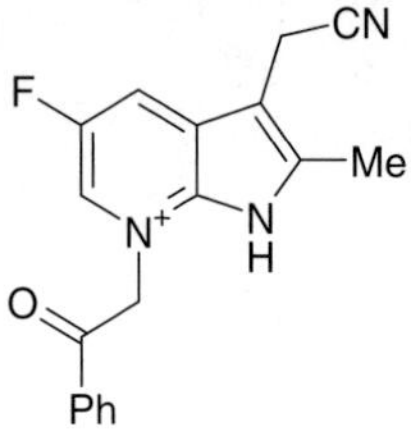

Pyrrolopyridine systems in which the pendant aryl-containing group is substituted from the nitrogen of the 5-membered ring have also been investigated. The Nisshin group described a series of compounds, exemplified by compound (121) [205]. This affords a modest 66% inhibition of H^+/K^+ ATPase activity at a concentration of 10 μg/mL. The same group also reported compounds exemplified by (122) [207].

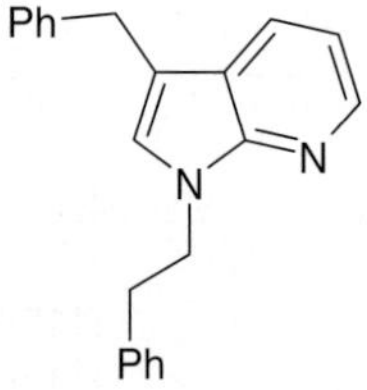

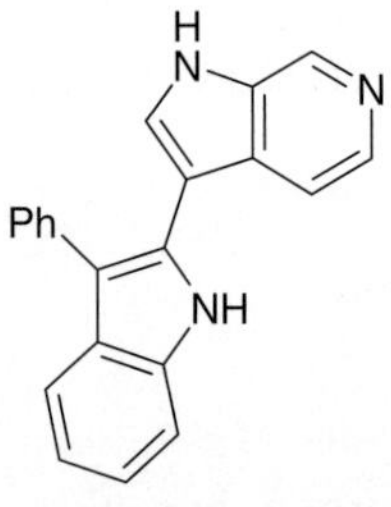

Certain heterobicyclic aromatic cores have received particular attention and these are reviewed systematically below.

BENZIMIDAZOLES

Early work on the benzimidazole template identified compounds such as (123) and (124) with IC_{50} values of 12.9 and 7.8 μM, respectively, for inhibition of acid secretion *in vitro* in isolated rabbit gastric gland [208–210]. The Astra/ AstraZeneca group showed that *ortho* substitution in the benzylamino group, as in compound (125), gave a significant increase in potency, with IC_{50} values <1.6 μM [210]. There is no information regarding activity of compounds bearing N1-methyl substitution (e.g., (126)), which are also claimed.

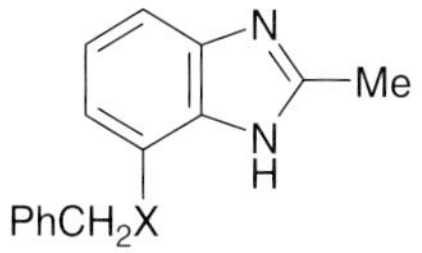

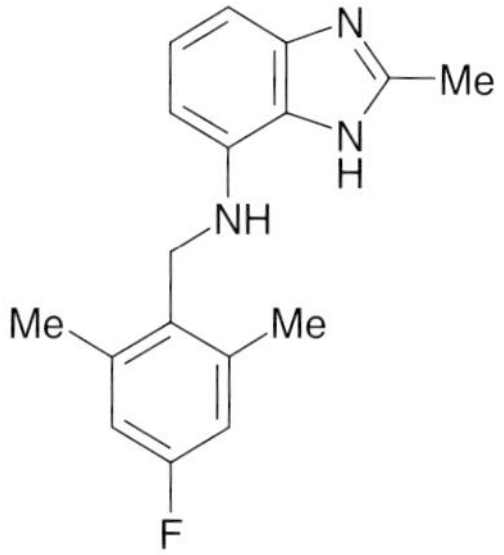

Similar substitution to that identified above for the imidazopyridine template has been applied to the benzimidazole template to afford compounds exemplified by (127) and (128), which are claimed to give at least 50% inhibition of pentagastrin-stimulated acid secretion of the perfused rat stomach after intraduodenal administration of 1 μmol/kg [211]. The N1 substituent has been further extended to give compounds such as (129) with a reported IC_{50} for inhibition of ion-tight vesicle porcine H^+/K^+ ATPase of 84 nM, and good inhibitory activity in the Heidenhain pouch model of histamine-stimulated gastric acid secretion [212].

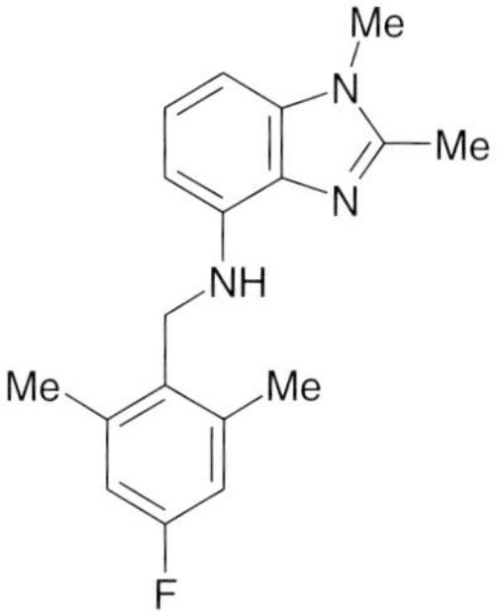

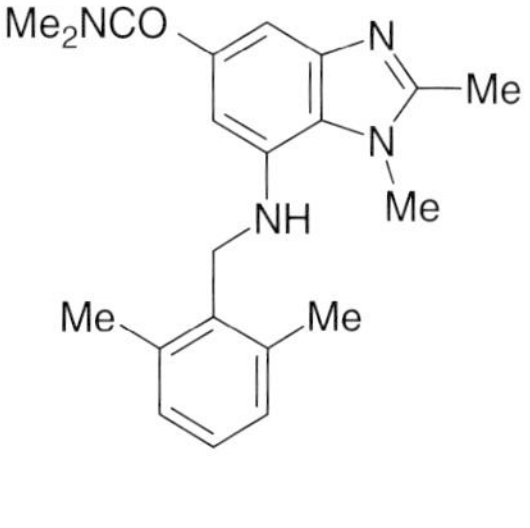

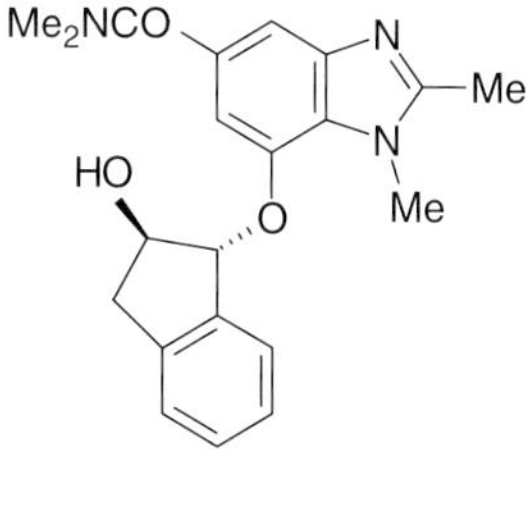

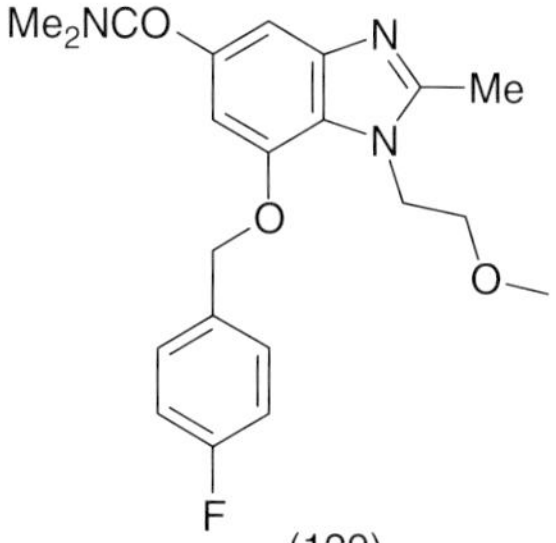

Table 3.10 EXAMPLES OF BENZIMIDAZOLE DERIVATIVES

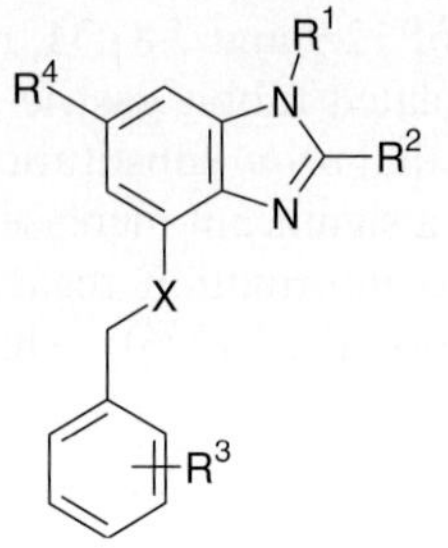

Compound	X	R^1	R^2	R^3	R^4
(130)	O	Me	Me	H	CH_2OMe
(131)	NH	Me	Me	2-Et, 6-Me	$CONH(CH_2)_2OH$
(132)	NH	Me	CH_2OMe	2-Et, 6-Me	$CONMe_2$
(133)	NH	H	Me	2-Et, 6-Me	$CONMe_2$

The isomeric benzimidazoles (130)–(133) (Table 3.10) and (134) are also reported to be active *in vivo* under the same assay conditions, but no distinction is made regarding relative potency for one isomer vs. the other [213]. Pfizer scientists have also applied their chromanyl-4-yl substituent in place of the benzylamino substituent at C8 to give analogues such as (135)–(140) (Table 3.11) [214] and (141)–(146) (Table 3.12) [215]. As with their imidazopyridine analogues described above, stereochemistry at the attachment point or the regioisomeric nature of the chromanyl substituent does not affect potency (this time in the ion-tight H^+/K^+ ATPase preparation). Compounds (139) and (141) are described as showing good inhibition of gastric acid secretion in the Heidenhain pouch dog model. They have extended this to replacement of the benzyl moiety with a tetrahydronaphthalen-2-yl or dihydroinden-2-yl moiety as in compounds (147) and (148), respectively [216]. Although the atom count between the benzimidazole core and the pendant benzylic phenyl ring is increased by one, potencies in the ion-tight vesicle assay are still 0.23 and 0.18 µM, respectively.

Table 3.11 FURTHER EXAMPLES OF BENZIMIDAZOLE DERIVATIVES

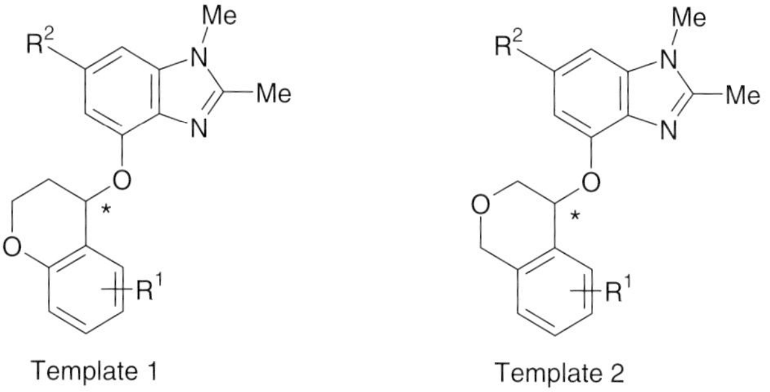

Compound	Template	R^1	R^2	Stereochemistry*	H^+/K^+ ATPase (ion-tight) IC_{50} μM
(135)	1	H	CONMe$_2$	Racemic	0.56
(136)	1	H	CON(Me)(CH$_2$)OH	Racemic	0.43
(137)	2	H	CONMe$_2$	Racemic	0.32
(138)	1	7-F	CONMe$_2$	Racemic	0.069
(139)	1	5,7-DiF	CONMe$_2$	(−)	0.21
(140)	1	5,7-DiF	CONMe$_2$	(+)	0.22

Furthermore, compound (147) is reported to inhibit gastric acid secretion in the Heidenhain pouch dog model with good activity (no specific data is presented) [216].

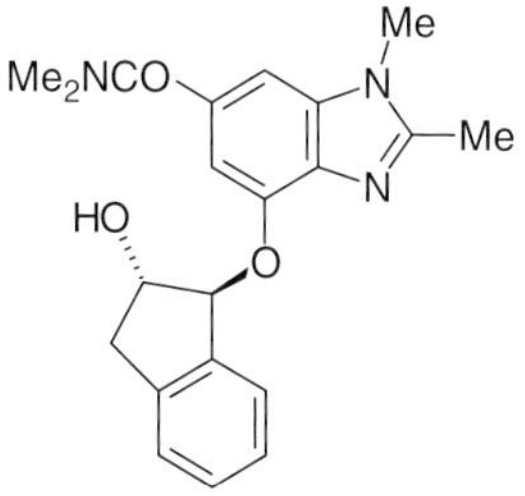

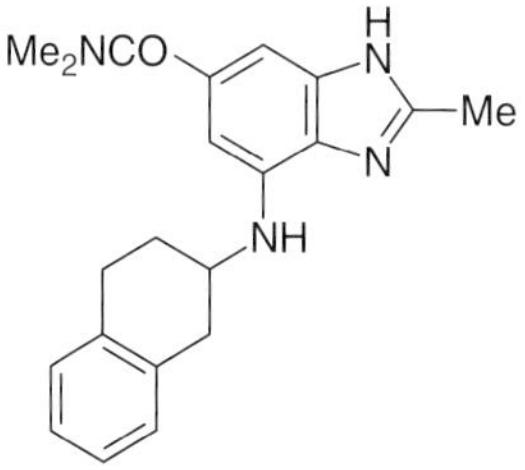

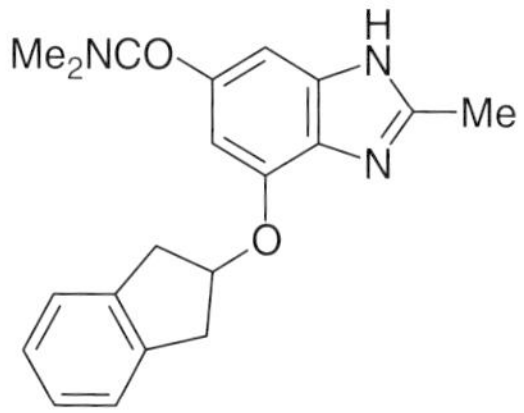

Table 3.12 FURTHER EXAMPLES OF BENZIMIDAZOLE DERIVATIVES

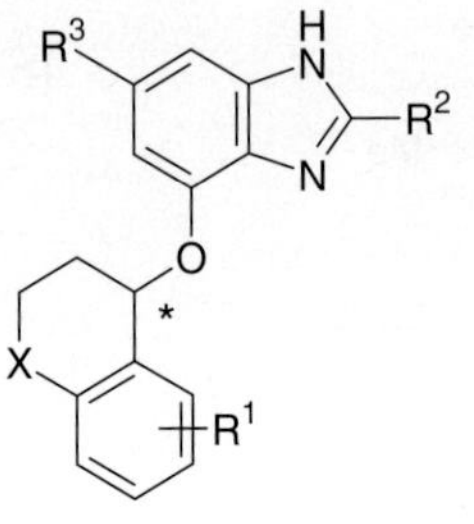

Compound	X	R^1	R^2	R^3	Stereochemistry*	H^+/K^+ ATPase (ion-tight) IC_{50} μM
(141)	O	5,7-DiF	Me	$CONMe_2$	(S)-	0.086
(142)	O	5,7-DiF	Me	$CONMe_2$	(R)-	0.098
(143)	O	5,7-DiF	Me	CO(N-pyrrolidine)	(−)-	0.030
(144)	O	5-Me	Me	$CONMe_2$	Racemic	0.068
(145)	O	5,7-DiF	Et	$CONMe_2$	Racemic	0.21
(146)	S	H	Me	$CONMe_2$	Racemic	0.16

Altana have also patented deuterated analogues based on the benzimidazole template, as discussed above for the imidazopyridine template, with the claimed benefit of improved metabolic stability, although no specific data is presented [217]. Compound (149) has an H^+/K^+ ATPase pIC_{50} of 6.12.

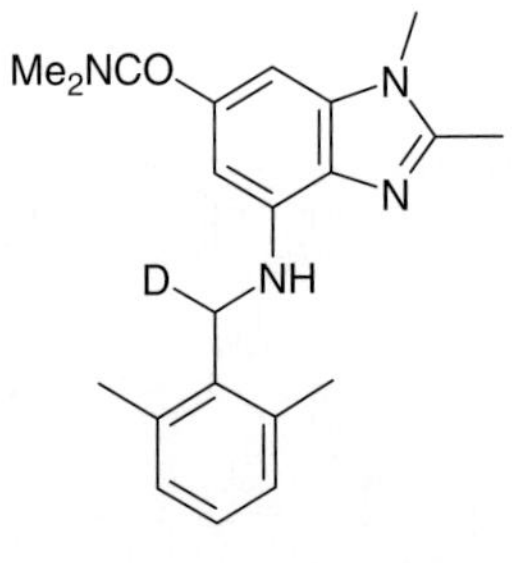

(149)

The corresponding 5-aza-benzimidazoles (150) and (151) are also claimed to afford inhibition of >30% of pentagastrin-stimulated acid secretion following 1 μmol/kg intraduodenal dose *in vivo* in the perfused rat stomach model [218].

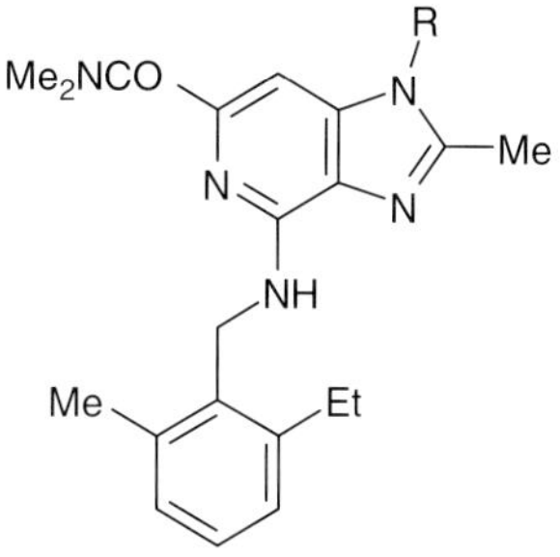

(150) R = Me
(151) R = H

TRICYCLIC BENZIMIDAZOLE ANALOGUES

The Altana Pharma group has applied the same tricyclic modification to the benzimidazole template as they have to the imidazopyridine template [213, 219]. Thus, the compounds of general structure (152) and (153) are claimed to have gastric protective activity and acid suppressive activity of at least 15% in the perfused rat stomach following intraduodenal administration of 3 μmol/kg. Again the preferred stereochemistry of the pendant phenyl group is in keeping with previous reports.

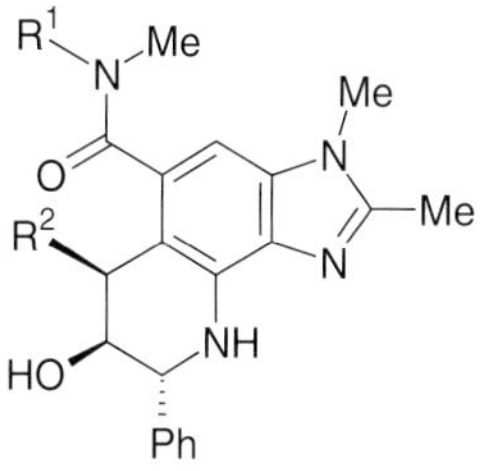

(152) R^1 = H, Me;
R^2 = H, OH, n-C_4H_9O, HC≡CCH_2O

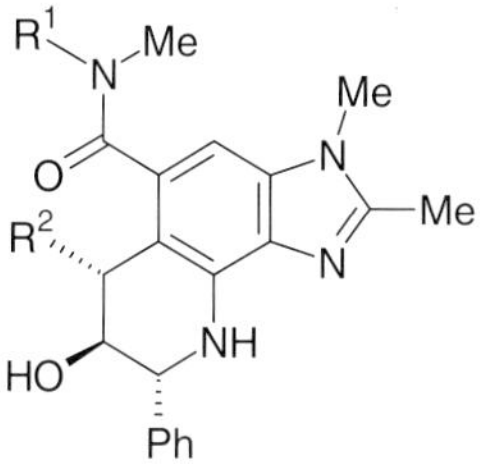

(153) R^1 = H, Me
R^2 = OH, n-C_4H_9O

It is not clear from these data whether the stereochemistry at the R^2-position is important for activity. However, data presented on the

Table 3.13 EXAMPLES OF TRICYCLIC BENZIMIDAZOLE DERIVATIVES

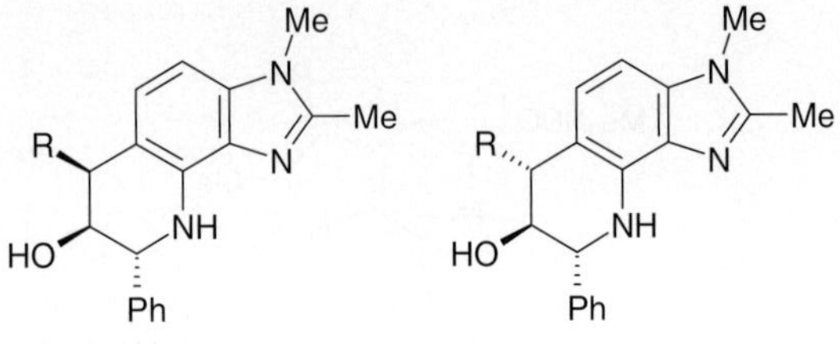

Template A

Template B

Compound	Template	R	% Inhibition[a]
(154)	B	MeOCH$_2$CH$_2$O	>50
(155)	A	MeOCH$_2$CH$_2$O	<50
(156)	B	HO	>50
(157)	A	HO	<50

[a]Pentagastrin-stimulated acid secretion in the perfused rat stomach following 2 μmol/kg intraduodenal administration *in vivo*.

C5-unsubstituted series (Table 3.13) suggests the (*R*)-stereochemistry at the R^2-position is critical [220].

Also, compounds (158)–(160) are reported to afford >50% inhibition of pentagastrin-induced acid secretion *in vivo* in the perfused rat stomach model at a dose of 1 μmol/kg intraduodenally [221]. Again, where distinction is made the focus is on the (8*S*)-stereochemistry. Both the chromene (X = O) and quinoline (X = N) versions of the template are reported to be active *in vivo*.

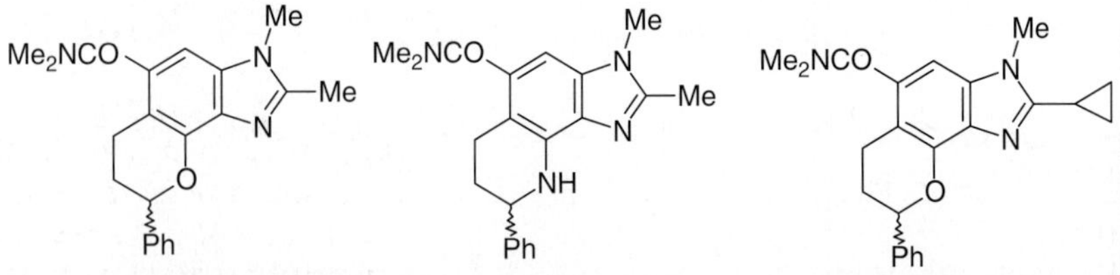

(158)   (159)   (160)

As for the imidazopyridine template, Altana have patented a series of spiro-tricyclicbenzimidazole analogues, exemplified by compound (161) which affords >80% inhibition of pentagastrin-stimulated acid secretion of the perfused rat stomach after 1 μmol/kg intraduodenal administration [222].

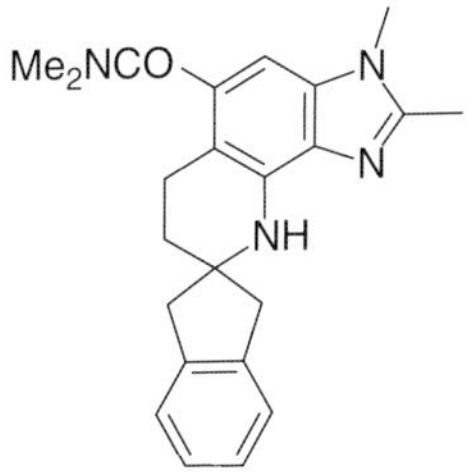

(161)

TRIAZOLOPYRIDINES

Although originally investigated by the Schering group [223], the triazolo[1,5-a]pyridines [224] and the triazolo[4,3-a]pyridines [225] have been explored by the Altana group. Data on (162)–(165) (Table 3.14)

Table 3.14 EXAMPLES OF TRIAZOLOPYRIDINE DERIVATIVES

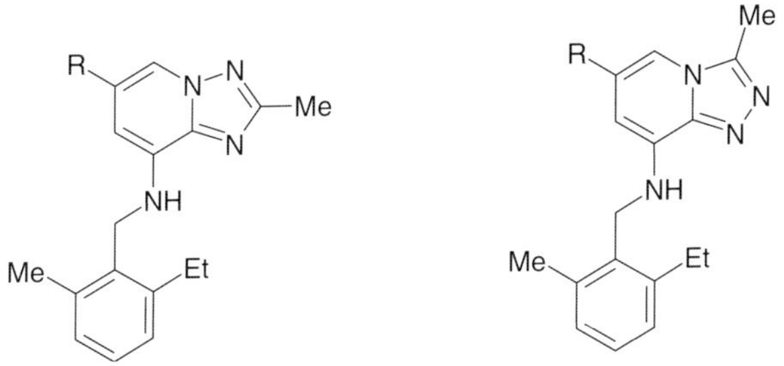

Compound	R	% Inhibition[a]
(162)	Me$_2$NCO	>25
(163)	MeOCH$_2$CH$_2$NHCO	<25
(164)	Me$_2$NCO	>30
(165)	MeOCH$_2$CH$_2$NHCO	>30

[a]Pentagastrin-stimulated acid secretion in the perfused rat stomach following 1 μmol/kg intraduodenal administration *in vivo*.

suggests that the latter provides the more potent template *in vivo*, but neither of these templates is extensively exemplified.

PYRROLO[2,3-D]PYRIDAZINES

The Sankyo/Ube group have developed a series of pyrrolo[2,3-d]pyridazines [187, 226, 227] for which there is an emphasis on the N-[(S,S)-2-methylcyclopropyl-methyl] substituent [228]. A multitude of different salt versions of compound CS-526 (166) has been disclosed [229]. Again, the position of the basic centre relative to the pendant aryl group will be different to that in the imidazopyridine series. Compound (166) is reported to inhibit hog gastric H^+/K^+ ATPase with an IC_{50} of 61 nM in a reversible, competitive manner [230]. It affords inhibition of gastric acid secretion in the pylorus-ligated rat with ID_{50} values of 0.7 mg/kg p.o. and 2.8 mg/kg i.d. It was also active following oral dosing in the dog Heidenhain pouch model. These observations were suggested [230] to imply that the compound may act on parietal cells through systemic availability, but also by direct permeation through the gastric mucosa. In a rat model of reflux oesophagitis (166) gave ID_{50} values of 5.4 mg/kg i.d. and 1.9 mg/kg p.o. for prevention of oesophageal lesions. It is also claimed [228] to have an excellent antibacterial effect against *H. pylori*, although no data are reported.

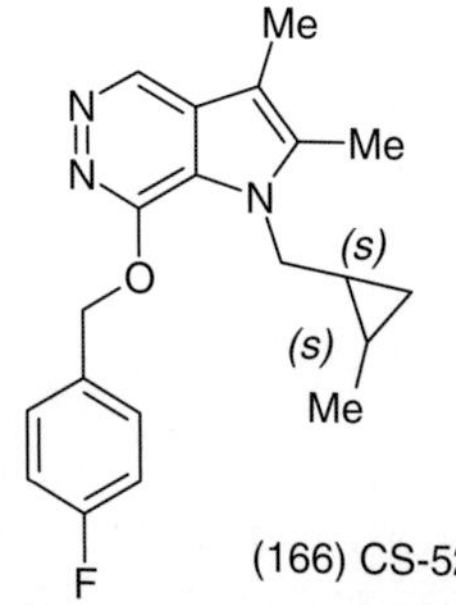

(166) CS-526, R105266

In 2001, Daiichi Sankyo and Ube initiated the co-development of (166) for the treatment of gastric ulcers and GERD. In 2003, Novartis obtained world-wide development rights to (166), as well as four other drug candidates, and initiated a Phase I European trial with (166) for GERD and gastric ulcers. Of all the alternative bicyclic heteroaryl templates investigated, this is the only one so far to have been reported to have progressed into clinical development. However, in 2005 Novartis announced that development of (166) had been terminated, although the reasons for this are not clear.

PYRROLOPYRIDINES

The Yuhan and Takeda groups have published patents disclosing various pyrrolopyridine isomers. Some of these appear to show excellent *in vitro* and *in vivo* potencies. Yuhan has reported on the pyrrolo[3,2-c]pyridine derivatives exemplified by (167)–(169) [231]. These derivatives inhibit pig H^+/K^+ ATPase (lyophilised vesicles) with IC_{50} values in the 0.1–1 μM range. This inhibition was reported to be fully reversible for (167). Substitution on N1 is tolerated, although increasing size of substituent appears to result in some loss in activity *in vitro*. Furthermore, (167) and (168) showed significant inhibition of acid output in the pylorus-ligated rat following oral dosing. Replacement of the tetrahydroisoquinoline (THIQ) moiety in (167) with a benzylamino substituent (169) gives similar levels of potency.

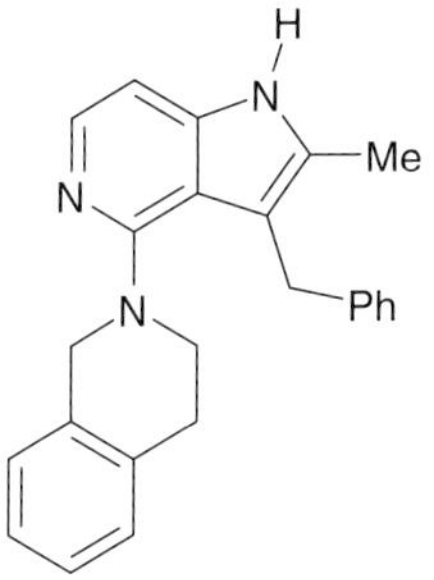

(167) 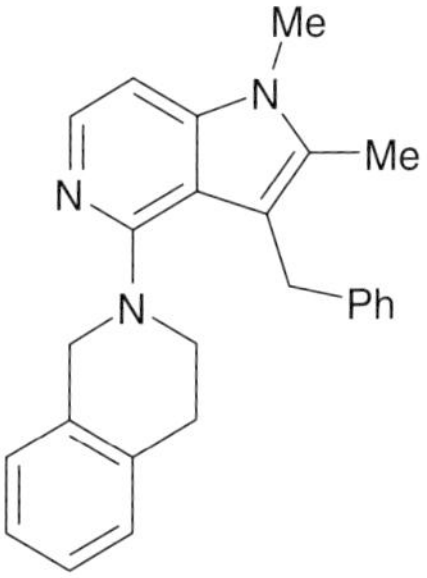 (168) 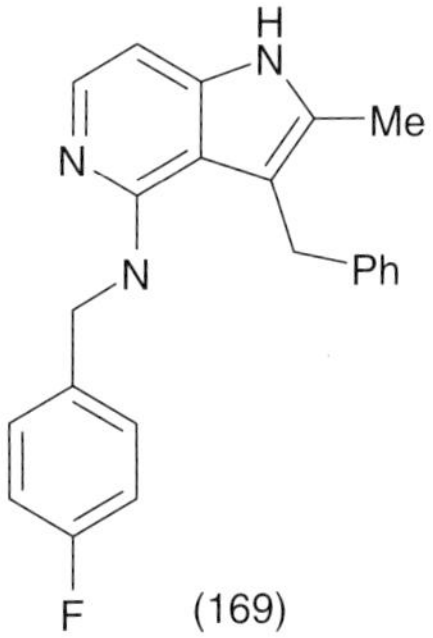 (169)

The isomeric pyrrolo[2,3-c]pyridine template was originally investigated by the Nisshin group [221], but the Yuhan group has more recently extended the investigation of this template to generate compounds such as (170)–(175) (Table 3.15) [232]. Disclosed data indicates H^+/K^+ ATPase IC_{50} values in the range of 0.03 to ~6 μM. Various lipophilic substituents are tolerated at N1. Substituents in the benxyloxy group are largely restricted to the *ortho-* and *para-*positions. *ortho-*Methyl and *para-*fluoro substituents are well tolerated, but chloro, and especially 2,4-dichloro are less well tolerated. The 5-carboxamide substituent in (175) appears to provide good H^+/K^+ ATPase-inhibitory activity. The Yuhan team also

Table 3.15 EXAMPLES OF YUHAN PYRROLOPYRIDINES

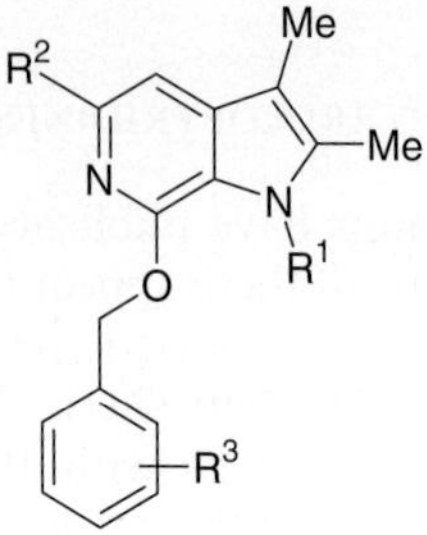

Compound	R^1	R^2	R^3	H^+/K^+ ATPase IC_{50} (μM)	Rat ED_{50} (mg/kg p.o.)[a]
(170)	MeO(CH$_2$)$_2$	H	4-F	0.047[b] (0.17)	
(171)	HC≡CCH$_2$	H	4-F	0.093[b] (0.18)	14
(172)	PhCH$_2$	H	4-F	0.39	
(173)	4-MeO-PhCH$_2$	H	4-F	1.84	
(174)	MeO(CH$_2$)$_2$	H	2,6-di-Cl	2.32	
(175)	PhCH$_2$	CONMe$_2$	4-F	0.028	

[a]Acid output in pylorus-ligated rat 6 h post p.o. dose.
[b]Tested as free base, value in brackets HCl salt.

claim and exemplify (but do not provide H^+/K^+ ATPase-inhibitory data for) other N1- and C7-substituent combinations in this template including some for which the C7-substituent is other than benzyloxy [232].

Compounds (176)–(179) are further examples of this type. *In vivo* data is provided for only (171). This suggests a poorer *in vivo* profile than the pyrrolo[3,2-c]pyridines (167)–(169) and the C7-amino analogues described in Table 3.16 (180–187).

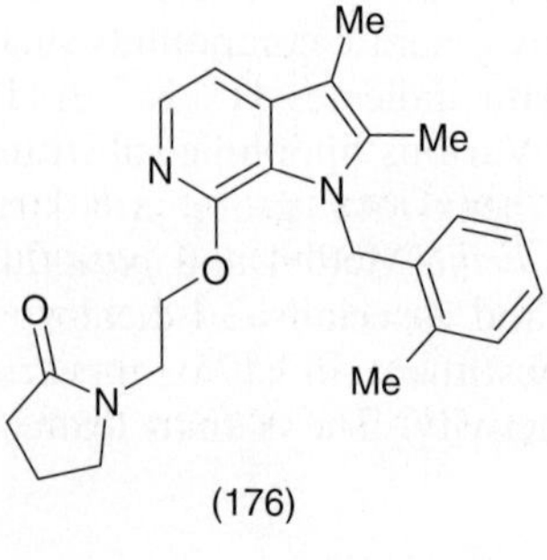

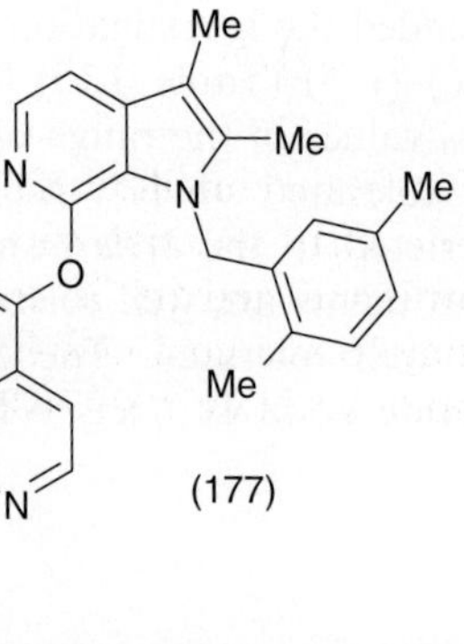

Table 3.16 FURTHER EXAMPLES OF YUHAN PYRROLOPYRIDINES

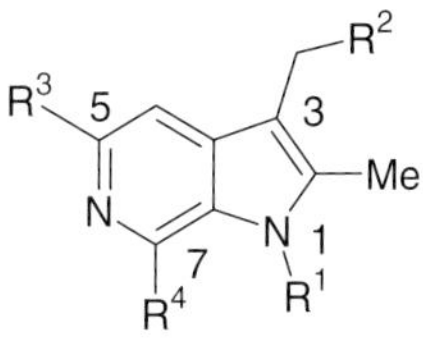

THIQ = (1,2,3,4-tetrahydroisoquinolin-2-yl)

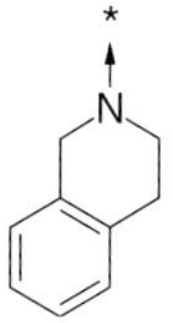

DHIQ = (3,4-dihydroisoquinolin-2-yl)

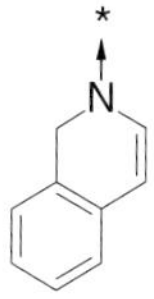

Compound	R^1	R^2	R^3	R^4	H^+/K^+ ATPase IC_{50} (μM)	Rat ED_{50} (mg/kg p.o.)[a]
(180)	iPrCH$_2$	H	H	4-F-PhCH$_2$NH	0.094	1.3
(181)	H	H	H	2-Naphthyl	0.063	11.2
(182)	CH$_2$:CHCH$_2$	H	H	THIQ	0.004	1.4
(183)	PhCH$_2$	CN	H	DHIQ	0.01	–
(184)	HC≡CCH$_2$	H	CONH$_2$	DHIQ	0.0015	–
(185)	CyPrCH$_2$	H	CONH$_2$	DHIQ	4.05	–
(186)	CH$_2$:CHCH$_2$	H	CO-(4-Me-piperazin-1-yl)	DHIQ	0.0007	–
(187)	(CH$_3$)$_2$C:CHCH$_2$	H	H	4-Cl-PhCH$_2$O	–	–

[a] Acid output in pylorus-ligated rat 6 h post p.o. dose.

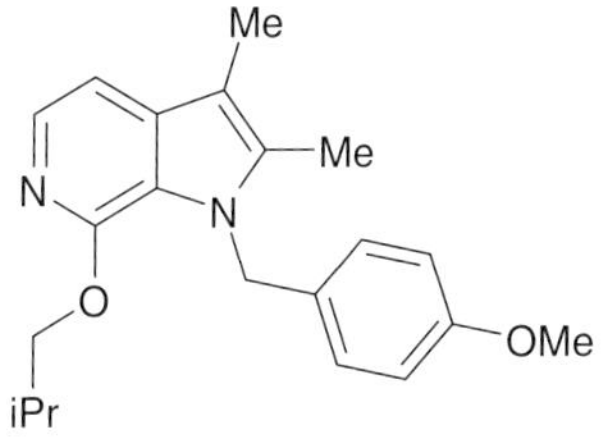

(178)

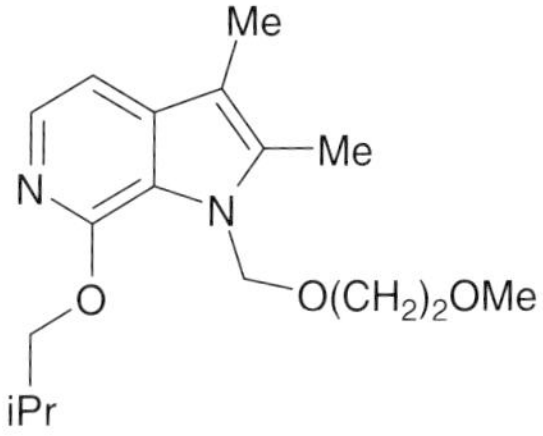

(179)

Further exemplification of this template by the Yuhan group is extensive, with a small selection shown in Table 3.16 [233]. A number of compounds show good activity in the pylorus-ligated rat model 6 h following oral dosing. Some compounds achieve *in vitro* potencies with IC_{50} in the low nanomolar or even sub-nanomolar region (e.g., (186)). However, the *in vivo* activity of these most potent compounds is not described. Emphasis is placed on the tetrahyro- and dihydro-isoquinoline moiety in place of the benzylamino group at C7. The 2-naphthyl moiety (181) at this position also retains some activity, but its *in vivo* performance is inferior. The length and nature of the linker from C7 to the pendant aryl moiety is critical. As in the imidazopyridine template, hydroxymethyl and cyanomethyl substituents are tolerated at the C3-position (e.g., (183)) and substitution at C5 affords compounds (e.g., (184)–(186)) with high H^+/K^+ ATPase-inhibitory potency.

The Yuhan group has extended the claimed utility of this template to include the potential treatment or prevention of cancer, presumably by a similar mechanism to that for PPIs discussed earlier. Thus, compound (187) showed IC_{50} values of 0.4–3.7 µM against the proliferation of a range of cancer cell lines.

The pyrrolo[2,3-c]pyridine template has also been described by the Takeda group at around the same time as the Yuhan team [234]. They focus more on the benzylamino substitutent at C7, exemplified by (188)–(190) in Table 3.17, rather than the THIQ or DHIQ groups, but again particularly potent compounds are achieved, the most potent of which is (190), with an IC_{50} of 3 nM.

Moving the nitrogen in the pyridine ring appears to have no detrimental effect on H^+/K^+ ATPase potency and may even be beneficial in some cases, since compounds such as (191), are reported to have an IC_{50} of 10 nM [233]. Again, SAR suggests that the pyrrolo-*N*-substituent contributes to an enhanced potency. The C7-THIQ group generally affords somewhat less

Table 3.17 EXAMPLES OF TAKEDA PYRROLOPYRIDINES

Compound	R^1	R^2	R^3	H^+/K^+ ATPase IC_{50} (µM)
(188)	H	Me	Ph	0.027
(189)	nPr	Me	4-F-2-Me-Ph	0.024
(190)	iBu	CH_2OH	4-F-2-Me-Ph	0.0032

potent analogues than the C7-(4-fluorobenzyloxy) group although (192) is reported to have an IC$_{50}$ of 30 nM.

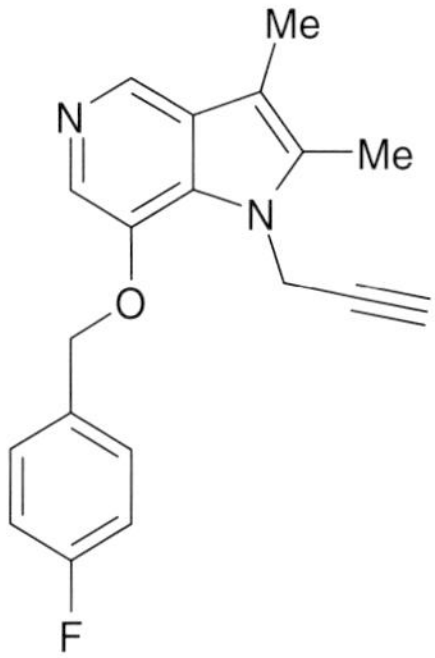

(191)

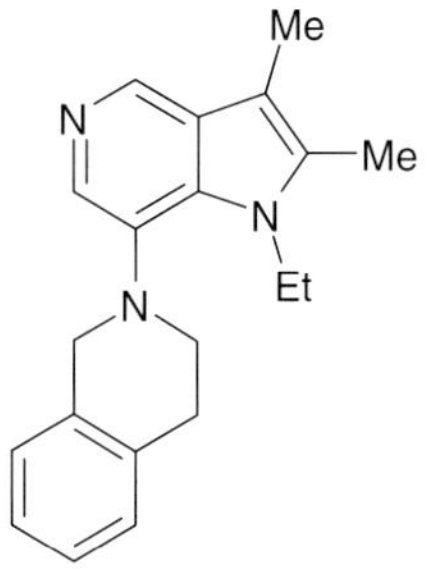

(192)

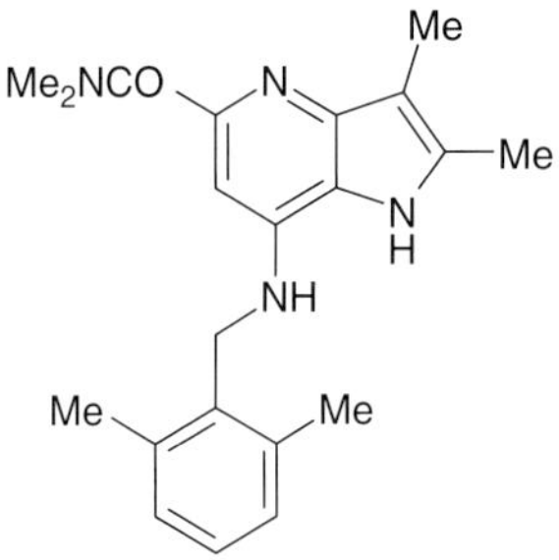

(193)

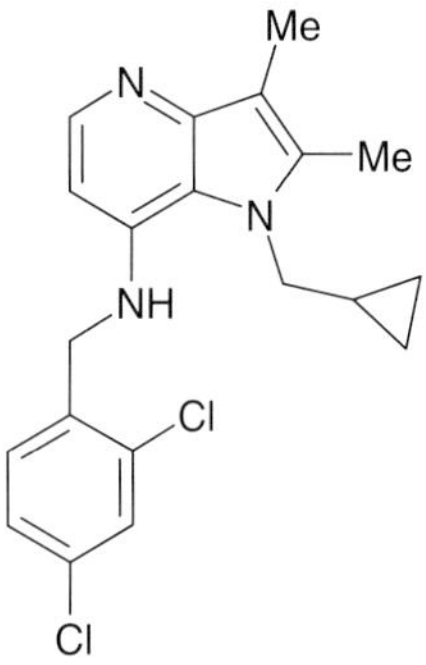

(194)

Moving the nitrogen to afford the pyrrolo[3,2-b]pyridine template has been reported by the Altana Pharma team [235]. For example, (193) gave >50% inhibition in the pylorus-ligated rat model 6 h following a dose of 1 μmol/kg i.d. These compounds lack an N1 substituent and there is a strong preference for the *ortho*-di-substituted benzyl group.

This same template is also reported by the Yuhan group, exemplified by (194), this time bearing the additional N1 substituent [236]. This appears to show somewhat different SAR at C7 – the *ortho* alkyl substituents are no longer required for potency. Compound (194) shows good H$^+$/K$^+$ ATPase-inhibitory activity (IC$_{50}$ = 20 nM) and is active *in vivo* by the oral route in the pylorus-ligated rat model 6 h post-dose ($ED_{50} = 1.6$ mg/kg).

4-AMINO-QUINOLINES

The discussion on reversible H^+/K^+ ATPase inhibitors so far in this review has centred on variations on a benzyloxy or benzylamino-substituted heterobicyclic aromatic. The 3-acyl-4-anilinoquinoline derivatives offer a somewhat different approach and have been a significant focus of efforts to identify reversible H^+/K^+ ATPase inhibitors [237–240]. The progenitor molecule for the series was compound (195). The key structural features of (195) were believed to be the intramolecular H-bond between the acyl group and exocyclic NH fixing the conformation of the aryl group relative to the mildly basic quinoline ring [237, 241]. The electron-withdrawing effect of the acyl group also contributed to this relative conformation as a result of increasing the conjugation between the N and quinoline ring [237]. This compound has a somewhat lower H^+/K^+ ATPase-inhibitory activity ($IC_{50} = 0.85 \mu M$) compared with that of SCH-28080 (28) ($IC_{50} = 0.1 \mu M$). Although the IC_{50} for accumulation of aminopyrine in intact pig gastric vesicles is more similar to that of (28) (0.037 μM vs. 0.013 μM, respectively), the *in vivo* activity in the lumen-perfused rat following compound administration by the i.v. route was significantly lower ($ED_{50} = 8.2$ μmol/kg vs. 0.31 μmol/kg for (28)). This is likely to be due to rapid metabolism to the inactive C2-acid derivative and is suggested also to account for the observed nephrotoxicity following 4 weeks dosing in the rat at 200 mg/kg/day p.o. [242].

The i-Pr analogue AHR-9294 (196) has also been explored and has a K_i of 0.28 μM for inhibition of H^+/K^+ ATPase activity. In the rat-perfused stomach model, (196) (0.25–4 mg/kg i.v.) achieved comparable inhibition to omeprazole (1) (0.25–1 mg/kg i.v.) at 1 h post-dose of pentagastrin-, histamine- or carbachol-stimulated acid secretion. In the Heidenhain pouch dog, histamine-stimulated acid secretion was rapidly and fully inhibited by a single dose of 3.2 mg/kg i.v. of (196) and this was maintained for at least 2 h post-dose. However, in the same model, meal-induced acid secretion was inhibited by 1 mg/kg i.v. of (196) for only about 2 h and then returned to control levels by 4 h, while (1), at the same dose, fully inhibited acid secretion throughout the duration of the experiment. Thus, while potency is achievable with this molecule its duration of action is somewhat compromised, presumably again due to metabolism of the ester group.

The SK&F group identified clinical candidate molecules from this series. The first of these, SKF 96067 (197) [242], affords inhibition of lyophilised pig gastric vesicle H^+/K^+ ATPase ($IC_{50} = 1.05\,\mu M$) and inhibition of aminopyrine accumulation in intact gastric vesicles with an IC_{50} of $0.06\,\mu M$, confirming accumulation of the mildly basic drug in the acidic compartment [243]. The compound is selective over Na^+/K^+ ATPase at pH 7.0 by at least 30-fold [241]. (197) showed inhibition of pentagastrin-stimulated acid secretion in the perfused rat lumen with ED_{50} of $2.6\,\mu mol/kg$ i.v. and showed none of the nephrotoxicity of (195), with no significant toxicological effects on dosing up to 1,000 mg/kg for 90 days in both rat and dog [242]. Oral dosing with (197) ($4\,\mu mol/kg$) afforded a longer duration of inhibition of histamine-stimulated acid secretion in the Heidenhain pouch dog than for the H_2RA cimetidine, with an ED_{50} of $1.6\,\mu mol/kg$ p.o. A significant dose-dependent reduction of 24 h intragastric pH was obtained after twice-daily oral dosing of SKF 96067 (300 mg) in healthy male volunteers [244], showing significant superiority to oral, twice-daily ranitidine (150 mg), and anti-secretory effects directly related to plasma levels. The compound had a T_{max} of 2.3 h and a half-life of 26 h [245]. Steady-state plasma concentrations were obtained by day 5 of a 300 mg twice-daily dosing study [244]. In the dose range 400–1200 mg, an oral suspension formulation of SKF 96067 (single dose) was generally well tolerated and afforded increasing systemic exposure with increasing dose, although less than dose-proportionality [246]. Following successful Phase I trials [16], the development of the compound was terminated after Phase II/III trials in Europe by SB and Byk Gulden (now Altana), although the reason for the termination has not been reported.

The SAR of this template has been extensively investigated. *ortho*-Methyl substitution in the pendant phenyl group is preferred over *ortho*-methoxy, which in turn is superior to an unsubstituted phenyl group. Small mesomerically donating substituents, for example, 4-OH, on this phenyl ring are also particularly favoured. The preferred substituents at the 3-position on the quinoline ring appear to be those which maintain its mildly basic nature, but which are also capable of intramolecular H-bonding to the C4-anilino-NH, and of enhancing the conjugation between the quinoline and phenyl rings. A fine balance of electronic effects needs to be maintained, with a high σ_p and relatively low σ_m [243]. Thus, replacement of the propionyl group with a strongly electron withdrawing group such as NO_2, CN or amide reduced potency, while alkyl groups, though affording less potent compounds, are surprisingly well tolerated in view of their lack of H-bonding capability, presumably due to their maintenance of an appropriate quinoline pK_a. The larger amide and alkyl substituents are further disfavoured. In spite of this, other authors have further investigated C3-amide derivatives, demonstrating reversible, potassium competitive H^+/K^+ ATPase inhibition and inhibition of histamine-induced acid secretion in rats

[247]. The aniline moiety in (195) can be replaced with a thienylamino moiety, for example, to give (198) [248]. (198) has an H^+/K^+ ATPase IC_{50} of 0.53 μM and inhibits pentagastrin-stimulated acid secretion in the rat following i.v. administration with an ED_{50} of 4.62 μmol/kg. At C6, a broad variety of O-, C- and N-linked substituents are tolerated, including electron-withdrawing substituents, but the most potent H^+/K^+ ATPase inhibition is again obtained with substituents enhancing the basicity of the molecule. The C8-unmodified compound, (199), at 10 μmol/kg i.v., was somewhat less potent than the 8-methoxy analogues and afforded only 61% inhibition of pentagastrin-stimulated acid secretion in the dog [249].

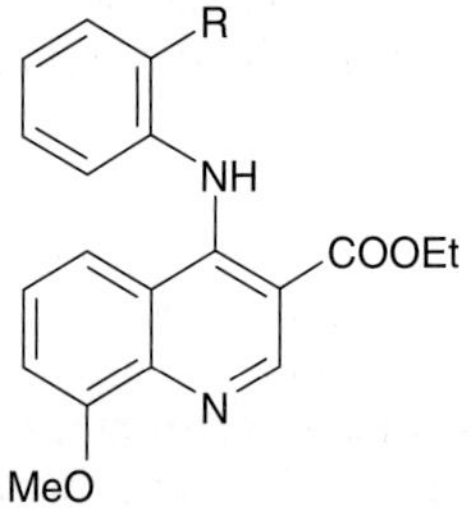

(195) R = Me
(196) AHR-9249 R = ^{i}Pr

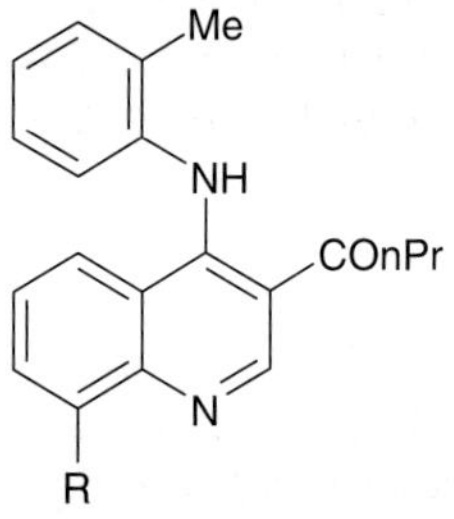

(197) SKF 96067 R = $O(CH_2)_2OH$
(200) SKF 97574 R = OMe
(201) R = CH_2OH

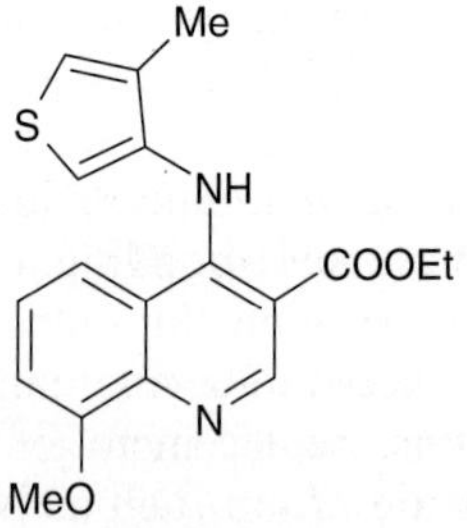

(198)

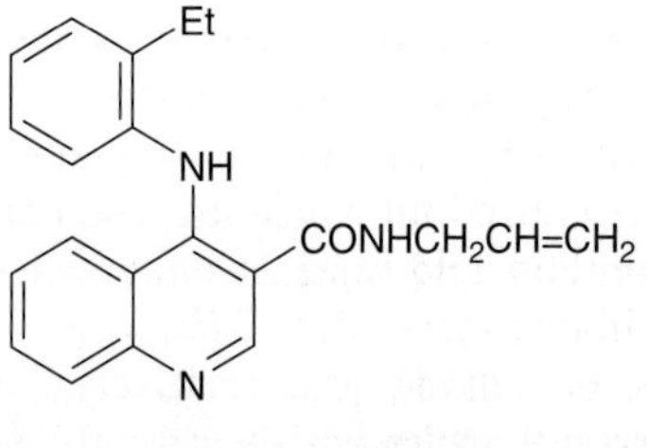

(199)

Extensive investigation of SAR, in particular of the C8-position and of the C4-aniline substituents, led to identification of SKF 97574 (200) [250]. This apparently has the best overall balance of potency, improved PK and *in vivo* potency. It was shown to be a potassium-competitive, reversible inhibitor of H^+/K^+ ATPase in lyophilised pig gastric membrane vesicles with an IC_{50} of 0.46 μM [238]. Although of a different structural type, the inhibition was shown to be competitive with that of SCH-28080 (28), suggesting a similar binding site. As with the imidazopyridine, it is again a weak base with a pK_a of 6.86, and accumulation in the acidic compartment of the parietal cell would again be expected. The H^+/K^+ ATPase-inhibitory activity is pH dependent, and consistent with the protonated form being the active species. (200) is selective over the Na^+/K^+ ATPase. In the Heidenhain pouch dog model, (200) inhibited histamine-stimulated acid secretion with an ED_{50} of 0.89 μmol/kg p.o. The duration of action in this model was found to be longer than that of (197), as well as that of the H_2RA cimetidine, but shorter than that of omeprazole (1). The development of (200) was terminated after Phase II trials, but again the reasons for this are not clear.

H^+/K^+ ATPase-inhibitory activity is retained when the quinoline is replaced with a thienopyridine [251] or thienopyrimidine [252] core. Compound (202) afforded an H^+/K^+ ATPase-inhibitory IC_{50} of 3 μM, but gave only a modest 32% inhibition of pentagastrin-stimulated acid secretion *in vivo* at 10 μmol/kg i.v. [251]. Compound (203) gave an H^+/K^+ ATPase IC_{50} of 19 nM, but still only 50% inhibition of rat gastric acid secretion following 10 μmol/kg i.v. dosing. Although in the histamine-stimulated Heidenhain pouch dog model, (203) gave 84% inhibition of acid secretion at 1 μmol/kg i.v., it afforded only 12% inhibition following an oral dose of 4 μmol/kg in the same model.

The quinoline has been replaced with a diaminoquinazoline moiety [252, 253]. Again, *in vivo* potency is modest, with (204) affording only 60% inhibition of pentagastrin-stimulated acid secretion in the rat at 10 μmol/kg i.v. [253]. Also while (205) gives an H^+/K^+ ATPase IC_{50} of 32 nM, it affords 94% inhibition of pentagastrin-stimulated acid secretion in the anaesthetised rat model at 10 μmol/kg i.v., and 91% inhibition of histamine-stimulated acid secretion in the Heidenhain pouch dog model at 1 μmol/kg i.v., it again gives only 32% inhibition when dosed orally at 4 μmol/kg in the same dog model [252]. The group from Yungjin Pharma Co. replaced the quinoline nucleus with the naphthyridine core to afford compounds with modest H^+/K^+ ATPase activity as well as moderate rat *in vivo* acid suppressive and anti-ulcer activity [254]. Compound (206), YJA20379-8 was singled out for further investigation and shown to be a reversible H^+/K^+ ATPase inhibitor, but with a short duration of action (less than 7 h at a dose of 30 mg/kg i.d.) [255].

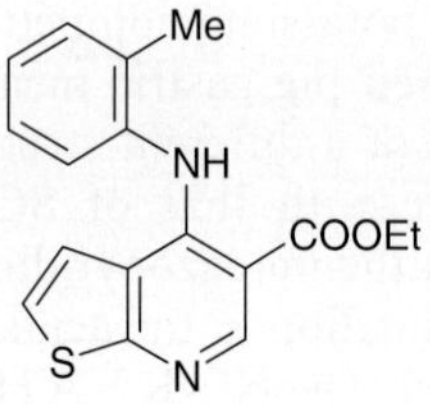

(202)

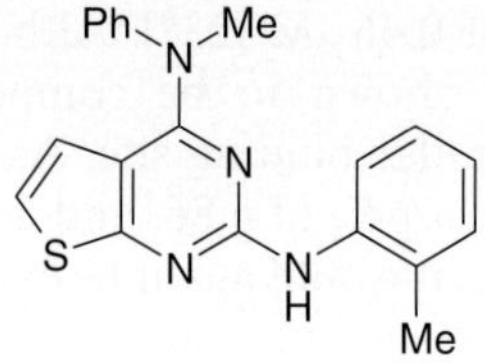

(203)

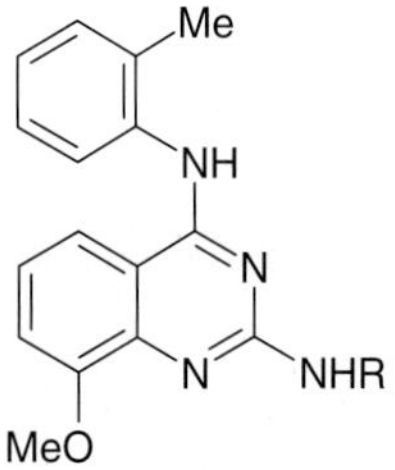

(204) R = H
(205) R = 4-F-2-Me-Ph

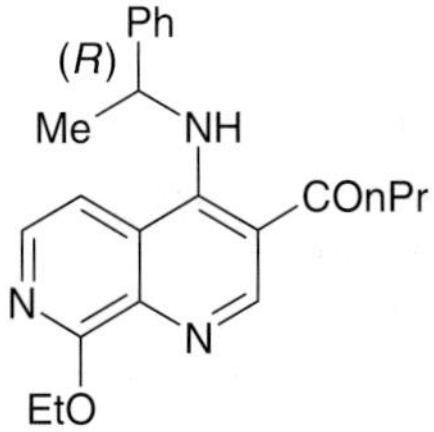

(206) YJA20379-8

The quinoline template has also been investigated by the Astra group [256, 257]. Thus, compounds such as (207) (H 335/25) gave 85% inhibition of pentagastrin/carbachol-induced acid secretion in gastric fistula rats following 6 μmol/kg i.d. The PK/PD relationship for H 335/25 has been studied in both dogs (histamine-stimulated Heidenhain pouch model) and human healthy volunteers (pentagastrin-stimulated). As with the imidazopyridines discussed above, the effect-time profile was delayed compared to the concentration-effect profile, for which modelling in both dog and man was consistent with a slow equilibration between free drug, free enzyme and drug–enzyme complex [258]. To achieve 100% inhibition of acid secretion in the clinic, doses of 500–800 mg were required.

Modification of the 8-position has been a focus for the Kyowa Hakko group, generating molecules such as (208), (209) and (210), which have H^+/K^+ ATPase IC_{50} values of 2.1, 0.67 and 1.1 μM, respectively [259].

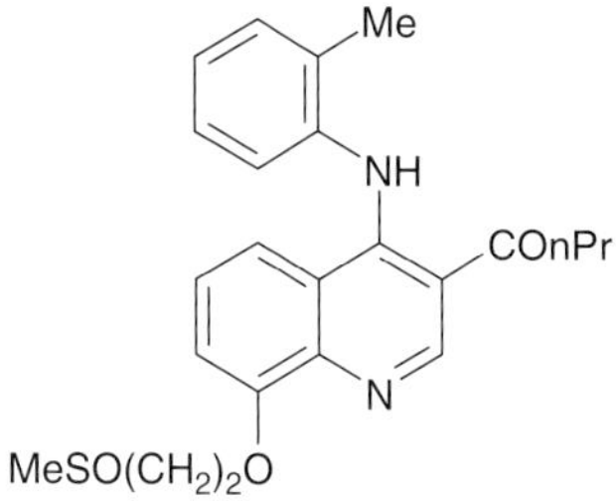

(207)

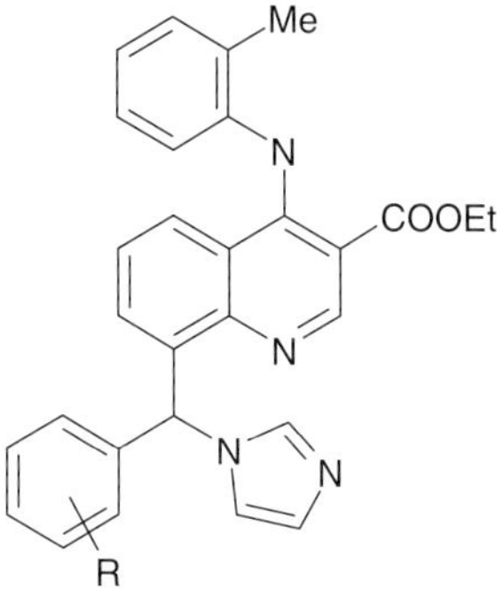

(208) R = 4-F
(209) R = 2,4,6-triMe
(210) R = 3-CF$_3$

Replacement of the intramolecular H-bond between the C4-NH and the C3-acyl group with a covalently cyclised system to give compounds such as (211), has been reported to retain ATPase-inhibitory activity (IC$_{50}$ = 0.42 μM), and inhibition of intact pig vesicle aminopyrine accumulation with an IC$_{50}$ of 28 nM, again confirming the concentrating effect in the acidic compartment [236]. However, at a dose of 10 μmol/kg i.v., only 43% inhibition of pentagastrin-stimulated acid secretion was observed in the lumen-perfused rat. Related compounds (212)–(214) were significantly less active, possibly due to reduced basicity. However, the aromatised-fused pyrrolo derivative (215) gave an almost identical activity profile to that of (211). The N-benzyl derivative (216) was also significantly less active. Both methyl and methoxy groups are tolerated at the 6-position, while both alkyl [260] and particularly substituted amino groups [261] at C4 give good activity. However, groups at C4 significantly affecting the basicity of the quinoline nitrogen greatly reduced H$^+$/K$^+$ ATPase-inhibitory activity; these include Cl, OH, alkoxy and ureido [261].

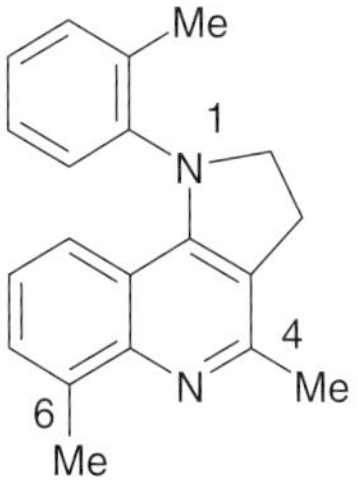

(211) 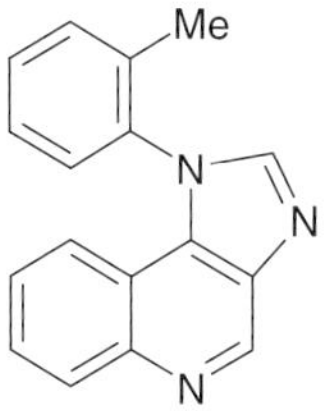　　　　　(212) 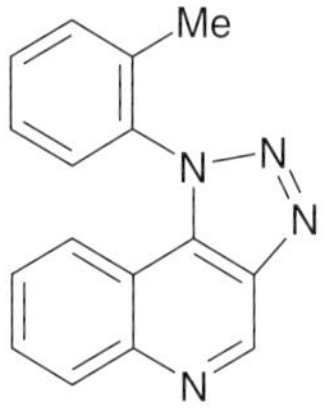　　　　　(213)

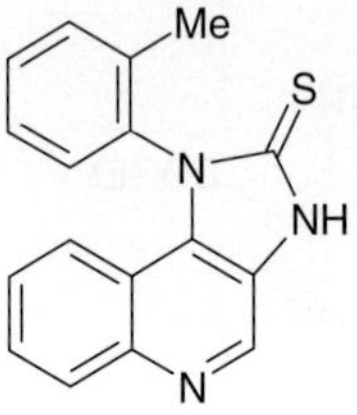

(214) 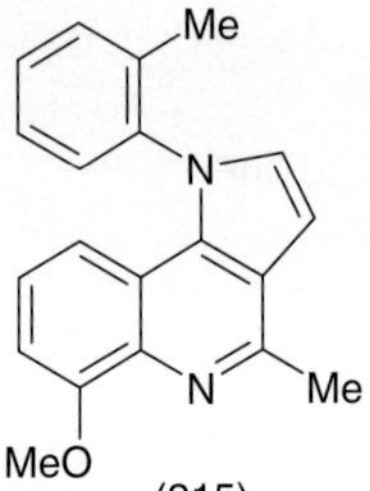   (215) 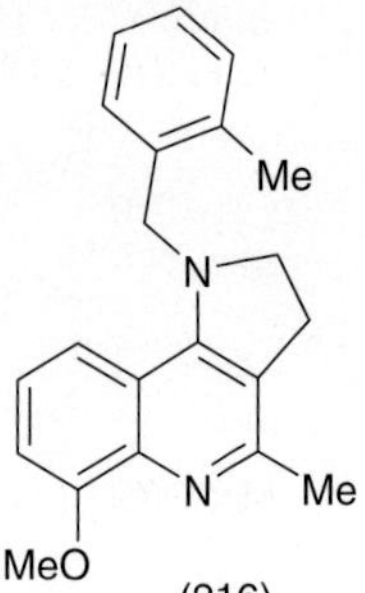   (216)

4-Amino-substituted analogues have also been shown to be active. Thus, MDPQ (217) inhibits K^+-stimulated H^+/K^+ ATPase (ATP hydrolysis) activity with a K_i of 0.22 µM [262]. This tricyclic system has been extended by the Korea Research Institute of Chemical Technology and DBM-819 (218) was identified as a potential drug candidate [263, 264]. (218) was shown to be a reversible, K^+-competitive inhibitor of rabbit gastric H^+/K^+ ATPase, with an IC_{50} of 5 µM [264]. This compound inhibited histamine- and pentagastrin-stimulated acid secretion in the perfused rat stomach model with ED_{50} values of 4.0 and 5.1 mg/kg i.d., respectively. In the chronic-fistulated rat, 10 mg/kg p.o. (218) gave significant acid inhibition for 10 h, compared with 48 h for the same dose of omeprazole (1) in the same model [263].

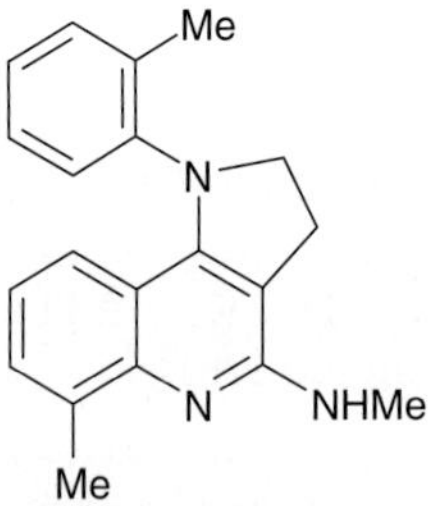

(217) MDPQ 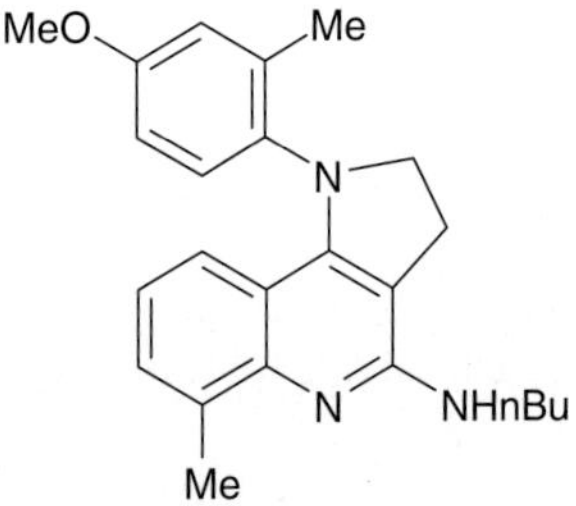   (218) DBM-819

Replacement of the 8-methoxy group with various fluoroalkoxy substituents in the pyrrolo or 2,3-dihydropyrroloquinoline template, to give reversible H^+/K^+ ATPase inhibitors (219–225, Table 3.18), is the focus of work by the Korea Research Institute of Chemical Technology [265]. One

Table 3.18 EXAMPLES OF KOREA RESEARCH INSTITUTE PYRROLOQUINOLINES AND DIHYDROPYRROLOQUINOLINES

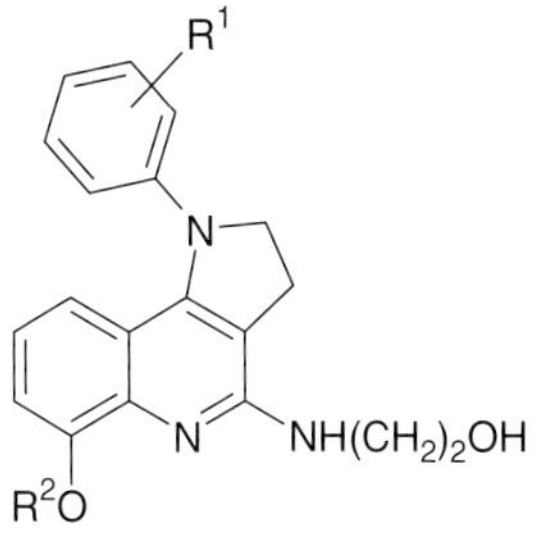

Template A

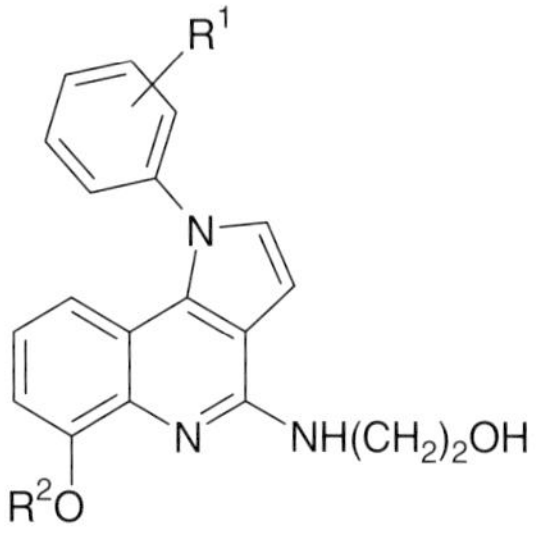

Template B

Compound	Template	R^1	R^2	Rat in vivo[a]
(219)	A	2-Et	CF_3	+ + +
(220)	A	2-OMe	CF_3	+
(221)	A	2-Me-4-OH	CF_3	+ + +
(222)	B	2-Me-4-OH	CH_2CF_3	+ + +
(223)	B	2-Me	CF_2CF_3	+ + +
(224)	A	2-Me	CF_2CF_3	+
(225)[b]	A	2-Me-4-MeO	CH_2CF_3	c

[a] Effect relative to SKF96067, + + + (strong), + + (similar), + (weak) on acid secretion in the pylorus-ligated rat 5 h following 20 mg/kg compound dosed intraduodenally.
[b] AU-461.
[c] ED_{50} = 9.7 mg/kg i.d. for histamine-induced acid secretion in lumen-perfused rat.

of the compounds (AU-461) (225) was singled out, however *in vitro* and *in vivo* potency was relatively poor [266]. (225) showed reversible, K^+-competitive IC_{50} values of 12 and 4.2 μM for inhibition of rabbit and pig gastric H^+/K^+ ATPase, respectively. In the lumen-perfused rat model, (225) afforded inhibition of histamine-stimulated acid suppression with an ED_{50} of 9.7 mg/kg i.d. Significant acid suppression was obtained at 6 h post 10 mg/kg p.o. dose in the chronic fistula rat model following pentagastrin stimulation.

The same group has also looked at 3-substituted analogues (226–229, Table 3.19), some of which afford superior activity to omeprazole (1) in the ethanol-induced lesion model in the rat, but data is only reported for the 1 h timepoint after fairly high doses [267].

Table 3.19 FURTHER EXAMPLES OF KOREA RESEARCH INSTITUTE PYRROLOQUINOLINES

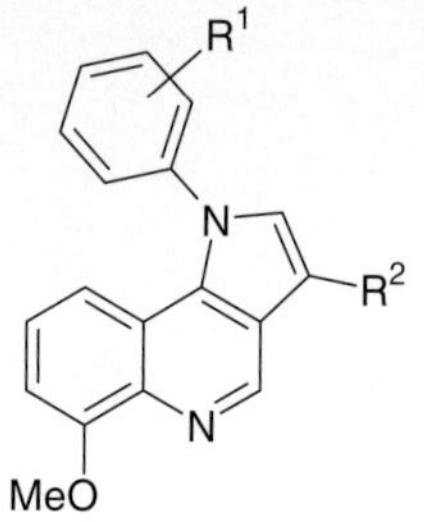

Compound	R^1	R^2	Rat in vivo[a]	Pig H^+/K^+ ATPase[**]
(226)	2-Me	Me	+++	+++
(227)	2-Me	i-Pr	+	++
(228)	2-Me-4-F	Et	+++	+++
(229)	2-Me-4-OH	Et	+	+++

[a]Effect relative to omeprazole, +++ (strong), ++ (similar), + (weak) on ethanol-induced ulcer in the rat 1 h following 30–300 mg/kg compound dosed orally.
[**]Effect relative to omeprazole, +++ (strong), ++ (similar), + (weak), in isolated pig H^+/K^+ ATPase inhibition.

Modifications to this template have included the generation of compounds exemplified by (230), which are reported to have IC_{50} values for inhibition of H^+/K^+ ATPase in the range 1.8–11.1 μM [268].

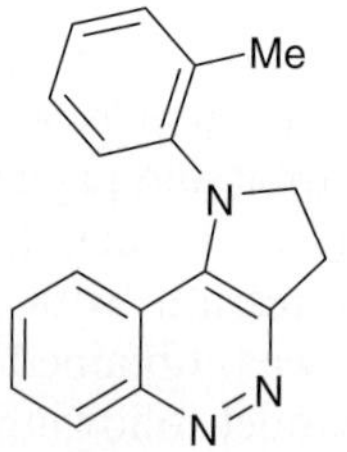

(230)

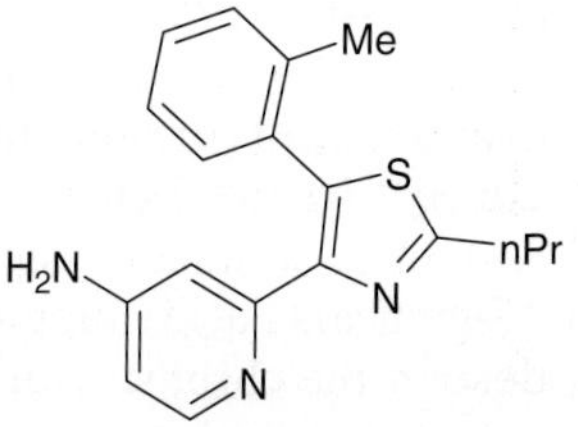

(231)

Ife et al. have reported compounds such as (231) [269]. (231) was shown to afford potassium-competitive inhibition of gastric H^+/K^+ ATPase with an IC_{50} of 2.2 μM, but only moderate selectivity over pig kidney Na^+/K^+ ATPase (IC_{50} = 44 μM). Its in vivo effect was also modest, affording only

36% inhibition of pentagastrin-stimulated acid secretion in the anaesthetised rat. The SAR indicates a variety of groups tolerated at the thiazole 2-position, significant reduction in activity on substituting the thiazole with an imidazole, significant reduction in activity on removal of the *ortho*-methyl susbtituent and a higher potency for the aminopyridyl analogues, compared to the corresponding unsubstituted pyridine derivative.

MONOCYCLIC DERIVATIVES

The Yuhan group has reported on a series of aminopyrimidine derivatives (232)–(238), which show modest H^+/K^+ ATPase-inhibitory potencies and activities in the pylorus-ligated rat at 5 h, following a 20 mg/kg i.d. dose, comparable to omeprazole (1, Table 3.20) [270].

There is little, if any, impact on potency of *ortho*-methyl substitution on the aniline aryl ring. The presence of the R^3 and R^4 substituents also impacts on potency. Compound (236) was identified as a clinical candidate molecule by the Yuhan group (revaprazan hydrochloride, YH-1885, SB-641257) [271]. In early 2001, GlaxoSmithKline had acquired rights to develop the compound for GERD, and was conducting Phase I/II studies, but the agreement was terminated in 2002. The reason for this has not been publicly disclosed. However, doses used in clinical trials were quite high (200–300 mg, cf. efficacy in Phase IIb studies at 25 mg for AZD-0865) (47), which is in line with relatively high IC_{50} values for H^+/K^+ ATPase inhibition. The pK_a of (236) is approximately 6.0 and it would therefore be expected to concentrate in the acidic compartment of the parietal cell canaliculus. Indeed, in radiolabelling studies, (236) was shown to reach peak levels in tissues at 4 h after oral dosing, but at 8 h the radioactivity was localised to the gastric wall [272]. The compound shows >100-fold selectivity for H^+/K^+ ATPase over Na^+/K^+ ATPase [272].

In rats and dogs, linear pharmacokinetics were observed following doses in the range 2–30 mg/kg p.o., with a corresponding oral bioavailability of 41–47% (rats) and 43–59% (dogs), and no change in PK was observed on repeat dosing [272]. However, above 200 mg/kg doses, dose proportionality was lost, which was attributed to poor aqueous solubility [273]. (236) was claimed to be more potent than omeprazole (1) in the inhibition of basal acid secretion in pylorus ligated and chronic fistula rats, as well as in histamine-stimulated acid secretion in the Heidenhain pouch dog, but not in pentagastrin-stimulated acid secretion in the lumen-perfused rat [272]. Peak plasma concentrations were reached in man within 1.3–2.5 h after a single oral dose, and PK remained constant on repeat dosing [274]. Single doses of 150 and 200 mg, but not 100 mg, afforded rapid increases in mean

Table 3.20 EXAMPLES OF YUHAN PYRIMIDINE DERIVATIVES

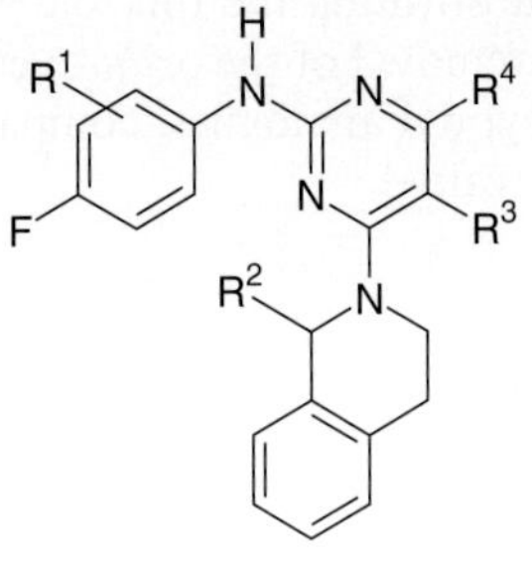

Template A

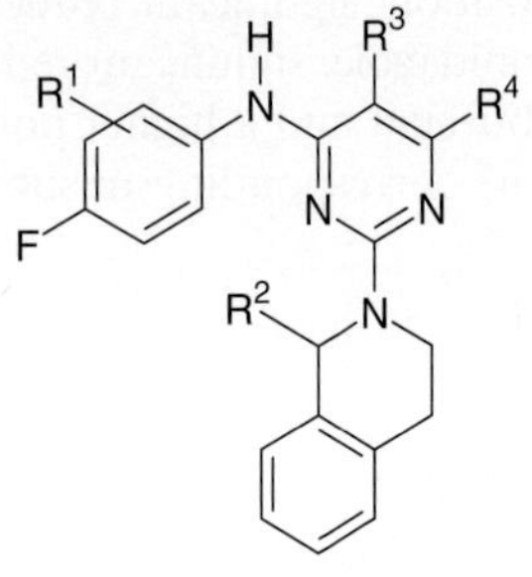

Template B

Compound	Template	R^1	R^2	R^3	R^4	H^+/K^+ ATPase IC_{50} (μM)	Relative inhibition of acid[a]
(232)	A	2-Me	Me	H	H	5.4	0.11
(233)	A	2-Me	Me	H	Me	0.9	0.89
(234)	A	2-Me	(R)-Me	H	Me	0.5	0.81
(235)	A	2-Me	(S)-Me	H	Me	0.7	0.72
(236)[b]	A	H	Me	Me	Me	1.6	0.99
(237)	B	2-Me	H	H	Et	>20	0.67
(238)	B	2-Me	Me	-(CH$_2$)$_3$-		0.4	0.68

[a]Inhibition of acid secretion by test compound relative to omeprazole (both at 20 mg/kg i.d. given 1 h before ligation) in the pylorus-ligated rat at 5 h post-ligation.
[b]Revazaprazan, YH-1885, SB-641257.

intragastric pH to 3.5 and 4.2, respectively [274], and effect was maintained from day 1 to day 7 on repeat dosing [275]. In September 2005, approval was obtained by Yuhan from the Korean FDA for use of revaprazan in the treatment of duodenal ulcer; it was launched as Revanex by the end of 2006.

Yuhan has extended the pyrimidine series to include fluorinated analogues, which show sub-micromolar H^+/K^+ ATPase IC_{50} values (239–242, Table 3.21) [276]. In the Shay pylorus-ligated rat, 5 h after pylorus ligation and administration of a 10 mg/kg intraduodenal dose, examples showed comparable inhibition of acid secretion to that of a similar dose of omeprazole (1).

Yuhan have also reported on alkylamino replacements for the aniline moiety [277]. For example, the allylamino-derivative (243) affords an IC_{50} of 7.85 μM against H^+/K^+ ATPase. In addition, the methyl groups on the

Table 3.21 FURTHER EXAMPLES OF YUHAN PYRIMIDINE DERIVATIVES

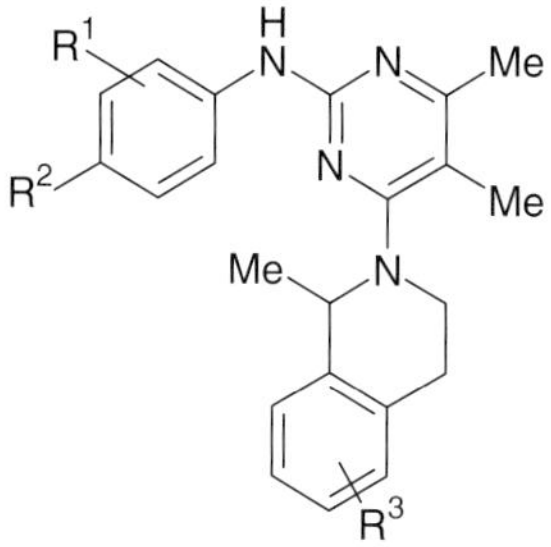

Compound	R^1	R^2	R^3	H^+/K^+ ATPase IC$_{50}$ (μM)
(239)	2-F	F	H	3.8
(240)	H	H	7-F	1.0
(241)	2-Me	F	7-F	0.3
(242)	2-Me	F	6-F	0.2

pyrimidine core can be replaced with a fused aryl to give quinazolines, such as (244), which afforded H^+/K^+ ATPase inhibition with an IC$_{50}$ of 2 μM [278].

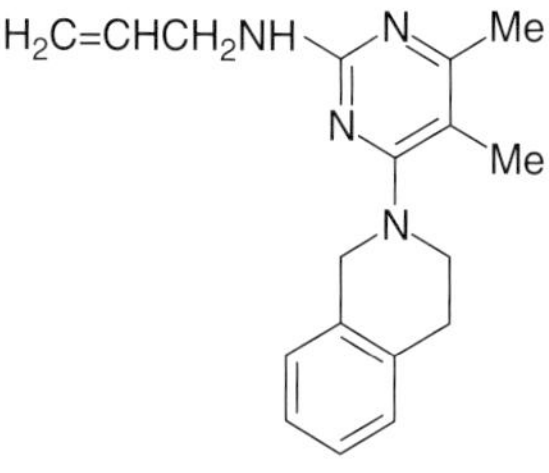

(243) 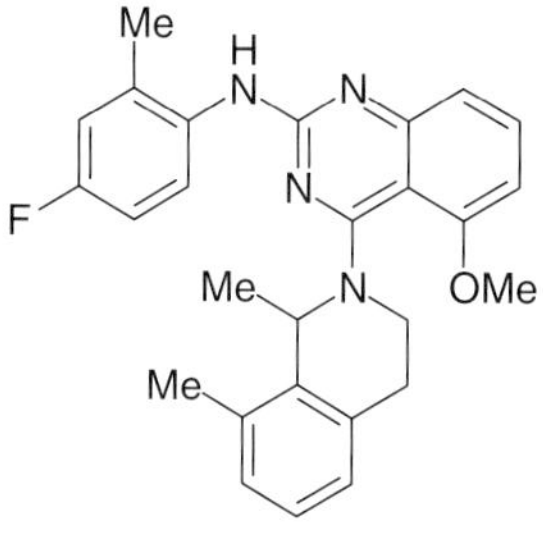   (244) 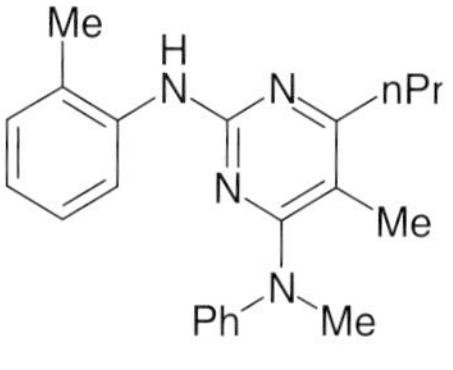   (245)

Bis-anilinopyrimidine derivatives (such as compound (245)) have also been disclosed in the patent literature, although no specific activities have been claimed [279]. The pyrimidine core in such compounds can be incorporated into a bicyclic heteroaryl moiety. The SK&F (now GSK) group demonstrated good activity in analogues with the quinazoline core (246–249, Table 3.22)

Table 3.22 EXAMPLES OF SKF QUINAZOLINE DERIVATIVES

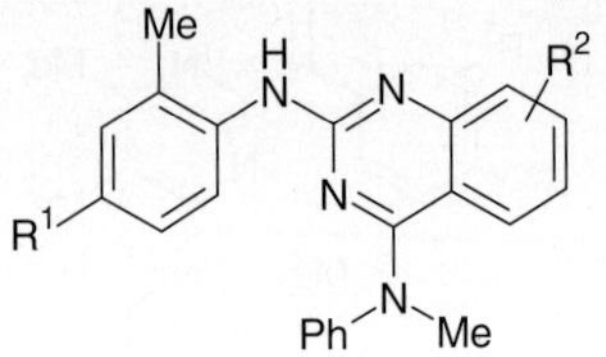

Compound	R^1	R^2
(246)	H	8-OH
(247)	F	6-OH
(248)	F	6-[Me$_2$N(CH$_2$)$_3$O]
(249)	H	8-[HO(CH$_2$)$_2$O]

Table 3.23 EXAMPLES OF PYRIDOPYRIMIDINE DERIVATIVES

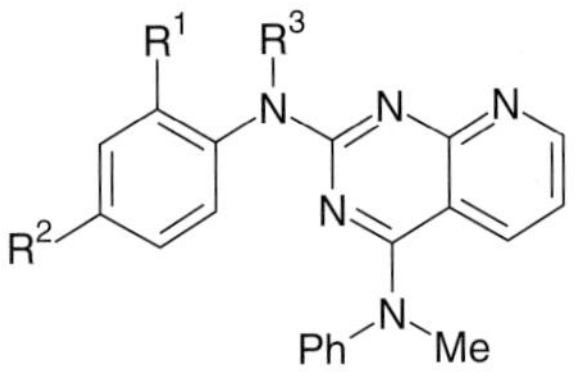

Compound	R^1	R^2	R^3
(250)	H	H	Me
(251)	Me	H	H
(252)	Me	F	Me

[280]. Examples in the table show H$^+$/K$^+$ ATPase-inhibitory IC$_{50}$ values of between 18 nM and 1.3 μM (although specific activities were not ascribed).

This work has been extended to the pyrido[2,3-d]pyrimidine core. Thus, (250–252) (Table 3.23) were claimed to have H$^+$/K$^+$ ATPase-inhibitory IC$_{50}$ values of 0.065–0.26 μM [281], although it is not clear which compound has which activity.

The core can be acyclic with the basic centre provided by an amidine or guanidine moiety, to give compounds such as (253) and (254) [282]. Although no specific data is provided, such compounds are claimed to have H^+/K^+ ATPase IC_{50} values <5.5 µM. The same group has also patented the analogous amidines and guanidines such as (255), although again, no specific data is provided [283]. If these compounds are fulfilling the same pharmacophore as other molecules discussed so far, it is not entirely clear which Ph group is satisfying the pendant benzylic interaction of the imidazopyridines. Compound (256) is representative of a further series of this type. Again no specific biological data is disclosed for such molecules (H^+/K^+ ATPase IC_{50} values <20 µM) [284].

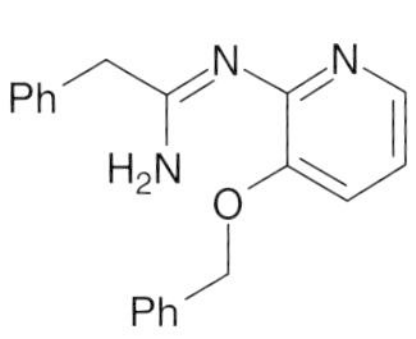

(253) 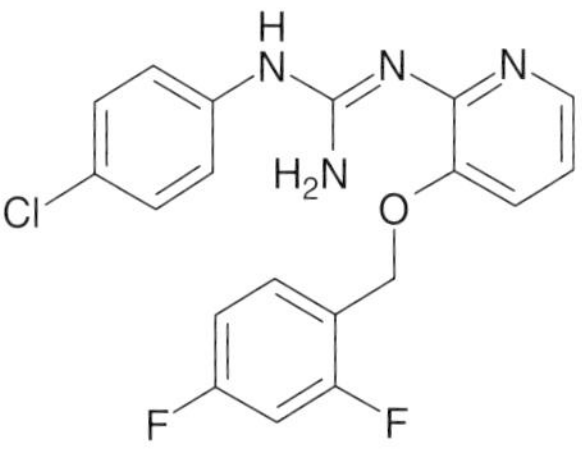 (254)

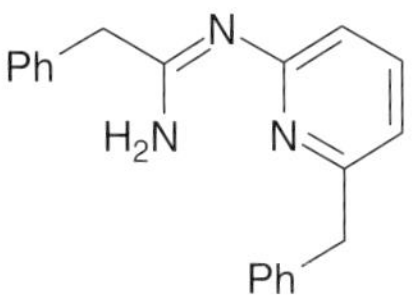

(255) 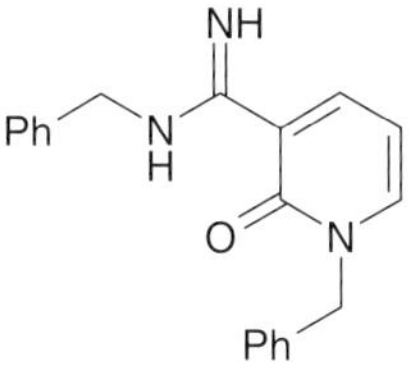 (256)

The Pfizer group have reported on a series of thiazolo-guanidine derivatives [285]. The chemical starting point came from a series of H_2RAs, including Zaltidine (257). Changing the imidazole of (257) to a pyrrole (258) eradicated the H_2 antagonist activity, but maintained gastric acid-inhibitory activity in vivo comparable to that of cimetidine ($ED_{50} = 26$ mg/kg i.d. in the pylorus-ligated rat model). This acid inhibition was found to be due to a reversible, potassium-dependent H^+/K^+ ATPase-inhibitory activity with an IC_{50} of 25 µM. Although the SAR in the series was not clear, the guanidine and pyrrole could be extensively substituted, for example, (259)

($IC_{50} = 2\,\mu M$), and the pyrrole could be exchanged for other aryl rings. The only replacement found for the thiazole moiety was a pyrimidine, for example, (260) ($IC_{50} = 8\,\mu M$).

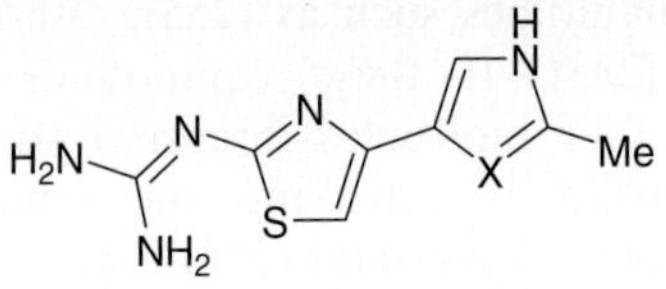

(257) X = N, Zaltidine
(258) X = CH

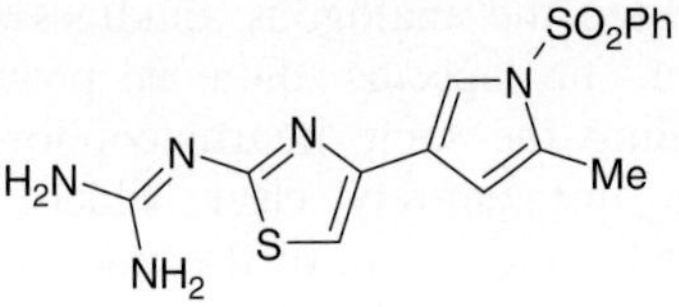

(259)

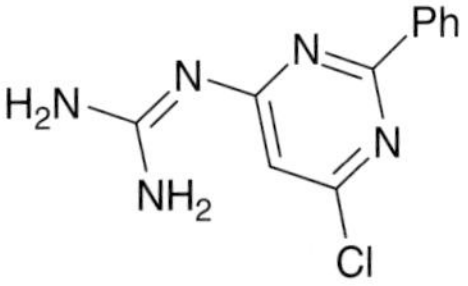

(260)

The pyridine derivative AU-1421 (261), reported by the Banyu group, inhibits the potassium-stimulated H^+/K^+ ATPase activity (hog gastric mucosal leaky membrane vesicles) in a reversible, potassium competitive and pH-dependent manner with a K_i of $0.1\,\mu M$ at pH 6.4. In the isolated rabbit gastric gland, histamine-stimulated [^{14}C]-aminopyrine uptake was inhibited with an IC_{50} of $\sim 1\,\mu M$. It also showed good potency in the acute fistula rat ($ED_{50} = 1.16\,mg/kg$) [286] and Heidenhain pouch dog models [287]. The compound was selective vs. the dog kidney Na^+/K^+ ATPase, for which the IC_{50} is $\sim 50\,uM$ [288]. AU-2064 (262) is a further compound which had been under development by Banyu [288–290]. This compound was as potent as cimetidine, affording potassium competitive H^+/K^+ ATPase-inhibitory activity with an IC_{50} of $0.5\,\mu M$ at pH 4, an ED_{50} of $20\,mg/kg$ i.d. in the pylorus-ligated rat model, 83% inhibition of tetragastrin-induced acid secretion in the acute fistula rat following a dose of $3\,mg/kg$ i.v. and suppression of histamine-stimulated acid secretion in the Heidenhain pouch dog with an ED_{50} of $1.2\,mg/kg$ i.d.

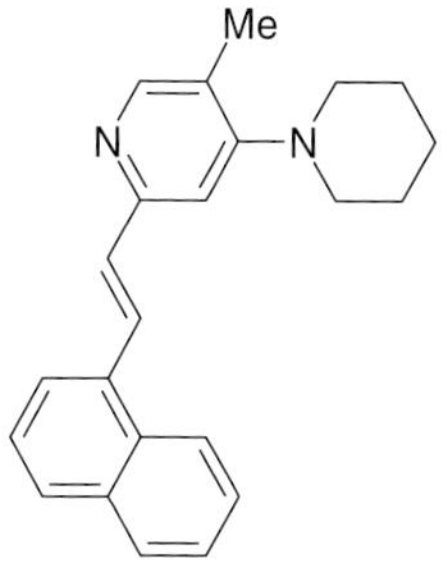

(261) AU-1421

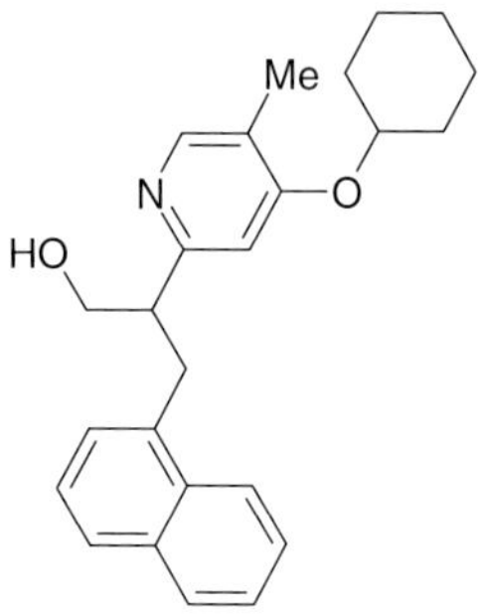

(262) AU-2064

Evaluation of SAR (263–266, Table 3.24) around the pyridine 4-position suggested that reduction of the basicity of the pyridyl nitrogen reduced the H^+/K^+ ATPase-inhibitory activity, in keeping with the potassium competitive mode of action [291]. Also reduction in the overall lipophilicity

Table 3.24 EXAMPLES OF BANYU PYRIDINE DERIVATIVES

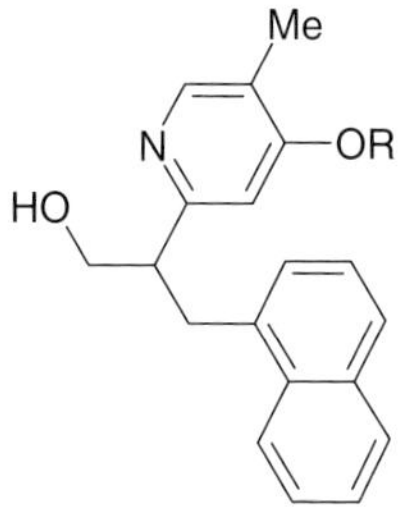

Template A

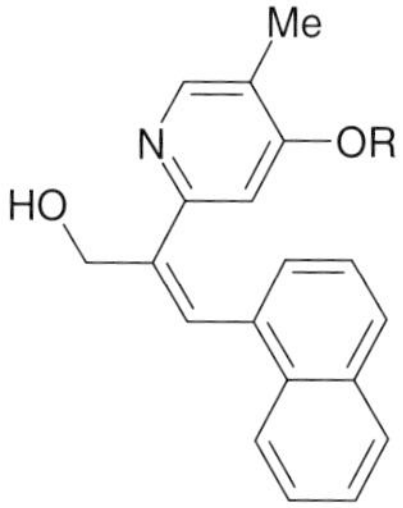

Template B

Compound	Template	R	H^+/K^+ ATPase IC_{50} (μM) pH 6.4	Acute fistula rat % inhibition at 3 mg/kg i.v.	Pylorus-ligated rats % inhibition at 10 mg/kg i.d.
(263)	A	CyHex	0.5	83	39
(264)	A	tBuCH$_2$	0.6	69	56
(265)	A	2-OHPhCH$_2$	1.8	77	33
(266)	B	2-OHPhCH$_2$	0.9	55	18

of the molecule improved the *in vivo* performance, especially following intraduodenal dosing, subject to first pass metabolism. The alkenyl derivatives showed a marginal improvement in H^+/K^+ ATPase-inhibitory activity (cf. (265) vs. (266)), suggesting the appropriate active conformation was achieved in this more constrained template. However, the *in vivo* performance of this template was inferior.

Imidazole has also been investigated as a monocyclic core [292–294]. However, at concentrations as high as 0.1 mM, compounds (267), (268) and (269) afforded only 79 (70% inhibition of H^+/K^+ ATPase), 71 and 64% inhibition of gastric acid secretion in guinea pigs, respectively.

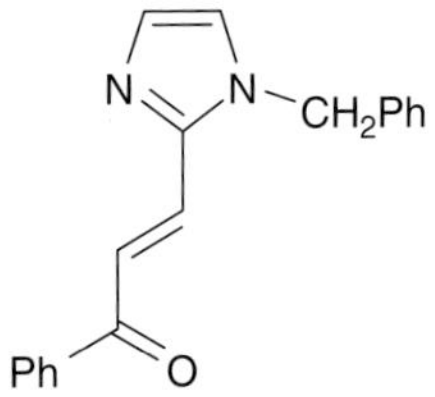

(267) 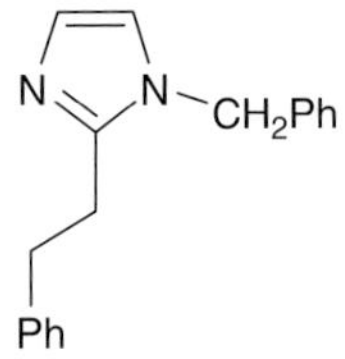 (268) 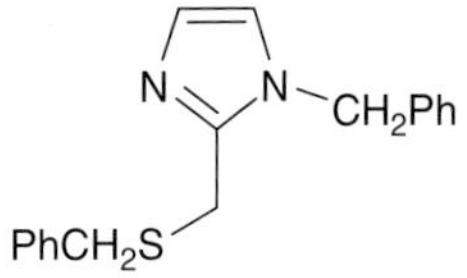 (269)

ALTERNATIVE TYPES OF H^+/K^+ ATPASE INHIBITORS

The inhibition of H^+/K^+ ATPase from porcine gastric mucosa by the imidazo[1,2-a]thieno[3,2-c]pyridine SPI-447 (270) was shown to be reversible, being attenuated by dilution, and not influenced by glutathione pre-treatment [295]. This inhibition was also pH dependent with an IC_{50} of 5.2 μM at pH 7.4 reducing to an IC_{50} of 1.05 μM at pH 6.8 [296]. This is comparable to values obtained for SCH-28080 (28) in the same assays (5.2 and 0.68 μM, respectively). Molecular Orbital calculations suggest that it is the N1 nitrogen which is the most basic. The activities of (28) and (270) are additive [296] and therefore SPI-447 may act at, or close to, the same binding site as that of (28).

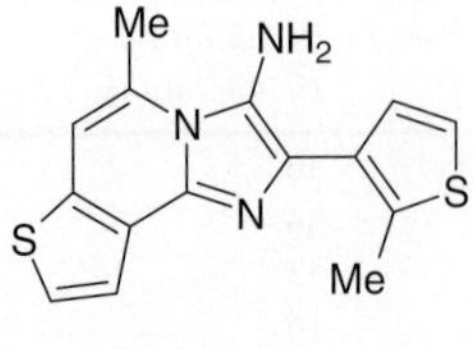

(270) SPI-447 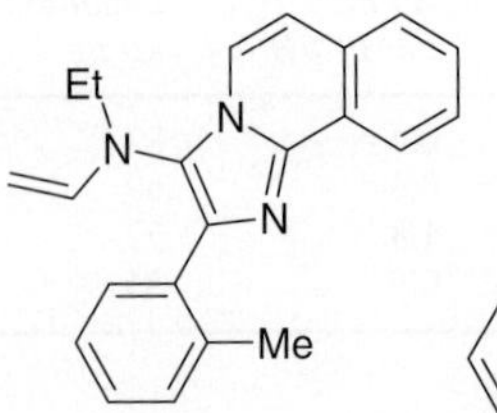   (271) 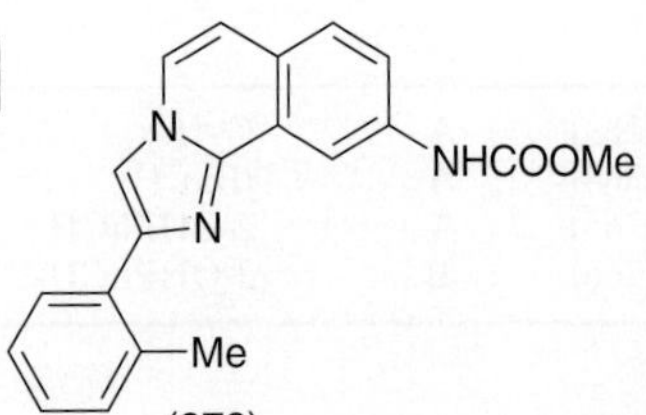   (272)

Table 3.25 EXAMPLES OF TAKEDA PYRROLE DERIVATIVES

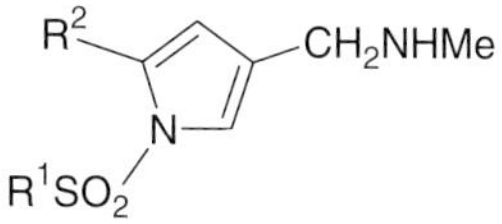

Compound	R^1	R^2	H^+/K^+ ATPase IC_{50} (nM)
(273)	Ph	3-Thienyl	4.2
(274)	3-(MeSO$_2$)-Ph	Ph	78
(275)	2-Benzothienyl	Ph	33
(276)	2-MeO-3-pyridyl	Ph	13

SPI-447 (270) was selective for H^+/K^+ ATPase over Na^+/K^+ ATPase [297]. However, no further development of this compound has been reported. Close analogues, such as compound (271), indicate that N-substitution is tolerated and is consistent with suppression of ethanol-induced stomach lesions in the rat. Further modifications generated a molecule (compound (272)) which is somewhat less potent (H^+/K^+ ATPase $IC_{50} = 22$ μM).

Takeda have recently patented a series of molecules (Table 3.25, 277–280) reported to give excellent inhibition of H^+/K^+ ATPase *in vitro*, and in which the moderately basic heteroaryl centre is replaced by a more basic alkylamine [296–300]. The mode of action of such compounds is yet to be demonstrated.

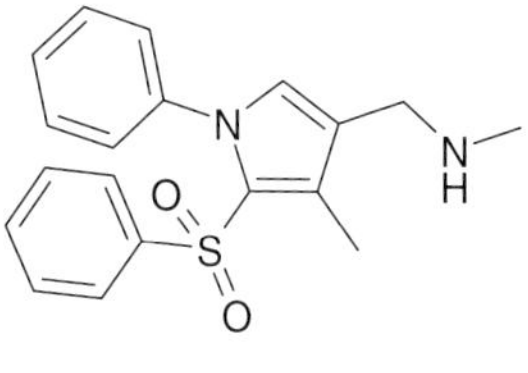

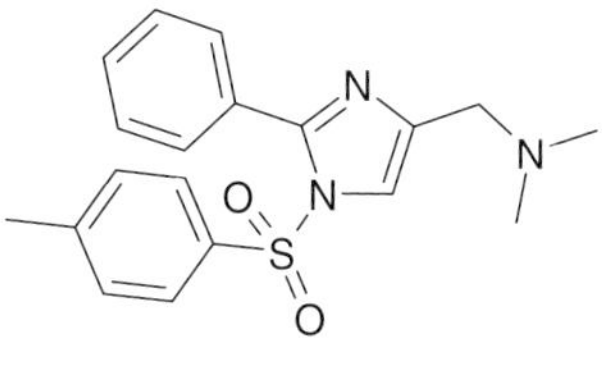

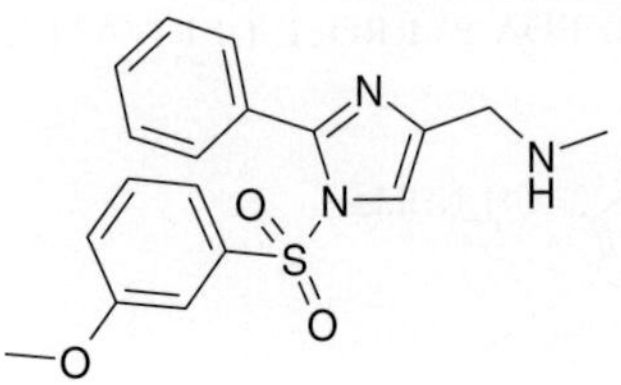

(279) 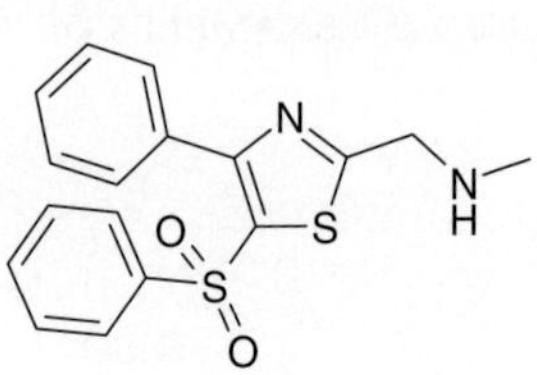   (280)

Other compounds such as (281), with an IC_{50} of 5 μM, show some activity as inhibitors of pig stomach H^+/K^+ ATPase. Such compounds afford inhibition of ethanol-induced erosion of rat stomach with an ED_{50} of between 1.2 and 3.9 mg/kg [301].

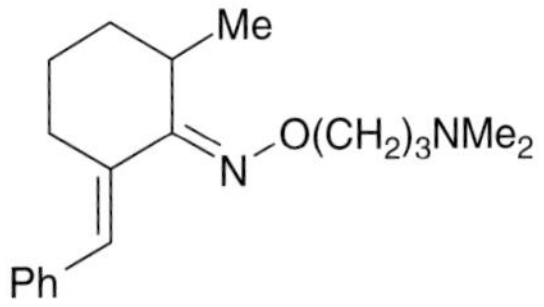

(281) 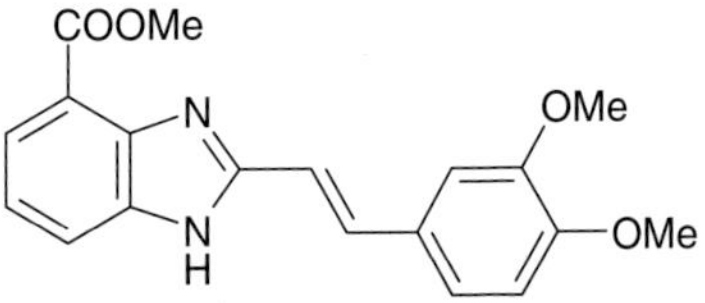   (282) ALE-36

A number of tertiary amine-containing compounds, including some marketed compounds such as verapamil and trifluoperazine, have been shown to inhibit H^+/K^+ ATPase activity in both permeabilised and intact gastric membranes, with the higher activity in the latter ascribed to the ability of the basic molecules to concentrate in the acidic compartment [302]. Again, no indication as to the mechanism is described. ALE-36 (282) was shown to have modest inhibitory activity against partially purified guinea pig H^+/K^+ ATPase, with an IC_{50} of between 10 and 30 μM which was not affected by co-incubation with β-mercaptoethanol indicating the likely reversible nature of the inhibition [303]. In the same experiment, SCH-28080 (28) provided an IC_{50} of between 0.3 and 1.0 μM. Both compounds were selective over Na^+/K^+ ATPase. In intact parietal cells, a moderate improvement in IC_{50} to approximately 0.3 μM was observed, with (28) showing an IC_{50} of approximately 30 nM.

There are a number of reports of tannins, catechins and other polyphenolic derivatives affording H^+/K^+ ATPase-inhibitory activity. For example, ellagic acid (283) [304], epigallocatechin gallate (284) [305] and tannic acid (285) [306] afford inhibition of pig gastric H^+/K^+ ATPase activity with IC_{50} values of 2.1 μM, 69 nM and 29 nM, respectively.

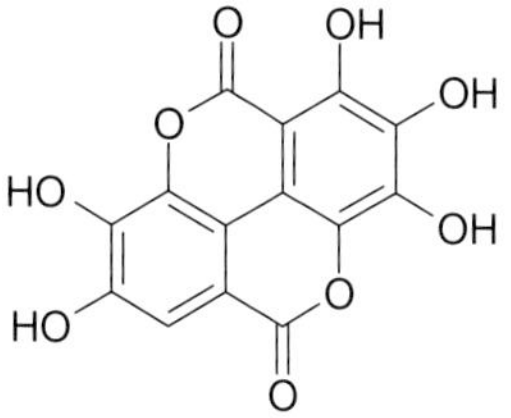

(283)

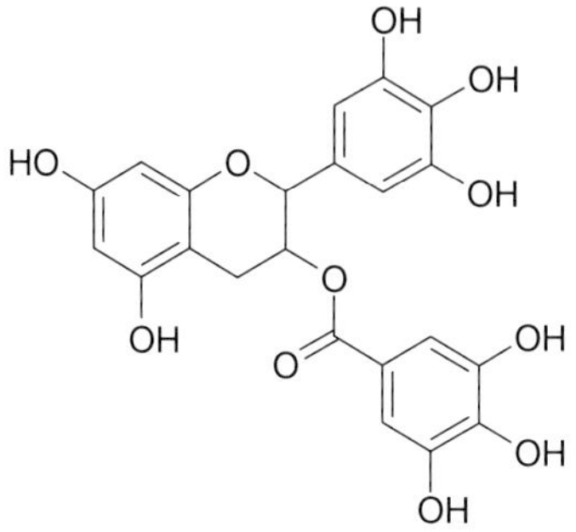

(284)

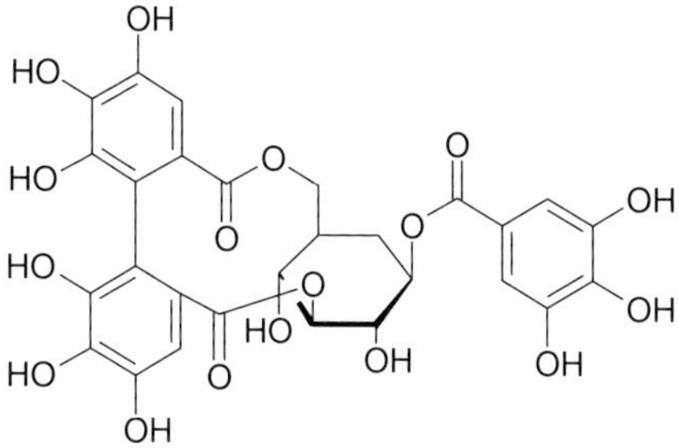

(285)

Importantly, the inhibition in each case was shown to be competitive with ATP and non-competitive with potassium. This is perhaps not surprising given the reported inhibitory effects of polyphenols on other proteins such as kinases. Indeed (283) is reported to have a broad spectrum of biological activities, and the strategy of targeting the ATP-binding site is very much a less attractive one for the inhibition of H^+/K^+ ATPase. In pylorus-ligated rats, (283) provided a modest, non-significant 30–35% inhibition of acid secretion at 5–10 mg/kg, but at similar doses provided approximately 85% inhibition of stress-induced gastric lesions in rats.

Similarly, the anti-ulcer agent sofalcone (286), its analogue sophoradin (287), and the analogous 4-hydroxyderricin (288) inhibit pig gastric H^+/K^+ ATPase with IC_{50} values of 15, 0.74 and 3.3 µM, respectively, at pH 7.4 [307, 308]. This inhibition was shown to be competitive with ATP. Sofalcone's anti-ulcer mode of action is primarily as an inhibitor of prostaglandin metabolism through its action on 15-hydroxyprostaglandin dehydrogenase, thereby strengthening gastric defensive factors. However, it also has a mild anti-secretory effect, with significant reduction of gastric juice volume and acidity when dosed at 50 mg/kg i.d. Compound (288) also showed anti-secretory activity at 100 mg/kg i.p.

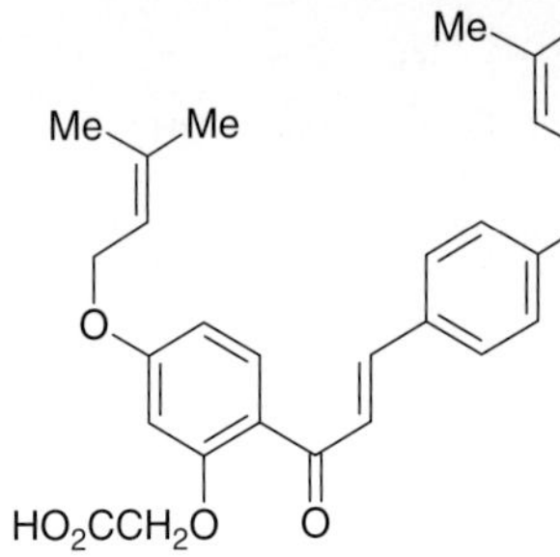

(286) Sofalcone

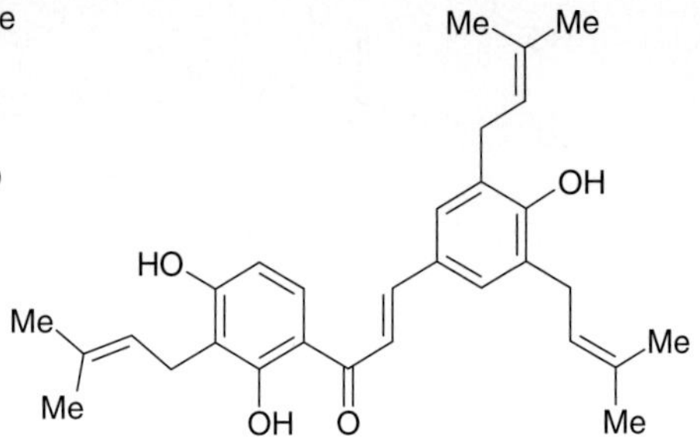

(287) Sophoradin

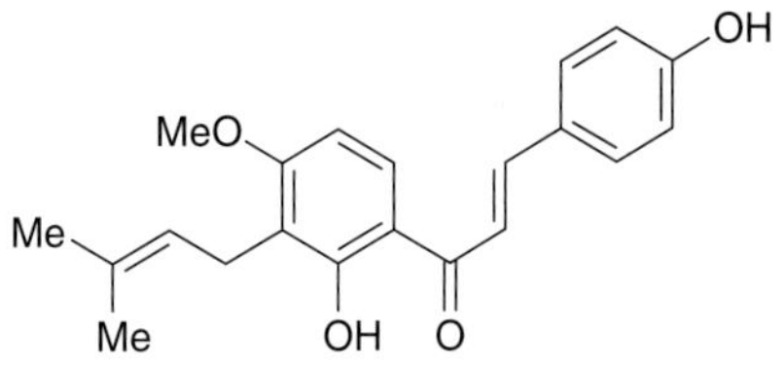

(288) 4-hydroxyderricin

A number of macrolide antibiotics have been shown to inhibit pig gastric H^+/K^+ ATPase [309]. IC_{50} values of 16–43 µg/mL were obtained, compared with 11.5 µg/mL for SCH-28080 (28). No activity was observed against the dog kidney Na^+/K^+ ATPase model. The mode of action is not at all clear. One such compound, copiamycin A, at 50–100 mg/kg i.p., gave 49–62% inhibition of pylorus-ligation-induced gastric ulcers in rats, compared with 91% inhibition by (28) at 20 mg/kg i.p. [309].

The tetracyclic diterpenoids scopadulcic acid B and scopadulciol, together with derivatives such as (289), have been shown to have H^+/K^+ ATPase-inhibitory activity [310]. (289) was shown to have an IC_{50} of 3 µM, which was claimed to be comparable to that of omeprazole (1) and SCH-28080 (28) in the same experiment. No mechanistic studies to provide an indication of mode of action are reported.

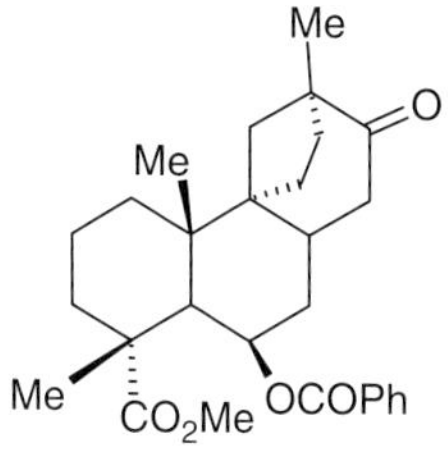

(289)

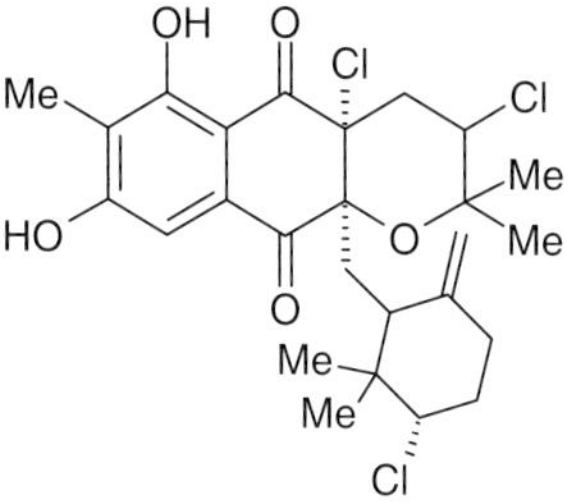

(290) A80915A

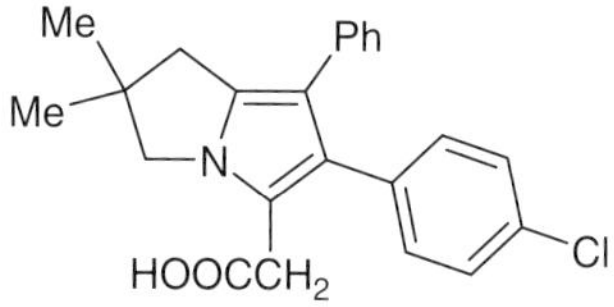

(291) ML3000

The semi-naphthoquinone natural product A80915A (290) was reported by Lilly to afford inhibition of pig gastric membrane vesicular H^+/K^+ ATPase at pH 7.4 with an IC_{50} of 2–3 µM [311]. The mode of action was shown to be distinct from that of (1) and (28), being neither dependent on irreversible thiol binding nor potassium competitive, and distinct binding sites on the E_1 enzyme conformation were invoked.

Reversible inhibition of pig gastric vesicular H^+/K^+ ATPase was reported for the pyrrolizine derivative ML3000 (291) with an IC_{50} of 16.4 µM [312]. This compound is of a similar structural type to non-steroidal anti-inflammatory drugs and has been shown to have 5-lipoxygenase and cyclooxygenase inhibitory activity. Indeed prostaglandin E_2 (PGE_2) ($IC_{50} = 15$ µM) and arachidonic acid ($IC_{50} = 16.7$ µM) have also been suggested to be inhibitors of H^+/K^+ ATPase activity. Whether these signalling pathways interact with the H^+/K^+ ATPase is not clear, but perhaps a more likely explanation for these relatively low-level inhibitory effects is that such anionic amphiphilic molecules exert inhibition indirectly through effects on the membrane lipids in which the protein resides. A similar mode of action could be inferred for a number of the natural products identified above.

THE BINDING SITE FOR POTASSIUM-COMPETITIVE INHIBITORS

Studies have been conducted to explore the site of interaction of potassium-competitive inhibitors with the protein, and thereby gain insight to the pharmacophoric requirements, to aid rational drug design. The binding site of the reversible inhibitors is believed to at least partly overlap with that of the irreversible inhibitors since pre-incubation with the imidazopyridine SCH-28080 (28) prevented covalent binding of omeprazole to Cys813 [313]. Furthermore, mutation of this residue (Cys813Thr) resulted in an approximately 6-fold decrease in affinity for (28) [314]. Scanning mutagenesis of this region of the protein has identified a number of other (mainly negatively charged) residues in TM4–6 and TM8, thought to play a part in binding the protonated imidazopyridine ring of (28), thereby blocking the ion-translocating pathway generated by these transmembrane helices [315, 316]. The hydrophobic phenyl ring is thought to interact with TM1–2 [316].

Although a structure of the porcine H^+/K^+ ATPase has been generated, the resolution is only 8 Å [317]. However, high-resolution structures of the Ca^{2+} ATPase have been generated in both the $[E_1]$ [318] and $[E_2]$ [319] conformers. Thus, the focus for gaining structural insight into the binding of ligands to H^+/K^+ ATPase has been on homology modelling [320].

The protein bound conformation of TMPIP (292) has been determined by NMR and agrees with the observation that constraint of the benzyloxymethyl side-chain (e.g., in (294) and other tricyclic imidazopyridine derivatives discussed above) affords potent compounds [321]. The bound conformation of TMPFPIP (293), an alkylated quaternary analogue of (28), has been determined by NMR [322]. The activity of the imidazopyridines is pH dependent and the N1-methyl quaternary imidazopyridine retains potency, corroborating the suggestion that the activity of the imidazopyridines such as (28) resides in the protonated form of these weakly basic compounds. (292) has been used in conjunction with the $[E_1]$ and $[E_2]$ Ca^{2+} ATPase structures to model the likely binding sites of the imidazopyridine class of reversible inhibitors. Thus, it is proposed that the inhibitor is first localised to a diffuse cluster of low-affinity (estimated K_i of 490 nM) ligand-binding sites on the extracellular face at the interface of TMs 5, 6 and 8 in the $[E_1]$ conformation [323]. On conversion to the phosphorylated $[E_2]$ conformer, a higher affinity, central binding pocket is generated, proximal to the cation transport region, with an estimated K_i of 22–99 nM (in good agreement with the experimentally determined K_i of 60–100 nM [324, 325]). In this way it is proposed that the protein is prevented from returning to the $[E_1]$ conformer through potassium-competitive, high-affinity binding to $[E_2]$ [323].

In this model, the imidazopyridine ring is proposed to be tightly bound by the lipophilic and aromatic residues Tyr928, Thr929, Phe932 (TM8), Ile820, Ile816 (TM6) and Tyr802 (TM5) (in agreement with SDM studies), while the benzylic moiety has more freedom and is directed to the extracellular space [323]. The only negatively charged residue in close proximity to the ligand in this model is Glu795, which is 5 Å from imidazopyridine N4. The role of the positive charge may be to provide a general ionised character for interaction in the more diffuse low-affinity states and in the transition to the high-affinity-binding conformation adjacent to the cation transport channel.

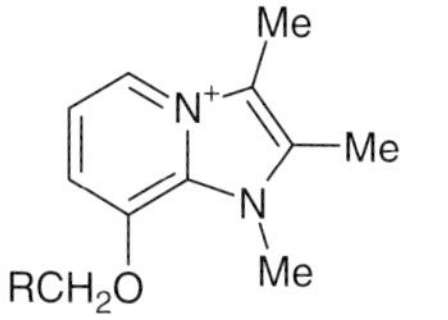

(292) R = Ph TMPIP
(293) R = F$_5$Ph TMFPIP

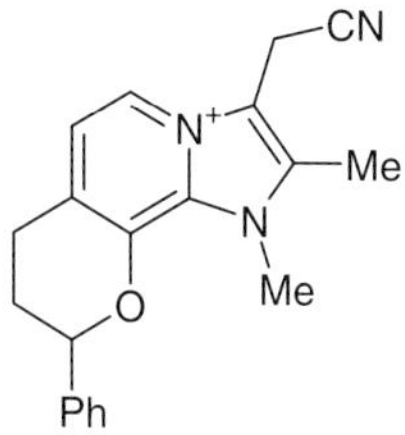

(294)

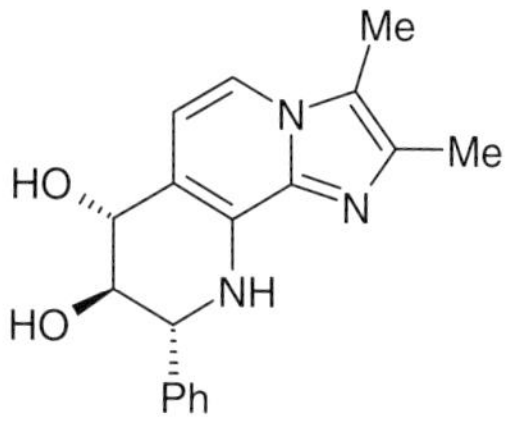

(295) Byk99

Asano *et al.* also invoke the importance of the tyrosine on TM5, referred to by this group as Tyr801 (care needs to be exercised when comparing the numbering of amino acid residues for H^+/K^+ ATPase by different authors) [311]. Mutation of Tyr801 to alanine and serine retained similar ^{86}Rb transport and potassium-dependent ATPase activity, but was 60–80-fold less sensitive to inhibition by (28). The inferred interaction of (28) with Tyr801 is likely to be an aromatic one, since the mutant Tyr801Phe resulted in only 2-fold lower sensitivity. A similar involvement of Tyr801 in the binding of SPI-447 (270) (see discussion of activity above) was also demonstrated. Using a protein homology model of H^+/K^+ ATPase based on the X-ray crystal structure of sarcoplasmic reticulum (SR) Ca^{2+} ATPase (which was comparable to the model previously defined by Sweadner and Donnet [326]), Asano *et al.* inferred the docking of (28) into a cavity distinct from the cation-binding sites formed in the [E$_2$] enzyme conformer by residues previously suggested to be involved in ligand binding by site-directed mutagenesis studies. These include Leu811, Cys815 and Met336. Furthermore, the model predicts the binding and stabilisation of the [E$_2$] enzyme conformer as reported in enzyme binding and kinetic studies. Cys815 is also reported to be involved in PPI binding and may explain the partial inhibition of omeprazole (1) binding by (28). In this model, the imidazopyridine ring lies perpendicular to the axes of the membrane helices.

Munson *et al.* derive a model based on the SR Ca^{2+} ATPase and the N-terminal domain of the Na^+/K^+ ATPase [320]. Mutagenesis studies direct them towards a binding site for (28) and the constrained analogue Byk99 (295) bounded by Phe332, Ala335 and Leu809. Of Phe332 it is the β-carbon which is expected to be in closest proximity to ligand since β-branched amino acid replacements have the biggest impact on binding. Nearby Met334 shows a similar effect. Ala335 is predicted to face directly into the middle of the imidazopyridine ring. The closest negatively charged residue to the proposed ligand-binding site is Gly795, which is within 5 Å of the proposed position of the ligand's N, thereby going some way to stabilising its positive charge. There is also a proposed open path to the ion-binding site, again consistent with the need for a protonated ligand. The *para*-position of the phenyl moiety of (28) and (295) is predicted to lie adjacent to Leu141, in accordance with photoaffinity labelling of this vicinity by *para*-N_3 substituted (28). The C2- and C3-positions of the ligand are predicted to be adjacent to Pro798 and Tyr799, accounting for the limited lipophilic substitution tolerated in these positions. Thus, in contrast to the Asano *et al.* model, the plane of the imidazopyridine ring is predicted to lie parallel to the axes of the transmembrane helices. Munson *et al.* have further refined this SR Ca^{2+} ATPase-based modelling approach to explain inhibitor or K^+ ion entry to the binding site, and the occlusion of the K^+ site by competitive inhibitors [327].

Despite these differences in proposed models, such studies afford useful insights into the mode of interaction of ligands with the target protein [328] and help in proposing new, potentially improved ligands. A more definitive model of the H^+/K^+ ATPase chemical space awaits the highly challenging goal of generating a ligand-bound X-ray crystal structure of this membrane-bound protein, more extensive studies of ligand interactions with site-directed mutants, or a more detailed and systematic structure–activity relationship study.

CONCLUDING REMARKS

The past 30 years have seen significant progress in the development of treatments for acid-related disorders. The PPIs have become a highly effective mainstay of the clinician's armoury. The reversible, potassium competitive acid blockers (PCABs) or APAs offer an alternative approach. Despite intense efforts over a number of years, no such compound has yet successfully rivalled the PPIs in clinical use.

REFERENCES

[1] American Digestive Health Foundation (http://www.gastro.org/adhf/ulcers.html).
[2] Hunt, R.H. (1995) *Aliment. Pharmacol. Ther.* **9**, 3–7.
[3] DiPalma, J.A. (2001) *J. Clin. Gastroenterol.* **32**, 19–26.
[4] Jones, R., Armstrong, D., Malfertheiner, P. and Ducrotté, P. (2006) *Curr. Med. Res. Opin.* **22**, 657–662.
[5] Fass, R., Shapiro, M., Dekel, R. and Sewell, J. (2005) *Aliment. Pharmacol. Ther.* **22**, 79–94.
[6] Geibel, J.P. and Wagner, C. (2006) *Rev. Physiol. Biochem. Pharmacol.* **156**, 46–60.
[7] Bristol, J.A. and Long, J.F. (1981) *Annu. Rep. Med. Chem.* **16**, 83–91.
[8] Caplan, M.J. (2007) *J. Clin. Gastroenterol.* **41**, S217.
[9] Walderhaug, M.O., Post, R.L., Saccomani, G., Leonard, R.T. and Briskin, D.P. (1985) *J. Biol. Chem.* **260**, 3852–3859.
[10] Sachs, G., Shin, J.M., Briving, C., Wallmark, B. and Hersey, S. (1995) *Annu. Rev. Pharmacol. Toxicol.* **35**, 277–305.
[11] Munson, K., Lambrecht, N. and Sachs, G. (2000) *J. Exp. Biol.* **203**, 161–170.
[12] Lee, J., Simpson, G. and Scholes, P. (1974) *Biochem. Biophys. Res. Commun.* **60**, 825–832.
[13] Wolosin, J.M. (1985) *Am. J. Physiol.* **248**, G595–G607.
[14] Hongo, T., Nojima, S. and Setaka, M. (1990) *Jpn. J. Pharmacol.* **52**, 295–305.
[15] Lindberg, P., Brandstrom, A., Wallmark, B., Mattson, H., Rikner, L. and Hoffmann, K. (1990) *Med. Res. Rev.* **10**, 1–54.
[16] Pope, A.J. and Parsons, M.E. (1993) *Trends Pharmacol. Sci.* **14**, 323.
[17] Hersey, S.J. and Sachs, G. (1995) *Physiol. Rev.* **75**, 155–189.
[18] Lindberg, P., Nordberg, P., Alminger, T., Braendstroem, A. and Wallmark, B. (1986) *J. Med. Chem.* **29**, 1327–1329.
[19] Shin, J.M., Cho, Y.M. and Sachs, G. (2004) *J. Am. Chem. Soc.* **126**, 7800.
[20] Besancon, M., Simon, A., Sachs, G. and Shin, J.M. (1997) *J. Biol. Chem.* **272**, 22438–22446.
[21] Tytgat, G.N. (2001) *Eur. J. Gastroenterol. Hepatol.* **13**, 29–33.
[22] Savarino, V., Mela, G.S., Zentilin, P., Cutela, P., Mele, M.R., Vigneri, S. and Celle, G. (1994) *Dig. Dis. Sci.* **39**, 161–168.
[23] Regardh, C.G., Langerstrom, P.O., Lundborg, P. and Skanberg, I. (1990) *Ther. Drug. Monit.* **12**, 163–171.
[24] Sohn, D.-R., Kobayashi, K., Chiba, K., Lee, K.-H., Shin, S.-G. and Ishizka, T. (1992) *J. Pharmacol. Exp. Ther.* **262**, 1195–1202.
[25] Furuta, T., Shirai, N., Sugimoto, M., Nakamura, A., Hishida, A. and Ishizaki, T. (2005) *Drug Metab. Pharmacokinet.* **20**, 153–167.
[26] Robinson, M. and Horn, J. (2003) *Drugs* **63**, 2739–2754.
[27] Vakil, N. and Fennerty, M.B. (2003) *Aliment. Pharmacol. Ther.* **18**, 559–568.
[28] Horn, J. (2006) *Aliment. Pharmacol. Ther. Symp. Ser.* **2**, 340.
[29] Espluges, J.V. (2005) *Drugs* **65**, 7–12.
[30] Maeda, M. (1994) *J. Biochem.* **115**, 6–14.
[31] Andersson, T., Rohss, K., Bredberg, E. and Hassan-Alin, M. (2001) *Aliment. Pharmacol. Ther.* **15**, 1563–1569.
[32] Kroemer, H.K., Klotz, U. and Schulz, M. (1995) *Pharm. Ztg.* **140**, 44, 46–48.
[33] Nagaya, H., Satoh, H. and Maki, Y. (1990) *J. Pharmacol. Exp. Ther.* **252**, 1289–1295.

[34] Nagaya, H., Satoh, H., Kubo, K. and Maki, Y. (1989) *J. Pharmacol. Exp.Ther.* **248**, 799–805.
[35] Sato, F., Kyotoku, T., Hori, Y., Hori, Y., Okano, S. and Inatomi, S. (2005) *Jpn. Pharmacol. Ther.* **33**, 113–122.
[36] Satoh, H., Inatomi, N., Nagaya, H., Inada, I., Nohara, A., Nakamura, N. and Maki, Y. (1989) *J. Pharmacol. Exp. Ther.* **248**, 806–815.
[37] Huber, R., Kohl, B., Sachs, G., Senn-Bilfinger, J., Simon, W.-A. and Sturm, E. (1995) *Aliment. Pharmacol. Ther.* **9**, 363–378.
[38] Jungnickel, P.W. (2000) *Clin. Ther.* **22**, 1268–1293.
[39] Kohl, B., Sturm, E., Senn-Bilfinger, J., Simon, W.A., Kruger, U., Schaefer, H., Rainer, G., Figala, V. and Klemm, K. (1992) *J. Med. Chem.* **35**, 1049–1057.
[40] Kromer, W., Kruger, U., Huber, R., Hartmann, M. and Steinijans, V.W. (1998) *Pharmacology* **56**, 57–70.
[41] Beil, W., Sewing, K.F. and Kromer, W. (1999) *Drugs Today* **35**, 753–764.
[42] Kromer, W., Postius, S., Riedel, R., Simon, W.A., Hanauer, G., Brand, U., Goenne, S. and Parsons, M.E. (1990) *J. Pharmacol. Exp. Ther.* **254**, 129–135.
[43] Garner, A. and Fadlallah, H. (1997) *Expert Opin. Investig. Drugs* **6**, 885–893.
[44] Steinijans, V.W., Huber, R., Hartmann, M., Zech, K., Bliesath, H., Wurst, W. and Radtke, H.W. (1996) *Int. J. Clin. Pharmacol. Ther.* **34**, 531–550.
[45] Simon, B., Mueller, P., Bliesath, H., Luehmann, R., Hartmann, M., Huber, R. and Wurst, W. (1990) *Aliment. Pharmacol. Ther.* **4**, 239–245.
[46] Langtry, H.D. and Markham, A. (1999) *Drugs* **58**, 725–742.
[47] King, F.D. and Joiner, K.A. (1986) *Eur. Pat Appl. Publ.* EP 178438.
[48] Uchiyama, K., Wakatsuki, D., Kakinoki, B., Takeuchi, Y., Araki, T. and Morinaka, Y. (1999) *J. Pharm. Pharmacol.* **51**, 457–464.
[49] Shin, J.M., Homerin, M., Domagala, F., Ficheux, H. and Sachs, G. (2006) *Biochem. Pharmacol.* **71**, 837–849.
[50] Digestive Disease Week (Los Angeles). (2006) Abstr. S1209.
[51] Zimmermann, P.J., Brehm, C., Palmer, A., Chiesa, M.V., Simon, W.-A., Postius, S., Kromer, W. and Sturm, E. (2005) *PCT Int. Appl.* WO2005123730.
[52] Digestive Disease Week (Los Angeles). (2006) Abstr. S-1200.
[53] Hirai, K., Koike, H., Ishiba, T., Ueda, S., Makino, I., Yamada, M., Ichihashi, T., Mizushima, Y. and Ishikawa, M. (1991) *Eur. J. Med. Chem.* **26**, 143–158.
[54] Sih, J.C., Im, W.B., Robert, A., Graber, D.R. and Blakeman, D.P. (1991) *J. Med. Chem.* **34**, 1049–1062.
[55] Hughes, P.M. (2005) *PCT Int. Appl.* WO200582337.
[56] Hughes, P.M. (2005) *PCT Int. Appl.* WO200582338.
[57] Shen, J., Welty, D.F. and Tang-Liu, D.D. (2005) *US Pat. Appl. Publ.* 2005075371.
[58] Garst, M.E., Sachs, G. and Shin, J.M. (2005) *PCT Int. Appl.* WO200009498.
[59] Hunt, R.H., Armstrong, D., Yaghoobi, M., James, C., Chen, Y., Leonard, J., Shin, J.M. and Sachs, G. (2006) *Am. J. Gastroenterol.* **101**, S92.
[60] Shen, J., Welty, D.F. and Tang-Liu, D.D. (2005) *PCT Int. Appl.* WO2005089758.
[61] Fang, X., Garvey, D.S. and Lette, L.G. (2004) *US Pat. Appl. Publ.* 200424014.
[62] Kinoshita, M., Saito, N., Noto, T. and Tamaki, H. (1996) *J. Pharmacol. Exp. Ther.* **277**, 28–33.
[63] Yamada, M., Yura, T., Morimoto, M., Harada, T., Yamada, K., Honma, Y., Kinoshita, M. and Sigiura, M. (1996) *J. Med. Chem.* **39**, 596–604.
[64] Kinoshita, M., Saito, N. and Tamaki, H. (1997) *Eur. J. Pharmacol.* **321**, 325–332.

[65] Karimian, K., Tam, T.F., Desilets, D., Lee, S. and Cappellotto, T. (1997) *PCT Int. Appl.* WO199731923.
[66] Karimian, K., Tam, T.F., Desilets, D., Lee, S., Cappellotto, T. and Li, W. (2000) *US Pat. Appl. Publ.* 5677302.
[67] Karimian, K., Tam, T.F., Desilets, D., Lee, S., Cappellotto, T. and Li, W. (2000) *US Pat. Appl. Publ.* 6093738.
[68] Sohn, S.-K., Chang, M.-S., Choi, W.-S., Kim, K.-B., Woo, T.-W., Lee, S.-B. and Chung, Y.-K. (1995) *Can. J. Physiol. Pharmacol.* **77**, 330–335.
[69] Kim, S.-H., Han, K.-S., Choi, W.-S., Chang, M.-S. and Lee, M.-G (1997) *Res. Commun. Mol. Pathol. Pharmacol.* **98**, 77–84.
[70] Yoo, H.-Y., Chung, K.-J., Chai, J.-P., Chang, M.-S. and Kim, S.-G. (1997) *PCT Int. Appl.* WO199703073.
[71] Yoo, H.-Y., Chung, K.-J., Chai, J.-P., Chang, M.-S. and Kim, S.-G. (1997) *PCT Int. Appl.* WO199703076.
[72] Chung, Y.K., Chang, M.S., Kim, K.B., Sohn, S.K., Woo, T.W., Lee, S.B. and Choi, W.S. (1998) *Can. J. Physiol. Pharmacol.* **76**, 921–929.
[73] Yoo, H.-Y., Chung, K.-J., Chang, M.-S., Kim, S.-G., Choi, W.-S. and Kang, D.-P. (1997) *PCT Int. Appl.* WO199703077.
[74] Loury, D.N., Beard, C.C., Fisher, P.E., Rosenkranz, R.P. and Waterbury, L.D. (1988) *FASEB J.* **2**(5), Abstr. 4394.
[75] Tabuchi, Y., Ogasawara, T. and Furuhama, K. (1994) *Arzneimittel-Forschung* **44**, 51–54.
[76] Esplugues, J.V. (2005) *Drugs* **65**(Suppl. 1), 7–12.
[77] Waldum, H.L., Gustafsson, B., Fossmark, R. and Qvigstad, G. (2005) *Dig. Dis. Sci.* **50**, S39–S44.
[78] Laine, L., Ahnen, D., McClain, C., Solcia, E. and Walsh, J.H. (2000) *Aliment. Pharmacol. Ther.* **14**, 651–668.
[79] Martin de Argila, C. (2005) *Drugs* **65**(Suppl. 1), 97–104.
[80] Martinsen, T.C., Bergh, K. and Waldum, H.L. (2005) *Basic Clin. Pharmacol. Toxicol.* **96**, 94–102.
[81] Wingate, D.L. (1990) *Lancet* **335**, 222.
[82] Larner, A.J. and Lendrum, R. (1992) *Gut* **33**, 860.
[83] Williams, C. and McColl, K.E.L. (2006) *Aliment. Pharmacol. Ther.* **23**, 3–10.
[84] Haakanson, R. and Sundler, F. (1990) *Eur. J. Clin. Invest.* **20**(Suppl. 1), S65–S71.
[85] Havu, N. (1986) *Digestion* **35**(Suppl. 1), 42–55.
[86] Ekman, L., Hansson, E., Havu, N., Carlsson, E. and Lundberg, C. (1985) *Scand. J. Gastroenterol.* **108**(Suppl.), 53–69.
[87] Carlsson, E., Larsson, H., Mattson, H., Ryberg, B. and Sundell, G.J. (1986) *Gastroenterology* **21**(Suppl. 118), 31.
[88] Herling, A.W. and Weidmann, K. (1994) *Prog. Med. Chem.* **31**, 233.
[89] Carlsson, E., Havo, N., Mattsson, H., Ekman, L. and Ryberg, B. (1991) *Fernstrom. Found. Ser.* **15**, 461–471.
[90] Judd, L.M., Andringa, A., Rubio, C.A., Spicer, Z., Shull, G.E. and Miller, M.L. (2005) *J. Gastroenterol. Hepatol.* **20**, 1266–1278.
[91] Lamberts, R., Creutzfeld, W., Struber, H.G., Brunner, G. and Solcia, E. (1993) *Gastroenterology* **104**, 1356.
[92] Freston, F.W., Borch, K., Brand, S.J., Carlsson, E., Creutzfeld, W., Hakanson, R., Olbe, L., Solcia, E., Walsh, J.H. and Wolfe, M.M. (1995) *Dig. Dis. Sci.* **40**(Suppl.), 50S.

[93] De Milito, A. and Fais, S. (2005) *Expert Opin. Pharmacother.* **6**, 1049–1054.
[94] De Milito, A., Iessi, E., Logozzi, M., Lozupone, F., Spada, M., Marino, M.L., et al. (2007) *Cancer Res.* **67**, 5408.
[95] Suzuki, H. and Hibi, F. (2005) *Expert Opin. Pharmacother.* **6**, 59–67.
[96] Yeo, M., Kwak, M.S., Kim, D.K., Chung, I.S., Moon, B.S., Song, K.S. and Hahm, K.-B. (2006) *J. Clin. Biochem. Nutr.* **38**, 1–8.
[97] Gillen, D. and McColl, K.E.L. (2001) *Best Pract. Res. Clin. Gastroenterol.* **15**, 487–495.
[98] Qvigstad, G. and Waldum, H. (2004) *Basic Clin. Pharmacol. Toxicol.* **94**, 202–208.
[99] Zacny, J., Zamakhshary, M., Sketris, I. and Van Zanten, S. (2005) *Alimentary Pharmacol. Ther.* **21**, 1299–1312.
[100] Johnson, D.A., Orr, W.C., Crawley, J.A., Traxler, B., McCullough, J., Brown, K.A. and Roth, T. (2005) *Am. J. Gastroenterol.* **100**, 1914–1922.
[101] Crawley, J.A. and Schmitt, C.M. (2000) *J. Clin. Outcomes Manag.* **7**, 29–34.
[102] Scarpignato, C., Pelosini, I. and Di Mario, F. (2006) *Dig. Dis.* **24**, 11–46.
[103] Bristol, J.A. and Puchalski, C. (1984) *US Pat. Appl. Publ.* 4450164.
[104] Kaminski, J.J., Bristol, J.A., Puchalski, C., Lovey, R.G., Elliott, A.J., Guzik, H., Solomon, D.M., Conn, D.J. and Domalski, M.S. (1985) *J. Med. Chem.* **28**, 876–892.
[105] Keeling, D.J., Taylor, A.G. and Schudt, C. (1989) *J. Biol. Chem.* **264**, 5545–5551.
[106] Munson, K.B. and Sachs, G. (1988) *Biochemistry* **27**, 3932–3938.
[107] Fallowfiled, C., Lawrie, K.W.M., Saunders, D., Ife, R.J. and Keeling, D.J. (1988) *Biochem. Soc. Trans.* **16**, 823.
[108] Briving, C., Andersson, B.M., Nordberg, P. and Wallmark, B. (1988) *Biochim. Biophys. Acta Biomembr.* **946**, 185–192.
[109] Keeling, D.J., Laing, S.M. and Senn-Bilfinger, J. (1988) *Prog. Clin. Biol. Res.* **273**, 255–260.
[110] Keeling, D.J., Laing, S.M. and Senn-Bilfinger, J. (1988) *Biochem. Pharmacol.* **37**, 2231–2236.
[111] Hersey, S.J., Steiner, L., Mendlein, J., Rabon, E. and Sachs, G. (1988) *Biochim. Biophys. Acta, Prot. Struct. Mol. Enzymol.* **956**, 49–57.
[112] Kaminski, J.J., Hilbert, J.M., Pramanik, B.N., Solomon, D.M., Conn, D.J., Rizvi, R.K., et al. (1987) *J. Med. Chem.* **30**, 2031–2046.
[113] Chiu, P.J.S., Casciano, C., Tetzloff, G., Long, J.F. and Barnett, A. (1983) *J. Pharmacol. Exp. Ther.* **226**, 121–125.
[114] Kaminski, J.J., Perkins, D.G., Frantz, J.D., Solomon, D.M., Elliott, A.J., Chiu, P.J.S. and Long, J.F. (1987) *J. Med. Chem.* **30**, 2047–2051.
[115] Long, J.F., Chiu, P.J.S., Derelanko, M.J. and Steinberg, M. (1983) *J. Pharmacol. Exp. Ther.* **226**, 114.
[116] Kaminski, J.J., Puchalski, C., Solomon, D.M., Rizvi, R.K., Conn, D.J., Elliott, A.J., Lovey, R.G., Guzik, H. and Chiu, P.J.S. (1989) *J. Med. Chem.* **32**, 1686–1700.
[117] Kaminski, J.J. and Doweyko, A.M. (1997) *J. Med. Chem.* **40**, 427–436.
[118] Yanagisawa, I., Ohta, M., Koide, T., Shikama, H. and Miyata, K. (1988) *Eur. Pat. Appl. Publ.* 0266890.
[119] Yuki, H., Kamato, T., Nishida, A., Ohta, M., Shikama, H., Yanagisawa, I. and Miyata, K. (1995) *Jpn. J. Pharmacol.* **67**, 59–67.
[120] 11th International Congress on Pharmacology (Amsterdam). (1990) Abstr. S.mo.06.5.
[121] Yamazaki, A., Ishii, Y. and Abe, M. (1990) *Jpn. Pat. Appl. Publ.* 02028177.
[122] Ueda, I., Shiokawa, Y., Take, K. and Itani, H. (1987) *Eur. Pat. Appl. Publ.* 0228006.

[123] Senn-Bilfinger, J., Grundler, G., Rainer, G., Postius, S., Riedel, R. and Simon, W.-A. (1994) *PCT Int. Appl.* WO199418199.
[124] Simon, W.-A., Riedel, R. and Postius, S. (1996) *PCT Int. Appl.* WO199603404.
[125] Simon, W.-A., Riedel, R., Postius, S. and Kley, H.-P. (1996) *PCT Int. Appl.* WO199603405.
[126] Riedel, R., Postius, S., Grundler, G., Senn-Bilfinger, J., Rainer, G. and Simon, W.A. (1996) *PCT Int. Appl.* WO199603402.
[127] Riedel, R., Postius, S., Simon, W.A., Rainer, G., Senn-Bilfinger, J. and Grundler, G. (1996) *PCT Int. Appl.* WO199603403.
[128] Senn-Bilfinger, J., Grundler, G., Riedel, R., Postius, S., Simon, W.-A. and Rainer, G. (1995) *PCT Int. Appl.* WO199510518.
[129] Riedel, R., Postius, S. and Simon, W.-A. (1996) *PCT Int. Appl.* WO199605199.
[130] Wurst, W., Hartmann, M. and Yale, J. (1996) *J. Biol. Med.* **69**, 233–243.
[131] Kromer, W., Postius, S. and Riedel, R. (2000) *Pharmacology* **60**, 179.
[132] Martinek, J., Blum, A.L., Stolte, M., Hartmann, M., Verdu, E.F., Luhmann, R., Dorta, G. and Wiesel, P. (1999) *Aliment. Pharmacol. Ther.* **13**, 27–34.
[133] Pfutzer et al. (1997) Dig. Dis. Week May 10–16, Washington D.C., Abstr. 3821.
[134] Fujisawa Pharma. Co. Ltd. (1990) *Jpn. Pat. Appl. Publ.* 02270873.
[135] Shiokawa, Y., Nagano, M. and Itani, H. (1989) *Eur. Pat. Appl. Publ.* 308917.
[136] Shiokawa, Y., Nagano, M. and Itani, H. (1991) *Jpn. Pat. Appl. Publ.* 03031280.
[137] Shiokawa, Y., Nagano, M. and Itani, H. (1988) *Eur. Pat. Appl. Publ.* 268989.
[138] Amin, K., Dahlström, M., Nordberg, P. and Starke, I. (1998) *PCT Int. Appl.* WO199837080.
[139] Amin, K., Dahlström, M., Nordberg, P. and Starke, I. (2000) *PCT Int. Appl.* WO200011000.
[140] Amin, K., Dahlström, M., Nordberg, P. and Starke, I. (2000) *PCT Int. Appl.* WO200010999.
[141] Amin, K., Dahlström, M., Nordberg, P. and Starke, I. (1995) *PCT Int. Appl.* WO199955706.
[142] Amin, K., Dahlström, M., Nordberg, P. and Starke, I. (1995) *PCT Int. Appl.* WO199955705.
[143] Lehmann, A. and Wrangstadh, M. (2004) *PCT Int. Appl.* WO200400856.
[144] Amin, K., Dahlstroem, M. and Nordberg, P. (2002) *PCT Int. Appl.* WO200260440.
[145] Nordberg, P. (2004) *PCT Int. Appl.* WO2004113339.
[146] Nordberg, P. (2004) *PCT Int. Appl.* WO2004113340.
[147] Nordberg, P. (2004) *PCT Int. Appl.* WO2004113338.
[148] Dahlstroem, M., Loevqvist, K. and Malm, B. (2002) *PCT Int. Appl.* WO200260442.
[149] Dahlstroem, M., Langkilde, F. and Loevqvist, K. (2002) *PCT Int. Appl.* WO200260441.
[150] Elman, B., Erback, S. and Thiemermann, E. (2002) *PCT Int. Appl.* WO200220523.
[151] Amin, K., Dahlström, M., Nordberg, P. and Starke, I. (2003) *PCT Int. Appl.* WO2003018582.
[152] Abelo, A., Andersson, M., Holmberg, A.A. and Karlsson, M.O. (2006) *Eur. J. Pharm. Sci.* **29**, 91.
[153] Andersson, K. (2006) Society for Medicines Research Symposium, September; Collingwood, S. and Witherington, J. (2007) *Drug News Perspect.* **20**, 139.
[154] Lilljequist, L., Lindkvist, M., Nordberg, P., Pettersson, U. and Sebhatu, T. (2005) *PCT Int. Appl.* WO200558895.

[155] Briving, C., Svensson, K., Maxvall, I. and Andersson, K. (2004) *Gastroenterology* **126**(Suppl. 2), Abstr. M1436.
[156] Andersson, K. and Carlsson, E. (2005) *Pharmacol. Ther.* **108**, 294–307.
[157] Briving, C., Svensson, K., Maxvall, I. and Andersson, K. (2004) *Gastroenterology* **126**, A-333 (M1436).
[158] Kirchhoff, P., Andersson, K., Socrates, T., Sidani, S., Kosiek, O. and Geibel, J.P. (2006) *Am. J. Physiol. Gastrointest. Liver Physiol.* **291**, G838.
[159] Holstein, B., Holmberg, A., Florentzson, M., Aurell Holmberg, A., Andersson, M. and Andersson, K. (2004) *Gastroenterology* **126**(Suppl. 2), Abstr. M1440.
[160] Holstein, B., Holmberg, A., Florentzson, M., Aurell Holmberg, A., Andersson, M. and Andersson, K. (2004) *Eur. J. Pharm. Sci.* **23**(Suppl. 1), S8.
[161] Andersson, K., Aurell Holmberg, A., Briving, C. and Holstein, B. (2004) *Gastroenterology* **126**, A-56.
[162] Kahrilas, P.J., Dent, J., Lauritsen, K., Malfertheiner, P., Denison, H., Franzen, S. and Lundborg, P. (2005) *Dig. Dis. Week, Chicago*, Abstr. T1671.
[163] Nilsson, C., Albrektson, E., Rydholm, H., Rohss, K., Hassan-Alin, M. and Hasselgren, G. (2005) *Dig. Dis. Week, Chicago*, Abstr. T1680.
[164] Holstein, B., Holmberg, A., Floretzson, M., Holmberg, A., Andersson, M. and Andersson, K. (2004) *Gastroenterology* **126**(Suppl. 2), Abstr. M1437.
[165] Kahrilas, P.J., Dent, J., Lauritsen, K., Malfertheiner, P., Denison, H., Franzen, S. and Hasselgren, G. (2006) *Dig. Dis. Week, Los Angeles*, Abstr. M1163.
[166] Dent, J., Kahrilas, P.J., Hatlebakk, J., Vakil, N., Denison, H., Franzen, S. and Hasselgren, G. (2006) *Dig. Dis. Week, Los Angeles*, Abstr. S1204.
[167] Simon, W.-A., Postius, S., Kromer, W., Buhr, W., Senn-Bilfinger, J. and Zimmermann, P.J. (2004) *PCT Int. Appl.* WO200446144.
[168] Senn-Bilfinger, J., Grundler, G., Schaeffer, H., Klemm, K., Rainer, G., Riedel, R., Schudt, C. and Simon, W. (1988) *Eur. Pat. Appl. Publ.* 290003.
[169] Zimmermann, P.J., Buhr, W., Brehm, C., Palmer, A.M., Feth, M.P., Senn-Bilfinger, J. and Simon, W.-A. (2007) *Bioorg. Med. Chem. Lett.* **17**, 5374.
[170] Kohl, B., Zimmermann, P.J., Zech, K., Buhr, W., Palmer, A., Brehm, C., et al. (2007) *PCT Int. Appl.* WO2007039464.
[171] Jinno, M., Shimokawa, H. and Yamagishi, T. (2006) *PCT Int. Appl.* WO200664339.
[172] Matsumoto, Y. and Shimokawa, H. (2007) *PCT Int. Appl.* WO2007026218.
[173] Matsumoto, Y., Shimokawa, H. and Yamagishi, T. (2007) *US Pat. Appl. Publ.* 219237.
[174] Bamford, M.J., Elliott, R.L., Giblin, G.M.P., Naylor, A., Witherington, J., Panchal, T.A. and Demont, E.H. (2006) *PCT Int. Appl.* WO2006100119.
[175] Bamford, M.J., Elliott, R.L., Giblin, G.M.P., Naylor, A., Panchal, T.A., Takle, A.K. and Witherington, J. (2007) *PCT Int. Appl.* WO2007003386.
[176] Gold, E.H., Kaminski, J.J. and Puchalski, C. (1984) *US Pat. Appl. Publ.* 4468400.
[177] Briving, C.B., Nordberg, P.M. and Starke, C.I. (1995) *PCT Int. Appl.* WO199527714.
[178] Grundler, G., Simon, W.-A., Postius, S., Riedel, R. and Thibaut, U. (1995) *PCT Int. Appl.* WO199854188.
[179] Mulder, G.J. (1992) *Annu. Rev. Pharmacol. Toxicol.* **32**, 25–49.
[180] Senn-Bilfinger, J. (2002) *US Pat. Appl. Publ.* 20020169320.
[181] Buhr, W., Kohl, B., Senn-Bilfinger, J. and Zimmermann, P. (2002) *PCT Int. Appl.* WO200234749.
[182] Simon, W.A., Postius, S., Huber, R., Kromer, W., Senn-Bilfinger, J. and Buhr, W. (2001) *PCT Int. Appl.* WO200172456.

[183] Simon, W.-A., Postius, S. and Kromer, W. (2001) *PCT Int. Appl.* WO200172755.
[184] Postius, S., Simon, W.-A., Grundler, G., Hanauer, G., Huber, R., Kromer, W. and Sturm, E. (2000) *PCT Int. Appl.* WO200017200.
[185] Senn-Bilfinger, J. (2000) *PCT Int. Appl.* WO200063211.
[186] Senn-Bilfinger, J. (2000) *PCT Int. Appl.* WO200026217.
[187] Simon, W.A., Herrmann, M., Klein, T., Shin, J.M., Huber, R., Senn-Bilfinger, J. and Postius, S. (2007) *J. Pharmacol. Exp. Ther.* **321**, 866.
[188] Iwabuchi, H., Hagihara, M., Shibakawa, N., Matsunobu, K. and Fujiwara, H. (2001) *PCT Int. Appl.* WO200158901.
[189] Buhr, W., Zimmermann, P.J., Brehm, C., Palmer, A., Chiesa, M.V., Simon, W.A., Postius, S., Kromer, W., Rast, G. and Doelling, U. (2006) *PCT Int. Appl.* WO200640338.
[190] Buhr, W., Simon, W.-A., Postius, S. and Kromer, W. (2003) *PCT Int. Appl.* WO2003014120.
[191] Buhr, W., Simon, W.-A., Postius, S., Kromer, W. and Sturm, E. (2003) *PCT Int. Appl.* WO2003016310.
[192] Simon, W.-A., Postius, S. and Kromer, W. (2001) *PCT Int. Appl.* WO200172757.
[193] Chiesa, M.V., Zimmermann, P.J., Brehm, C., Simon, W. and Kromer, W. (2005) *PCT Int. Appl.* WO2005090358.
[194] Buhr, W., Chiesa, M.V., Zimmermann, P.J., Brehm, C., Simon, W.-A., Kromer, W. and Postius, S. (2005) *PCT Int. Appl.* WO2005058325.
[195] Palmer, A.M., Grobbel, B., Jecke, C., Brehm, C., Zimmermann, P.J., Buhr, W., Feth, M.P., Simon, W.-A. and Kromer, W. (2007) *J. Med. Chem.* **50**, 6240.
[196] Buhr, W., Zimmermann, P.J., Brehm, C., Palmer, A., Kromer, W., Postius, S. and Simon, W.-A. (2005) *PCT Int. Appl.* WO2005077949.
[197] Brehm, C., Chiesa, M.V., Zimmermann, P.J., Buhr, W., Simon, W.-A., Kromer, W., Postius, S. and Palmer, A. (2005) *PCT Int. Appl.* WO2005090346.
[198] Zimmermann, P.J., Senn-Bilfinger, C., Buhr, J., Brehm, W., Chiesa, M.V., Palmer, A., Simon, W.-A., Postius, S. and Kromer, W. (2006) *PCT Int. Appl.* WO2006134112.
[199] Zimmermann, P.J., Brehm, C., Buhr, W., Palmer, A.M., Volz, J. and Simon, W.-A. (2008) *Bioorg. Med. Chem.* **16**, 536.
[200] Amin, K., Dahlström, M., Nordberg, P. and Starke, I. (1999) *PCT Int. Appl.* WO199928322.
[201] Bristol, J.A. and Lovey, R.G. (1984) *US Pat. Appl. Publ.* 4464372.
[202] Chiesa, M., Palmer, A., Brehm, C., Grundler, G., Senn-Bilfinger, J., Simon, W.-A., Postius, S. and Kromer, W. (2004) *PCT Int. Appl.* WO2004074289.
[203] Romero, R.S., Franco, F., Castaneda, A. and Muchowski, J.M. (1991) *US Pat. Appl. Publ.* 5041442.
[204] Bristol, J.A. and Lovey, R.G. (1983) *US Pat. Appl. Publ.* 4409226.
[205] Takehashi, T., Horigome, M., Momose, K., Nagai, S., Sugita, M., Katsuyama, K., Suzuki, C. and Nakamura, K. (1994) *Jpn. Pat. Appl. Publ.* 06247966.
[206] Briving, C., Elebring, M., Carlsson, S., Carter, R., Kuehler, T., Nordberg, P., Starke, I. and Svensson, A. (1992) *Eur. Pat. Appl. Publ.* 509974.
[207] Takahashi, T., Horigome, M., Momose, K., Nagai, S., Oshida, N., Sugita, M., Katsuyama, K., Suzuki, C. and Nakamura, K. (1994) *Jpn. Pat. Appl. Publ.* 06247967.
[208] Briving, C.B., Carlsson, S.A.I., Lindberg, P.L., Mattson, A.H., Nordberg, M.P. and Wallmark, B.M.G. (1988) *Eur. Pat. Appl. Publ.* 0266326.
[209] Briving, C.B., Carlsson, S.A.I., Lindberg, P.L., Mattson, A.H., Nordberg, M.P. and Wallmark, B.M.G. (1988) *US Pat. Appl. Publ.* 5106862.

[210] Amin, K., Dahlström, M., Nordberg, P. and Starke, I. (1997) *PCT Int. Appl.* WO199747603.

[211] Chiesa, M.V., Palmer, A., Brehm, C., Simon, W.-A., Postius, S., Kromer, W., Zimmermann, P.J. and Buhr, W. (2005) *PCT Int. Appl.* WO2005111000.

[212] Iwamuro, Y., Koike, H., Morita, M. and Sakakibara, S. (2007) *PCT Int. Appl.* WO2007072142.

[213] Zimmermann, P.J., Brehm, C., Palmer, A., Chiesa, M.V., Simon, W.-A., Postius, S., Kromer, W. and Senn-Bilfinger, J. (2005) *PCT Int. Appl.* WO2005058893.

[214] Koike, H. and Sakakibara, S. (2006) *PCT Int. Appl.* WO2006134460.

[215] Hanazawa, T. and Koike, H. (2007) *US Pat. Appl. Publ.* 142448.

[216] Hanazawa, T., Koike, H. and Sakakibara, S. (2007) *PCT Int. Appl.* WO2007031860.

[217] Kohl, B., Zimmermann, P.J., Zech, K., Zimmerman, P., Palmer, A., Brehm, C., et al. (2007) *PCT Int. Appl.* WO2007023135.

[218] Buhr, W., Zimmermann, P.J., Brehm, C., Palmer, A., Simon, W.-A., Postius, S. and Kromer, W. (2005) *PCT Int. Appl.* WO2005026164.

[219] Zimmermann, P.J., Buhr, W., Chiesa, M.V., Palmer, A., Postius, S., Kromer, W. and Simon, W.-A. (2005) *PCT Int. Appl.* WO2005121139.

[220] Zimmermann, P.J., Brehm, C., Palmer, A., Chiesa, M.V., Simon, W.A., Postius, S., Kromer, W., Senn-Bilfinger, J. and Buhr, W. (2005) *PCT Int. Appl.* WO2005103057.

[221] Takahashi, T., Horigome, M., Momose, K., Nagai, S., Oshida, N., Sugita, M., Katsuyama, K., Suzuki, C. and Nakamura, K. (1994) *Jpn. Pat. Appl. Publ.* 06247967.

[222] Zimmermann, P.J., Senn-Bilfinger, J., Brehm, C., Buhr, W., Chiesa, M.V., Palmer, A., Simon, W.-A., Postius, S. and Kromer, W. (2006) *PCT Int. Appl.* WO2006134111.

[223] Bristol, J.A. and Lovey, R.G. (1982) *US Pat. Appl. Publ.* 4358453.

[224] Zimmermann, P.J., Brehm, C., Palmer, A., Chiesa, M.V., Simon, W.-A., Postius, S. and Kromer, W. (2005) *PCT Int. Appl.* WO2005070927.

[225] Zimmermann, P.J., Brehm, C., Palmer, A., Chiesa, M.V., Simon, W.-A., Postius, S. and Kromer, W. (2005) *PCT Int. Appl.* WO2005077947.

[226] Kimura, T., Fujihara, Y., Shibakawa, N., Fujiwara, H., Itoh, E., Matsunobu, K., Tabata, K. and Yasuda, H. (1995) *PCT Int. Appl.* WO199519980.

[227] Hagihara, M., Shibakawa, N., Matsunobu, K., Fujiwara, H. and Ito, K. (2001) *Jpn. Pat. Appl. Publ.* 2001058993.

[228] Hagihara, M., Shibakawa, N., Matsunobu, K., Fujiwara, H. and Ito, K. (2000) *PCT Int. Appl.* WO200077003.

[229] Ito, K., Hagihara, M., Shibakawa, N. and Shimizu, M. (2004) *PCT Int. Appl.* WO2004058768.

[230] Ito, K., Kinoshita, K., Tomizawa, A., Inaba, F., Morikawa-Inomata, Y., Makino, M., Tabata, K. and Shibaka, N. (2007) *J. Pharmacol. Exp. Ther.* **323**, 308.

[231] Choi, R., Kim, J.-G., Ahn, B.-N., Lee, H.-W., Yoon, Y.-A., Kim, D.-H., Keum, S.-H., Shin, Y.-A. and Kang, H.-I. (2006) *PCT Int. Appl.* WO2006025714.

[232] Choi, R., Kim, J.-G., Ahn, B.-N., Lee, H.W., Yoon, S.-W., Yoon, Y.-A., et al. (2006) *PCT Int. Appl.* WO2006025717.

[233] Kim, J.-G., Ahn, B.-N., Lee, H.-W., Yoon, S.-W., Yoon, Y.-A., Choi, H.-H. and Kang, H.-I. (2006) *PCT Int. Appl.* WO2006025715.

[234] Hasuoka, A. and Arikawa, Y. (2006) *PCT Int. Appl.* WO2006011670.

[235] Buhr, W., Zimmermann, P.J., Brehm, C., Chiesa, M.V., Grundler, G., Simon, W.-A., Kromer, W. and Postius, S. (2006) *PCT Int. Appl.* WO200613195.

[236] Choi, R., Kim, J.-G., Ahn, B.-N., Lee, H.-W., Yoon, S.-W., Yoon, Y.-A., Lee, C.-H., Cha, M.-H. and Kang, H.-I. (2006) *PCT Int. Appl.* WO2006038773.
[237] Brown, T.H., Ife, R.J., Keeling, D.J., Laing, S.M., Leach, C.A., Parsons, M.E., Price, C.A., Reavill, D.A. and Wiggall, K.J. (1990) *J. Med. Chem.* **33**, 527–533.
[238] Grundler, G., Rainer, G., Ife, R.J., Leach, C.A., Postius, S., Riedel, R., Schaeffer, H., Senn-Bilfinger, J. and Simon, W.A. (1993) *PCT Int. Appl.* WO199312090.
[239] Pope, A.J., Boehm, M.K., Leach, C., Ife, R.J., Keeling, D. and Parsons, M.E. (1995) *Biochem. Pharmacol.* **50**, 1543–1549.
[240] Parsons, M.E., Rushant, B., Rasmussen, T.C., Leach, C., Ife, R.J., Postius, S. and Pope, A.J. (1995) *Biochem. Pharmacol.* **50**, 1551–1556.
[241] Munson, H.R., Jr. and Reevis, S.A. (1982) *US Pat. Appl. Publ.* 4343804.
[242] Ife, R.J., Brown, T.H., Keeling, D.J., Leach, C.A., Meeson, M.L., Parsons, M.E., Reavill, D.R., Theobald, C.J. and Wiggall, K.J. (1992) *J. Med. Chem.* **35**, 3413–3422.
[243] Keeling, D.J., Malcolm, R.C., Laing, S.M., Ife, R.J. and Leach, C.A. (1991) *Biochem. Pharmacol.* **42**, 123–130.
[244] Broom, C., Eagle, S., Steel, S., Pue, M. and Laroche, J. (1993) *Gastroenterology* **104**, A46.
[245] Ife, R.J. (1992) *Drugs Future* **17**, 796–798.
[246] Eagle, S., Gill, C., Acton, G., Horton, J., Writer, D., Meineke, I., De Mey, C. and Broom, C. (1993) *Proceedings BPS* (April), 176P.
[247] Uchida, M., Otsubo, K., Matsubara, J., Ohtani, T., Morita, S. and Yamasaki, K. (1995) *Chem. Pharm. Bull.* **43**, 693–698.
[248] Ife, R.J., Leach, C.A. and Brown, T.H. (1989) *Eur. Pat. Appl. Publ.* 0336544.
[249] Ife, R.J. and Leach, C.A. (1995) *US Pat. Appl. Publ.* 5432182.
[250] Leach, C.A., Brown, T.H., Ife, R.J., Keeling, D.J., Parsons, M.E., Theobald, C.J. and Wiggall, K.J. (1995) *J. Med. Chem.* **38**, 2748–2762.
[251] Ife, R.J., Brown, T.H. and Leach, C.A. (1989) *PCT Int. Appl.* WO198908112.
[252] Ife, R.J., Brown, T.H., Blurton, P., Keeling, D.J., Leach, C.A., Meeson, M.L., Parsons, M.E. and Theobald, C.J. (1995) *J. Med. Chem.* **38**, 2763–2773.
[253] Ife, R.J., Brown, T.H. and Leach, C.A. (1989) *PCT Int. Appl.* WO198905297.
[254] Yoo, H.Y., Chung, K.J., Chang, M.S., Kim, S.G., Choi, W.S., Kang, D.P., et al. (1997) *PCT Int. Appl.* WO199703074.
[255] Sohn, S.K., Chang, M.S., Choi, W.S., Chung, Y.K., Kim, K.B., Woo, T.W., Lee, S.B. and Park, C.J. (1999) *J. Pharm. Pharmacol.* **51**, 1359–1365.
[256] Starke, I. (1996) *PCT Int. Appl.* WO199617830.
[257] Starke, I. (1994) *PCT Int. Appl.* WO199429274.
[258] Abelo, A., Gabrielsson, J., Holstein, B., Eriksson, U.G., Holmberg, J. and Karlsson, M.O. (2001) *Eur. J. Pharm. Sci.* **14**, 339.
[259] Onoda, Y., Nomoto, J., Takai, H., Seo, N., Kase, H., Yokoyama, S. and Ishii, A. (1995) *Jap. Pat. Appl. Publ.* 07173138.
[260] Ife, R.J., Brown, T.H. and Leach, C.A. (1989) *Eur. Pat. Appl. Publ.* 307268.
[261] Leach, C.A., Brown, T.H., Ife, R.J., Keeling, D.J., Laing, S.M., Parsons, M.E., Price, C.A. and Wiggall, K.J. (1992) *J. Med. Chem.* **35**, 1845–1852.
[262] Rabon, E.C., Sachs, G., Leach, C.A. and Keeling, D. (1992) *Acta Phys. Scand. Suppl.* **607**, 269–273.
[263] Cheon, H.G., Lee, S.S., Lim, H. and Lee, D.H. (2001) *Eur. J. Pharmacol.* **411**, 187–192.
[264] Cheon, H.G., Lim, H. and Lee, D.H. (2001) *Eur. J. Pharmacol.* **411**, 181–186.

[265] Choi, J.-K., Kim, S.-S., Yum, E.-K., Kang, S.K., Yoo, Y.K., Cheon, H.G. and Kim, H.J. (1999) *PCT Int. Appl.* WO199909029.
[266] Cheon, H.G., Kim, H.J., Mo, H.K., Lee, B.H. and Choi, J.K. (2000) *J. Pharmacol.* **60**, 161–168.
[267] Choi, J.K., Yum, E.K., Kim, S.S., Kang, S.K., Cheon, H.G. and Kim, H.J. (2000) *PCT Int. Appl.* WO200001696.
[268] Brown, T.H. and Blurton, P. (1990) *Eur. Pat. Appl. Publ.* 380257.
[269] Ife, R.J., Leach, C.A., Pope, A.J. and Theobald, C.J. (1995) *Bioorg. Med. Chem. Lett.* **5**, 543–546.
[270] Lee, J.W. (1996) *PCT Int. Appl.* WO 9605177.
[271] Yuhan Corp. (2004) *Drugs Future* **29**, 455–459.
[272] Park, S., Ahn, B., Lee, B., Kang, H. and Song, K. (2003) *Gut* **52**(Suppl. VI), Abstr. MON-G-008.
[273] Han, K.S., Kim, Y.G., Yoo, J.K., Lee, J.W. and Lee, M.G. (1998) *Biopharm. Drug Dispos.* **19**, 493–500.
[274] Yu, K.S., Bae, K.S., Shon, J.H., Cho, J.Y., Yi, S.Y. and Chung, J.Y. (2004) *J. Clin. Pharmacol.* **44**, 73–82.
[275] Park, S., Song, K., Moon, B.S., Joo, S., Choi, M. and Chung, I. (2002) *Gut* **51**(Suppl. 3), Abstr. TUE-G-312.
[276] Lee, J.W., Lee, B.Y., Kim, C.S., Lee, S.K. and Lee, S.J. (2000) *PCT Int. Appl.* WO200029403.
[277] Lee, J.W., Lee, B.Y., Kim, C.S., Lee, S.K., Son, K.S., Lee, S.J., Shim, W.J. and Hwang, M.S. (1998) *PCT Int. Appl.* WO199843968.
[278] Lee, J.W., Chai, J.S., Kim, C.S., Kim, J.K., Lim, D.S., Lee, J.W., Shon, M.K. and Jo, D.W. (1994) *PCT Int. Appl.* WO199414795.
[279] Ife, R.J., Brown, T.H. and Leach, C.A. (1991) *PCT Int. Appl.* WO199118887.
[280] Brown, T.H., Ife, R.J., Leach, C.A. and Keeling, D.J. (1990) *Eur. Pat. Appl. Publ.* 0404322.
[281] Brown, T.H., Ife, R.J. and Leach, C.A. (1990) *Eur. Pat. Appl. Publ.* 0404355.
[282] Ife, R.J., Leach, C.A. and Dhanak, D. (1993) *PCT Int. Appl.* WO199315055.
[283] Ife, R.J. and Leach, C.A. (1994) *PCT Int. Appl.* WO199426715.
[284] Ife, R.J., Leach, C.A. and Dhanak, D. (1993) *PCT Int. Appl.* WO199315056.
[285] LaMattina, J.L., McCarthy, P.A., Reiter, L.A., Holt, W.F. and Yeh, L.A. (1990) *J. Med. Chem.* **33**, 543–552.
[286] Hosoi, M., Nishioka, R., Hioki, Y., Iida, Y., Takeshita, H., Niiyama, K. and Hidaka, Y. (1988) *Eur. Pat. Appl. Publ.* 264883.
[287] Hioki, Y., Takada, J., Hidaka, Y., Takeshita, H., Hosoi, M. and Yanao, M. (1990) *Arch. Int. Pharm. Ther.* **305**, 32–44.
[288] Takada, J. (1990) *Biochem. Pharmacol.* **40**, 1527–1531.
[289] Takeshita, H., Hirayama, Y., Amano, H., Hioki, Y., Takada, J., Ishikawa, J. and Yano, M. (1990) *Jpn. J. Pharmacol.* **52**(Suppl. 1), 242.
[290] Niiyama, K., Nagase, T., Fukami, T., Takezawa, Y., Takezawa, H., Hioki, Y., Takeshita, H. and Ishikawa, K. (1997) *Bioorg. Med. Chem. Lett.* **7**, 527–532.
[291] Ishikaw, K., Fukami, T., Niiyama, K., Nagase, T., Hioki, Y. and Takeshita, H. (2003) *Jpn. Pat. Appl. Publ.* 2003072458.
[292] Nagasawa, H., Toyofuku, H., Sawaki, S. and Edenami, K. (1988) *Jpn. Pat. Appl. Publ.* 63270666.

[293] Nagasawa, H., Toyofuku, H., Sawaki, S. and Edenami, K. (1988) *Jpn. Pat. Appl. Publ.* 63270664.
[294] Toyofuku, H., Nagasawa, H., Sawaki, S. and Edenami, K. (1988) *Jpn. Pat. Appl. Publ.* 63270667.
[295] Tanaka, H., Fukuzumi, K., Togawa, T., Banno, K., Ushiro, T., Morii, M. and Nakatani, T. (1996) *PCT Int. Appl.* WO199633195.
[296] Tsukimi, Y., Ushiro, T., Yamazaki, T., Ishikawa, H., Hirase, J., Narita, M., Nishigaito, T., Banno, K., Ichihara, T. and Tanaka, H. (2000) *Jpn. J. Pharmacol.* **82**, 21–28.
[297] Fukizumi, K. and Morii, M. (1999) *Jpn. Pat. Appl. Publ.* 11343289.
[298] Kajino, M., Hasuoka, A., Tarui, N. and Takagi, T. (2006) *PCT Int. Appl.* WO200636024.
[299] Kajino, M., Hasuoka, A. and Nishida, H. (2007) *PCT Int. Appl.* WO2007026916.
[300] Hasuoka, A., Takagi, T., Kajino, M. and Nishida, H. (2007) *PCT Int. Appl.* WO2007114338.
[301] Bajnogel, J., Blasko, G., Budai, Z., Egyed, A., Feketa, M., Karaffa, E., Mezei, T., Reiter, K. and Simig, G. (1994) *Eur. Pat. Appl. Publ.* 619298.
[302] Im, W.B., Blakeman, D.P., Mendlein, J. and Sachs, G. (1984) *Biochim. Biophys. Acta* **770**, 65–72.
[303] Yamamoto, O., Tanaka, H., Ueda, F. and Kimura, K. (1989) *Scand. J. Gastroenterol. Suppl.* **162**, 178–181.
[304] Erdei, L., Szabo-Nagy, A. and Laszlavik, M. (1994) *J. Plant Physiol.* **144**, 49–52.
[305] Murukami, S., Muramatsu, M. and Otomo, S. (1992) *J. Pharm. Pharmacol.* **44**, 926–928.
[306] Murukami, S., Muramatsu, M. and Otomo, S. (1992) *J. Nat. Prod.* **55**, 513–516.
[307] Murukami, S., Muramatsu, M., Aihara, H. and Otomo, S. (1991) *Biochem. Pharmacol.* **42**, 1447–1451.
[308] Murakami, S., Kijima, H., Isobe, Y., Muramatsu, M., Aihara, H., Otomo, S., Baba, K. and Kozawa, M. (1990) *J. Pharm. Pharmacol.* **42**, 723–726.
[309] Hamagishi, Y., Kawano, K., Kamei, H. and Oki, T. (1991) *Jpn. J. Pharmacol.* **55**, 283–286.
[310] Hayashi, T., Asanao, S., Mizutani, M., Takeguchi, N., Kojima, T., Okamura, K. and Morita, N. (1991) *J. Nat. Prod.* **54**, 802–809.
[311] Asano, S., Yoshida, A., Yashiro, H., Kobayashi, Y., Morisato, A., Ogawa, H., Takeguchi, N. and Morii, M. (2004) *J. Biol. Chem.* **279**, 13968–13975.
[312] Smolka, A.J., Goldenring, J.R., Gupta, S. and Hammond, C.E. (2004) *BMC Gastroenterol.* **4**, Online computer file.
[313] Hersey, S.J., Steiner, L., Mendlein, J., Rabon, E. and Sachs, G. (1988) *Biochim. Biophys. Acta* **956**, 49–57.
[314] Lambrecht, N., Munson, K., Vagin, O. and Sachs, G. (2000) *J. Biol. Chem.* **275**, 4041–4048.
[315] Swarts, H.G., Klaasse, C.H., de Boer, M., Fransen, A.M. and De Pont, J.J. (1996) *J. Biol. Chem.* **271**, 29764–29772.
[316] Vagin, O., Munson, K., Lambrecht, N., Karlish, S.J. and Sachs, G. (2001) *Biochemistry* **40**, 7480–7490.
[317] Xian, Y. and Herbert, H. (1997) *J. Struct. Biol.* **118**, 169–177.
[318] Toyoshima, C., Nakasako, M., Nomura, H. and Ogawa, H. (2000) *Nature (London)* **405**, 647–655.

[319] Toyoshima, C. and Nomura, H. (2002) *Nature (London)* **418**, 605–611.
[320] Munson, K., Garcia, R. and Sachs, G. (2005) *Biochemistry* **44**, 5267–5284.
[321] Middleton, D.A., Robins, R., Feng, X., Levitt, M.H., Spiers, I.D., Schwalbe, C.H., Reid, D.G. and Watts, A. (1997) *FEBS Lett.* **410**, 269–274.
[322] Watts, J.A., Watts, A. and Middleton, D.A. (2001) *J. Biol. Chem.* **276**, 43197–43204.
[323] Kim, C.G., Watts, J.A. and Watts, A. (2005) *J. Med. Chem.* **48**, 7145–7152.
[324] Vagin, O., Denevich, S., Munson, K. and Sachs, G. (2002) *Biochemistry* **41**, 12755–12762.
[325] Munson, K., Lambrecht, N. and Sachs, G. (2000) *Biochemistry* **39**, 2997–3004.
[326] Sweadner, K.J. and Donnet, C. (2001) *Biochem. J.* **356**, 685–704.
[327] Munson, K., Law, R.J. and Sachs, G. (2007) *Biochemistry* **46**, 5398.
[328] Sachs, G., Shin, J.M., Vagin, O., Lambrecht, N., Yakubov, I. and Munson, K. (2007) *J. Clin. Gastroenterol.* **41**, S226.

4 The Adenosine A_1 Receptor and its Ligands

PETER G. NELL[1] and BARBARA ALBRECHT-KÜPPER[2]

[1]*Global Drug Discovery – Operations, Bayer HealthCare AG, Bayer Schering Pharma, Müllerstraße 178, 13353 Berlin, Germany*

[2]*Global Drug Discovery – Department of Cardiology Research, Bayer HealthCare AG, Bayer Schering Pharma, Aprather Weg 18a, 42096 Wuppertal, Germany*

INTRODUCTION	163
Adenosine and Adenosine Receptor Subtypes	163
PHARMACOLOGY OF ADENOSINE A_1 RECEPTORS	165
Structure and Function of Adenosine A_1 Receptors	165
Species Differences	167
Therapeutic Application of A_1 Receptor Agonists	167
CHEMISTRY OF ADENOSINE A_1 RECEPTOR AGONISTS	171
Adenosine-Derived A_1 Agonists	171
Agonists with Non-Adenosine-Derived Structures	189
CONCLUSIONS	195
REFERENCES	195

INTRODUCTION

ADENOSINE AND ADENOSINE RECEPTOR SUBTYPES

Both adenosine and its action on the heart were discovered nearly 80 years ago [1]. Adenosine is a ubiquitous purine nucleoside and it is released under a variety of different physiological and pathophysiological circumstances. It is synthesized extra- and intra-cellularly by the enzymes 5′-nucleotidase and S-adenosylhomocysteine hydrolase [2, 3] and is released into the extra-cellular space, where it is cleared rapidly by an energy-dependent nucleotide transport system [4] (half-life of about 10 s [5]). Adenosine exerts its action by binding to specific adenosine receptors, which results in the activation or inhibition of second messenger pathways leading to the alteration of cellular functions. Following the administration of adenosine in mammals, pronounced physiological effects are observed, for example, on cardiovascular function. Depending on the region of the heart, different effects are evoked by adenosine. The most prominent effects are found in the sinus node, the atrium and the atrio-ventricular (A-V) node. In the sinus and A-V nodes adenosine leads to the reduction and shortening of the action potential. Minor effects of adenosine are seen in ventricle or nodal-His bundle stimulation [6].

The clinically relevant use of adenosine is limited to medication in which a short-acting response is sufficient to achieve the desired pharmacological effect. The short half-life of adenosine allows selective action at the site of administration. Adenosine, as Adenocard™ (Medco), is used for the diagnosis of coronary stenosis, as a hypotensive agent during aneurysm surgery and for the termination of paroxysmal supraventricular tachycardia [7–9].

The effects of adenosine are mediated by stimulation of four distinct G-protein-coupled cell surface receptors A_1, A_{2A}, A_{2B} and A_3. Adenosine receptors were first classified as P1 receptors based on their preference of binding adenosine [10]. The P1 receptors were divided into A_1 and A_2 receptors depending on their ability to reduce or increase cellular cAMP levels, respectively [11–13]. A_1 receptors, which have a high affinity for adenosine, have been cloned from several species including mouse, rat, rabbit, dog, bovine and human, and have been found to have a high overall sequence homology of 87% [14–24]. They are expressed in brain, testis, white adipose tissue, heart and kidney [17, 18, 22, 25–28].

A_2 receptors are defined by their activation of the adenylate cyclase system and thereby increasing cellular cAMP levels. A_2 receptors have been divided into A_{2A} and A_{2B} receptors which have high (nM range) and low (μM range) affinities for adenosine, respectively [29, 30]. The pharmacology, signalling pathways, organ specific activity and therapeutic opportunities of A_2 receptors have been extensively reviewed [31–34]. A_{2A} and A_{2B} receptors have been cloned from many species, including mouse, rat, guinea pig, dog and human [14, 15, 35–42].

Adenosine A_3 receptors were first cloned in 1991 [43, 44]. As with A_1 receptors, A_3 receptors inhibit the adenylate cyclase system. Further characteristics of A_3 receptors have been reviewed [45–47].

PHARMACOLOGY OF ADENOSINE A_1 RECEPTORS

STRUCTURE AND FUNCTION OF ADENOSINE A_1 RECEPTORS

The adenosine A_1 receptor is a G-protein-coupled receptor of about 36 kDa with a seven transmembrane domain structure (Figure 4.1). The N-terminus and loops E1–E3 are located on the extra-cellular side of the plasma membrane, whereas the C-terminus and loops C1–C3 are intra-cellular. The transmembrane domains consist of seven right-handed α-helices (Figure 4.1) [48].

A_1 receptors are expressed in highest density in the brain and adipose tissue, followed by kidney and atria, and with lower density in ventricles, lung, pancreas, liver and GI tract [49]. In the heart, A_1 receptors are expressed on cardiomyocytes [50], smooth muscle cells [50] and in supraventricular tissue (atria), where the receptor is more abundant than in the ventricles [51].

Adenosine A_1 receptors couple to $G_{i(1-3)}$ and G_o proteins inhibiting adenylate cyclase and thus reducing cellular cAMP levels. This has been

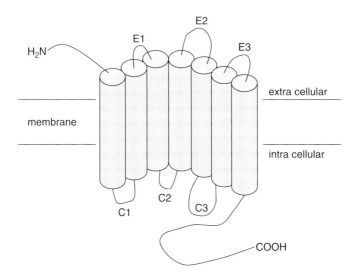

Fig. 4.1 Seven transmembrane domain structure of adenosine A_1 receptors. The extra-cellular part of the A_1 receptor contains the N-terminus and three loops (E1–E3). The C-terminus and the remaining loops (C1–C3) are intra-cellular.

analyzed in several tissues and cellular systems [52–54]. The activated receptor has only minor effects on adenylate cyclase activity in the absence of stimulation by, for example, catecholamine, but it does inhibit the activation of adenylate cyclase by stimulators [55–57]. Additionally, noradrenalin release from sympathetic nerve endings in the heart is inhibited by the A_1 receptor-dependent adenylate cyclase modulation, resulting in cardio protective effects [58]. Furthermore, release of the cardiac hormone ANP is increased by A_1 receptor activation-dependent inhibition of the adenylate cyclase system, again resulting in cardioprotective effects [59].

A_1 receptors also modulate phospholipase C, thereby influencing inositol triphosphate and Ca^{2+} release from internal stores, and act on potassium and calcium channels [52–54, 60–67]. A_1 receptors have only minor inhibitory effects on basal L-type calcium current in atrial myocytes [68], but reduce the calcium current in A-V nodal cells through an NO-dependent pathway [69].

Receptor desensitization *in vitro* has been reported after extended exposure to A_1 receptor agonists, for example, at the level of cultured atrial myocytes or in adipocytes [70, 71], and *in vivo* [72, 73]. Receptor desensitization is a mechanism known for many G-protein-coupled receptors. Chronic exposure and activation of A_1 receptors results in a slow (hours to days) down-regulation of response [74, 75]. There are several serine and threonine residues in the cytoplasmatic domains of the receptor which are potential sites for protein kinase-A, protein kinase-C and β-adrenoceptor kinase-mediated phosphorylation, and which may play a role in receptor desensitization [48]. Desensitization might be a challenge for the use of full A_1 agonists in chronic therapeutic applications. Partial A_1 agonists open up the opportunity for reduced desensitization during chronic treatment [76].

Human recombinant A_1 receptors have been expressed in Chinese hamster ovary cells [77] to allow convenient testing of A_1 receptor agonists and antagonists. Several agonists activate A_1 receptors with high (0.3–3 nM: N^6-(cyclopentyl)adenosine (CPA), N^6-(cyclohexyl)adenosine (CHA), N^6-(R-phenylisopropyl)adenosine (R-PIA), adenosine amine congener (ADAC), medium (3–30 nM: 5'-N-ethylcarboxamidoadenosine (NECA), 2-chloroadenosine (2-CADO), adenosine (ADO)) and low (30–350 nM: N^6-(S-phenylisopropyl)adenosine (S-PIA)), N^6-[2-(3,5-dimethoxyphenyl)-2-(2-methylphenyl) ethyl]adenosine (DPMA)) potency [78]. Additionally A_1 antagonists have been characterized with high (0.5–2 nM: 1,3-dipropyl-8-cyclo-pentylxanthine (DPCPX), xanthine amine congener (XAC)), medium (2–200 nM: 8-cyclophenyltheophylline (CPT), 9-chloro-2-(2-furanyl)-5,6-dihydro-[1,2,4]-triazolo[1,5] quinazolin-5-imine monomethanesulfonate (CGS15943)) and low (1– >20 µM: theophylline, caffeine, 3-isobutyl-1-methylxanthine (IBMX)) potency [78].

Adenosine and adenosine A_1 receptors are involved in the modulation of several physiological effects in heart, kidney, CNS and lipid metabolism [79].

SPECIES DIFFERENCES

An analysis of the binding characteristics of the A_1 antagonist DPCPX in rat, rabbit, guinea pig and pig hearts showed that the affinity of the A_1 receptors in atria and ventricles is similar in these species, but that the receptor density is higher in atrial compared to ventricular tissue [51]. The similarity of A_1 receptor-binding affinity in atria and ventricles has also been shown in human tissue [80].

Guinea pig hearts have a lower affinity for the antagonist DPCPX when compared with other species. Nevertheless, the guinea pig heart is much more sensitive to A-V nodal conduction block by A_1 receptor agonists than rat or rabbit hearts. This might be due to variations in receptor reserve in the atrial tissue and A-V node of these species [81]. Differences in binding characteristics between species might be due to small variations in the A_1 receptor amino acid sequences [82]. Additionally, species differences in G-protein coupling and A_1 receptor density in the same tissue have been reported [51, 54].

These binding and activity differences of A_1 receptor agonists and antagonists in different species have to be taken into consideration when the effects of A_1 receptor modulation are assumed to transfer between species or taken as predictive for humans.

THERAPEUTIC APPLICATION OF A_1 RECEPTOR AGONISTS

Adenosine A_1 receptor activation leads to: bradycardia, vasoconstriction, negative cardiac inotropy and dromotropy and preconditioning in heart; decreased glomerular filtration rate, inhibition of renin release, antidiuresis and vasoconstriction of afferent arteries in the kidney; decreased transmitter release, decreased locomotor activity, sedation and anti-convulsant effects in the CNS; anti-lipolytic effects and an increase of insulin sensitivity in lipid metabolism [79].

There are several options for new therapies involving A_1 receptor activation in cardiovascular diseases such as angina pectoris, control of cardiac rhythm and ischemic injury during an acute coronary syndrome (ACS) or coronary artery bypass graft (CABG) surgery.

Angina pectoris

Coronary heart disease (CHD) is the most significant cause of death in industrial countries. Angina pectoris is a discomfort in the chest caused by myocardial ischemia brought on by exertion. The episode usually lasts between 5 and 15 min and is relieved by rest or nitroglycerin.

Myocardial ischemia is caused by a transient imbalance between myocardial oxygen demand and supply resulting from a stenosis due to an atherosclerotic plaque. Exertion-induced angina can appear when a stenosis of a coronary vessel is >50%.

Standard therapies are mainly β-blockers which improve oxygen utilization in the heart by reducing heart rate. However, β-blockers also decrease the contractility of the heart which can result in reduced physical performance of patients [83].

Activation of A_1 receptors by adenosine itself or by A_1 receptor agonists induces bradycardia by modulating the action potential (AP) in the heart [6]. Bradycardia, and only a slight decrease in contractility by A_1 receptor activation, causes a reduction in oxygen consumption which is a very useful approach for the treatment of angina as oxygen consumption is linearly related to the number of beats per minute. Furthermore, stimulation of A_1 receptors leads to improvement in myocardial perfusion since perfusion is also directly related to the duration of the diastole which is inversely related to the heart rate.

Adenosine A_1 receptor agonists therefore offer a new therapeutic opportunity for the treatment of angina pectoris without a pronounced reduction of contractility. The efficacy of an A_1 agonist in angina pectoris treatment has been proven in a first clinical trial with Capadenoson [84].

Atrial fibrillation

Atrial fibrillation is the most common form of atrial arrhythmias [85]. Atrial fibrillation is characterized by the coexistence of multiple activation waves within the atria, which result in rotary atrial excitations leading to an irregular and increased heartbeat and thereby to a reduced pumping efficacy of the heart.

There are two different forms of therapy of atrial fibrillation: rate control which reduces heart rate without restoring the sinus rhythm to the heart and rhythm control in which AF is reverted in an attempt to restore sinus rhythm [86]. The effectiveness of both approaches is comparable, as clinical trials such as AFFIRM (Atrial Fibrillation Follow-up Investigation of Rhythm Management) have shown [87].

In atrial fibrillation, activation of adenosine A_1 receptors decreases heart rate by modulating the AP in the A-V node. At least two mechanisms are involved in this effect. One mechanism is the direct activation of the acetylcholine-activated potassium current (IK_{ACH}) reducing the excitability of nodal cells [88]. IK_{ACH} activation decreases membrane excitability by hyperpolarizing the membrane potential and slowing membrane depolarization [89]. This accounts for functions that slow automaticity. Activation of A_1 receptors in the A-V node results in a conduction delay which may cause a complete third degree A-V block [26].

In addition, the increase in L-type calcium current caused by catecholamine is attenuated by A_1 receptor activation (the "indirect" or anti-β-adrenergic effect of adenosine) [90, 91]. A_1 receptors antagonize the β-adrenergic system, which functions by Gs protein induced activation of adenylate cyclase and increasing cAMP concentrations.

Both effects on automaticity of the heart can be used in rate control therapy of atrial fibrillation. Animal and clinical studies with full A_1 agonists, such as Tecadenoson (CVT-510, an N^6-substituted derivative of adenosine, in clinical development for treatment of paroxysmal supraventricular tachyarrhythmias (PSVT)) and Selodenoson (in development for atrial fibrillation), have shown significant reduction of heart rate in animal models, for example, the goat atrial fibrillation model [92] and in human patients [93].

One issue with full A_1 receptor agonists is the induction of third degree A-V block in rate control therapy [94]. Complete A-V block means that the heart rate depends only on the ventricular frequency (about 40 beats per minute), a life threatening condition in at-risk patients.

Partial A_1 agonists have the opportunity of slowing A-V conduction only to a certain degree, thereby reducing the heart rate without generating a third degree A-V block [95].

Cardio protection: Ischemic preconditioning

ACS and myocardial infarction (MI) are common and severe diseases. About 9% of patients with unstable angina die or suffer from MI within 30 days and >20% of MI patients die during the first 24 h in hospital. Two mechanisms are responsible for the generation of tissue injury during ACS and MI. Firstly ischemia results in tissue necrosis and activation of inflammatory processes in the occluded area of the heart. Secondly reperfusion which, although beneficial by itself, increases the injured area by enhanced necrosis, apoptosis and inflammation [96].

Cardioprotection from ischemia-induced tissue injury by preconditioning is the most powerful endogenous mechanism for protecting the heart during prolonged ischemic episodes [97]. Preconditioning can be induced with short episodes of recurrent sub-lethal ischemia followed by reperfusion phases, which protects the myocardium from tissue damage during more prolonged ischemic episodes. The cardio protective effect of preconditioning has been shown in several animal models of acute MI, in which infarct size could be reduced by >75% by this mechanism.

Two time windows for preconditioning are known; a short and transient phase lasting less than 2 h and a late phase occurring 24 h after the preconditioning stimulus and lasting for 48 h [98].

Endogenous activation of adenosine A_1 receptors by adenosine, or by A_1 receptor agonists, has been shown in several animal studies to be involved in this protective effect and in the reduction of infarct size [99–102]. Activation of A_1 receptors results in the subsequent modulation of several downstream signalling cascades [98, 103]. Furthermore, repeated application of A_1 agonists leads to a chronic preconditioned state [104].

The importance of developing a clinically useful drug for cardioprotection by pharmacologically induced preconditioning is well accepted. A_1 receptor agonists are promising candidates to fulfil this need.

Central nervous system

A_1 receptor activation is involved in sedative and anti-convulsant effects [105, 106]. It modulates the dopaminergic control of movement and can reduce neuropathic pain [107, 108] in the brain. Furthermore, neuroprotection and preconditioning have been reported to be mediated by A_1 receptors [109], which might offer a new therapeutic opportunity for A_1 agonists in patients with an increased risk of stroke.

In multiple sclerosis, a chronic disease characterized by neuroinflammation that leads to demyelination and axonal/neuronal injury [110], A_1 agonists might have therapeutically beneficial effects. A_1 receptors are expressed on microglia/macrophages and neurons. Reduced expression of A_1 receptors in these cells leads to a severe progressing-relapsing form of experimental allergic encephalomyelitis in A_1 receptor knockout mice when compared to WT mice. This is believed to occur by enhancement of proinflammatory responses and the release of cytokines [111].

Heteromer formation between G-protein-coupled receptors has been described for A_1 receptors in the brain. A_1–A_{2A} heteromers show different modulations on glutamate release compared to A_1 receptors [112]. Furthermore, heteromers of A_1 and the dopamine D_1 receptor have been described [113].

Lipid metabolism

A_1 receptors are highly expressed in adipocytes. Activation of the receptor inhibits lipolysis and lowers circulating free fatty acid (FFA) concentrations [114]. Elevated plasma levels of FFA are characteristic of non-insulin-dependent diabetes [115] and worsen insulin resistance and hyperglycemia in patients [116, 117].

Over-expression of A_1 receptors in the adipose tissue of transgenic mice lowered plasma FFA levels and improved insulin sensitivity compared to WT mice on a high fat diet [118]. The A_1 receptor agonist RPR 749 reduced FFA in a first double-blind, single increasing-dose, placebo-controlled, randomized study in humans [119]. Limitations of using full A_1 agonists in dyslipidemia treatment might include their cardiac effects. Partial A_1 agonists open an opportunity to avoid these effects. Partial A_1 agonists, such as CVT-3619, have anti-lipolytic effects at doses which have no, or only slight, cardiac activity [114].

CHEMISTRY OF ADENOSINE A_1 RECEPTOR AGONISTS

ADENOSINE-DERIVED A_1 AGONISTS

The vast majority of adenosine receptor agonists are closely related in structure to the endogenous ligand adenosine (1). It has been shown that the ribose moiety plays an important role in the agonistic effects of adenosine. Changes of either structural properties, including stereochemistry, of the ribose moiety in most cases lead to a loss of receptor-binding potency and intrinsic activity. An exception to this concept of adenosine analogues has been introduced by Bayer HealthCare AG using aminodicyanopyridine derivatives. These compounds will be described in more detail in the following section.

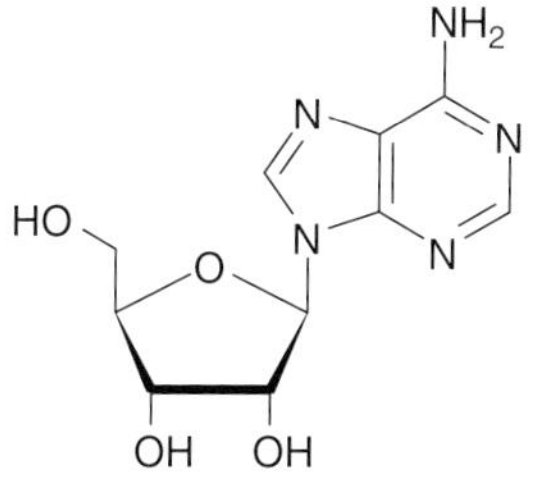

(1) Adenosine

Most of the adenosine-related compounds are modified in the N^6- or C^2-position of the adenine moiety. In addition, modifications in the 3'-, 4'- or 5'-position of the ribose unit are in some cases tolerated. The nature of the attached groups in N^6- and 5'-positions exerts a profound influence on receptor affinity and subtype selectivity.

N^6-Cycloalkyl-substituted adenosine derivatives

Adenosine analogues such as CPA (2) and CCPA (3) emerged by simple substitution at the N^6-position with a cyclopentyl ring system and introduction of a chlorine atom at the C^2-position of the adenine ring system. Both CPA and CCPA are highly selective for the rat adenosine A_1 receptor, but less selective for the human receptors. A team from Monash University, the University of Melbourne and the University of Florida [120] evaluated two series of N^6-substituted adenosines with monocyclic and bicyclic ring systems containing a heteroatom. All compounds were prepared starting from 6-chloro-9-[2,3,5-tris-*O*-(*tert*-butyldimethylsilyl)-β-D-ribofuranosyl]-9*H*-purine [121] with substitution of the chlorine atom under standard S_N Ar conditions and subsequent deprotection of the *tert*-butyldimethylsilyl groups. Within the five-membered ring series it was found that the seleno derivative (Table 4.1, entry (6)). In the bicyclic series the unsubstituted 7-azabicyclo[2.2.1]heptan-2-yl derivative (10) showed reasonable potency as an A_1 agonist N-substitution yielded slightly more active and selective compounds (Table 4.1, entries (11) and (14)).

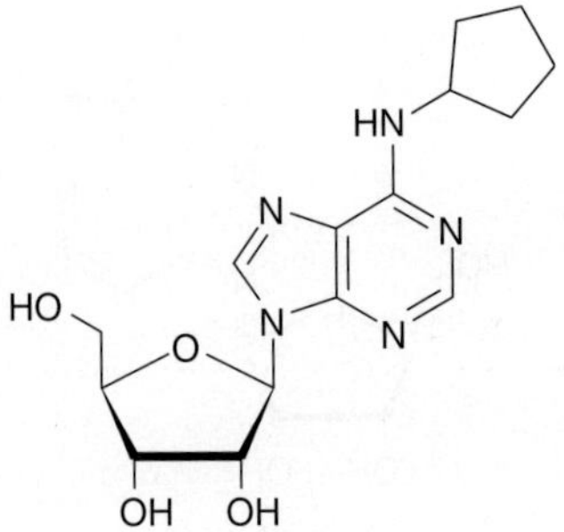

(2) CPA

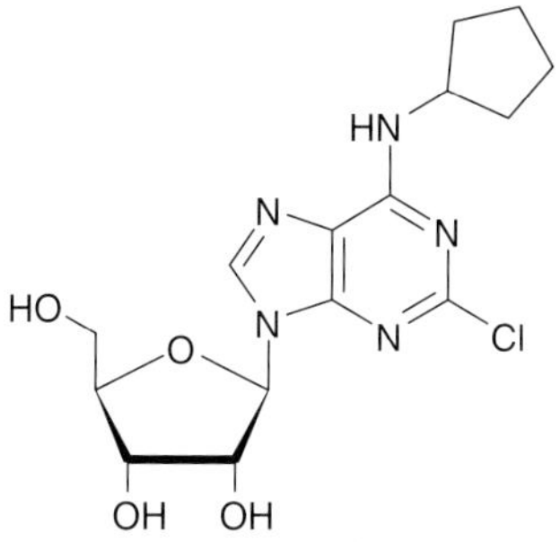

(3) CCPA

Table 4.1 K_I AND IC_{50} OF N^6-CYCLIC ADENOSINE ANALOGUES

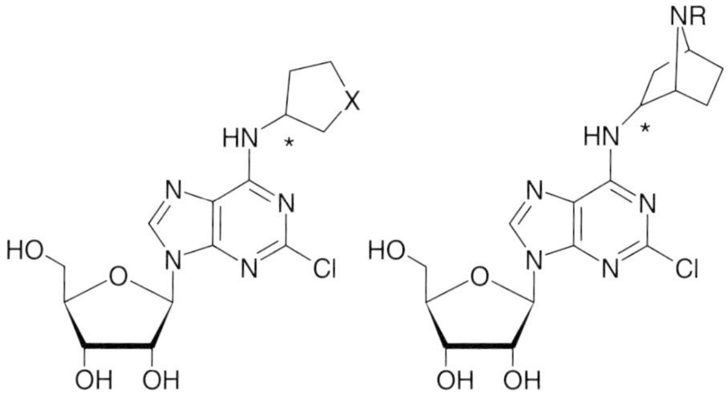

Compound	X	R	A_1 AR^a		A_2 AR^b
			K_i (nM)	IC_{50} (nM)	K_i (nM)
(4) (R)	O		65 ± 16	8.2 ± 1.2	>100,000
(5) (R)	S		70 ± 9	5.3 ± 2	9,700 ± 1,200
(6) (R)	Se		8 ± 2	1.9 ± 0.1	5,500 ± 500
(7) (R)	NBoc		929 ± 397	380 ± 84	>100,000
(8) (R)	NAc		803 ± 341	85 ± 24	>100,000
(9) (R)	NCbz		288 ± 127	3.2 ± 0.4	>100,000
(10) (endo)		H	51 ± 16	35 ± 9	7,800 ± 1,500
(11) (endo)		Boc	32 ± 8	2.5 ± 0.5	34,700 ± 3,900
(12) (endo)		Cbz	164 ± 9	68 ± 13	>100,000
(13) (endo)		2-Cl-Cbz	783 ± 73	254 ± 58	>100,000
(14) (endo)		2-Br-Cbz	21 ± 4	9.0 ± 2.7	>100,000

[a] K_i values calculated from the concentration of the compounds that initiated [^{3}H]CPX binding by 50%, IC_{50} values are concentrations of compounds that inhibited (−)-isoproterenol (1 μM)-stimulated cAMP accumulation by 50% in DDT cells.
[b] K_i values calculated from the concentration of the compounds that inhibited [^{3}H]ZM241385 binding by 50% had the highest affinity, potency and selectivity.

Table 4.2 AFFINITIES IN A_1 AND A_2 ADENOSINE RECEPTOR BINDING OF N^6-BICYCLOALKYLADENOSINES

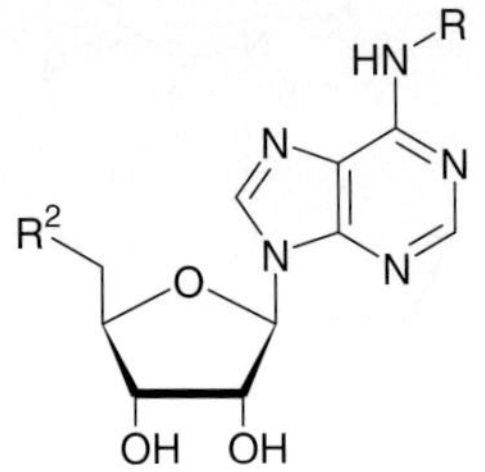

Compound	R	R″	K_i (nM) A_1	K_i (nM) A_2
(15)	Cyclopentyl	OH	0.59	462
(16)	2-endo-Norbornyl	OH	0.42	750
(17)	2-exo-Norbornyl	OH	0.91	970
(18)	1R, 2S, 4S isomer of (16)	OH	0.30	1,390
(19)	1S, 2R, 4R isomer of (16)	OH	1.65	610
(20)	Cyclopentyl	Cl	0.72	1,870
(21)	2-endo-norbornyl	Cl	0.42	2,100
(22)	1R, 2S, 4S isomer of (21)	Cl	0.24	3,900
(23)	1S, 2R, 4R isomer of (21)	Cl	1.86	1,600

Note: A_1 binding: [^{3}H]CHA in rat brain membranes; A_2 binding: [^{3}H]-NECA in presence of 50 nM CPA in rat striatal membranes.

Similar work had previously been published by a research group from Warner-Lambert [122] for purely aliphatic non-heterocyclic ring systems (Table 4.2). When the 5′-hydroxy group was replaced by a chlorine atom, affinity remained in the same range; however, selectivity versus the A_2 receptor was increased by a factor of 3–4. Another team from Warner-Lambert [123] investigated the effect of ring size of N^6-substituted cycloalkyl systems on the affinity. A comparison of all ring sizes ranging from three- to twelve-membered rings revealed the five-membered ring (CPA) as having the highest affinity and selectivity for the A_1 receptor over the A_2 receptor (Table 4.3, rat values) [124].

Another compound of the N^6-cyclopentyl class, GR-79236 (52), originated from the GlaxoSmithKline labs [125], was initially evaluated for pain [126].

Table 4.3 AFFINITIES OF N^6-CYCLOALKYLADENOSINES

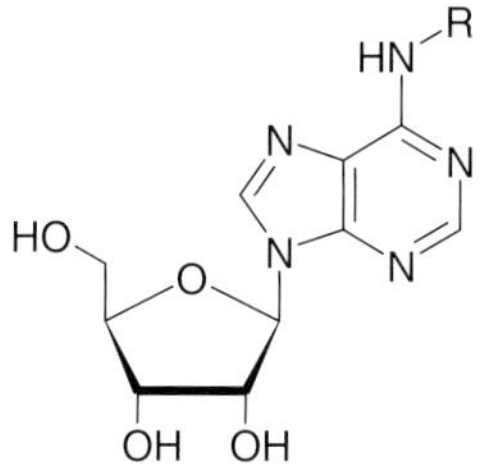

Compound	R	K_i (nM) A_1	A_2
(24)	Cyclopropyl	319 ± 0.49	1,240 ± 130
(25)	Cyclobutyl	0.777 ± 0.028	263 ± 5
(15)	Cyclopentyl (CPA)	0.590 ± 0.023	462 ± 15
(26)	Cyclohexyl (CHA)	1.31 ± 0.009	363 ± 46
(27)	Cycloheptyl	2.79 ± 0.36	1,690 ± 190
(28)	Cyclooctyl	4.03 ± 0.16	2,450 ± 500
(29)	Cyclodecyl	100 ± 11	5,190 ± 50
(30)	Cyclododecyl	5,390 ± 351	47,700 ± 8,500

The binding affinities of GR-79236 gave K_i values at the rat A_1 receptor of 3.1 nM and for the A_2 receptor, 1,300 nM.

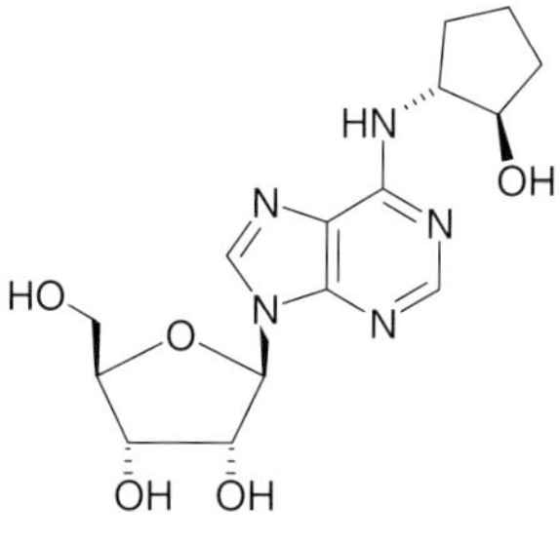

(52) GR-79236

As described earlier, cyclopentyl and norbornenyl substitution at the N^6-position are very favourable. Many studies have therefore been related to

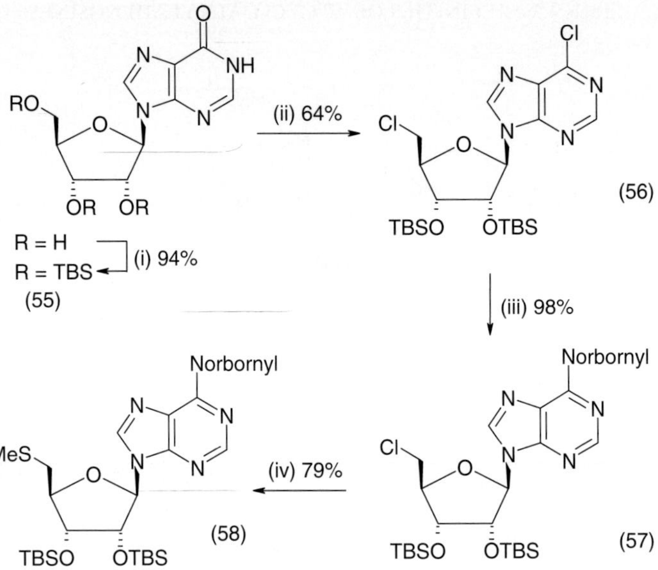

Scheme 4.1 Synthesis of N^6-substituted-5′-modified adenosines. Reagents and conditions: (i) TBSCl, imidazole, DMF, rt; (ii) cat. DMF, SOCl$_2$, 79 °C; (iii) (±)-endo-norborn-2-yl amine · HCl, N(i-Pr)$_2$Et, t-BuOH, 83 °C and (iv) NaSMe, DMF, rt. Followed by deprotection of TBS groups using NH$_4$F in MeOH at 50–60 °C.

this variation with the goal of identifying other sites of substitution for optimisation. Research teams from the Monash University [127, 128] identified a new synthetic route to N^6-substituted 5′-modified adenosines starting from relatively inexpensive inosine (Scheme 4.1).

In a recent publication researchers from the same groups [129] evaluated several compounds, applying this new route to generate analogues with different N^6-substitution based on cyclopentyl amine, endo-norbornyl amine and 2S-endo-amino-5,6-exo-epithiobicyclo[2.2.1]heptane (Table 4.4). This study showed that modification of the 5′-position can have relatively large effects on affinity and potency. All derivatives tested were full agonists for cAMP inhibition as compared to CPA. Not surprisingly, the compound in this study with the highest affinity was the one with R = C(O)NHEt (38), as it is a close derivative of Seledenoson/RG 14202 (53) [130]. Seledenoson has been in clinical Phase II trials for the intravenous and oral treatment of atrial fibrillation. Half-maximal binding affinities of Seledenoson to the rat brain

Table 4.4 AFFINITIES FOR THE A_1 ADENOSINE RECEPTOR IN DDT_1 MF-2 CELLS

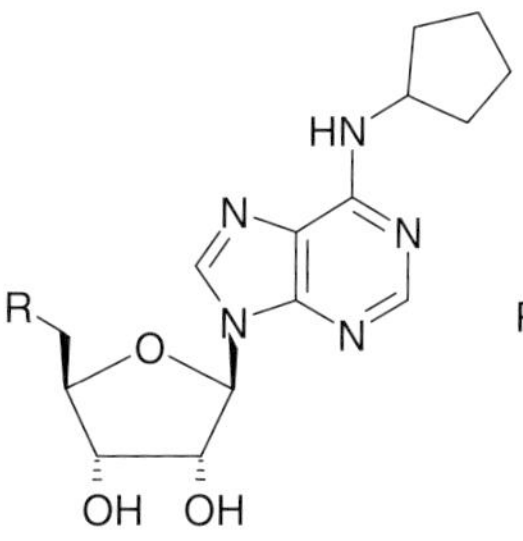

Compound	R	K_i (nM) A_1	IC_{50} (nM)
(31)	OH	21.5±5.7	2.7±0.6
(32)	SEt	173±35	21±8
(33)	F	17±6	7±3
(34)	Cl	9.8±3.4	2.6±1.6
(35)	SEt	122±42	28±15
(36)	NH_2	3,933±1,008	351±135
(37)	OC(O)NHMe	590±110	410±50
(38)	C(O)NHEt	0.9±0.2	0.12±0.03
(39)	CH_2Cl	2.8±0.8	0.3±0.1
(40)	CH_2SMe	5.0±0.2	0.9±0.3
(41)	CH_2SeMe	30±3	12±3

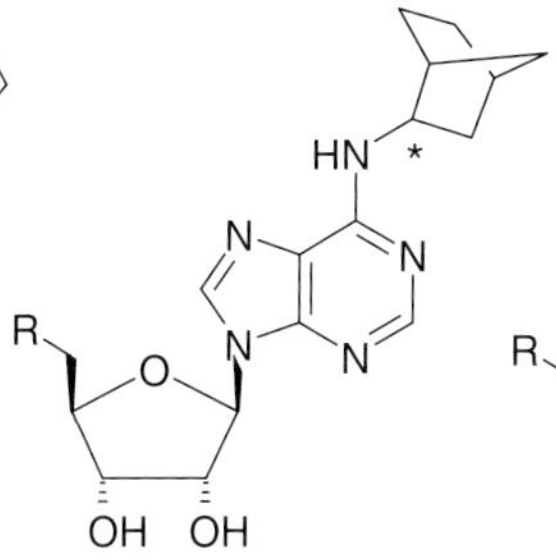

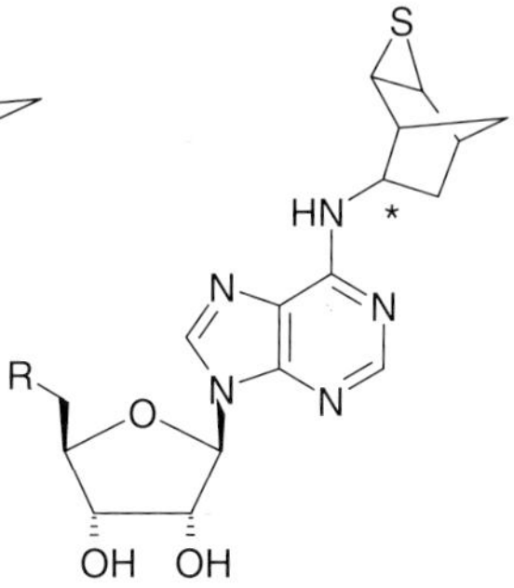

Note: K_i values calculated from concentrations of compounds that initiated [^{3}H]CPX binding by 50%; IC_{50} values are concentration of compounds that inhibited (–)-isoproterenol (1 μM) stimulated cAMP accumulation by 50% in DDT cells.

A_1 and A_2 receptor were determined to be 1.11±0.32 and 306±49 nM, respectively.

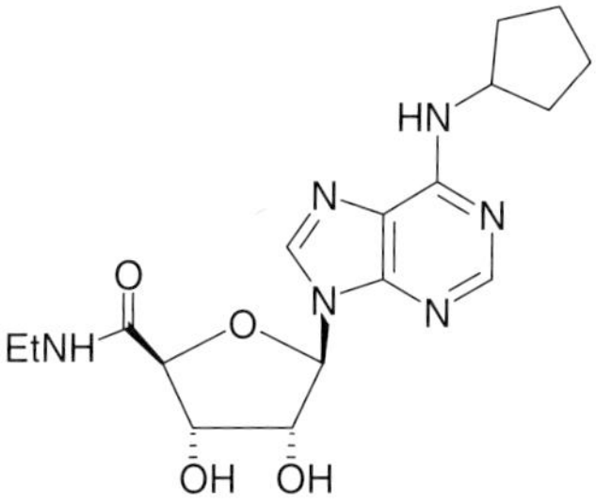

(53) Seledenoson

Another interesting compound from CV-Therapeutics, based on N^6-substitution with a cyclopentyl ring system, is CVT-3619 (54) [131, 132]. This compound, a potent and selective adenosine A_1 partial agonist with a binding affinity K_i of 1.1 nM, is currently in Phase I clinical trials for dyslipidemia/diabetes. The compound dose-dependently reduced free fatty acid and triglyceride levels in normal, overnight-fasted rats (at 2.5–10 mg/kg p.o.). In Zucker diabetic fatty rats dosed s.c., it effectively reduced plasma levels of glucose, insulin, free fatty acids and triglycerides at 10 mg/kg. b.i.d. over five days. These effects were seen without significant changes in atrial rates in isolated rat and guinea pig hearts over a dose range of 0.01–30 μM, and with only a small increase of coronary conductance. The compound did not cause significant cardiovascular or CNS-related adverse effects in rats (1–50 mg/kg p.o.) and did not induce any A-V blockade. This might be correlated to the fact that the compound is a partial agonist. By comparison, the full agonists described above, for example, CPA, when given in similar experiments to reduce free fatty acid levels, also cause significant bradycardia [133]. CVT-3619 shows approximately 25% oral bioavailability in rats and dogs.

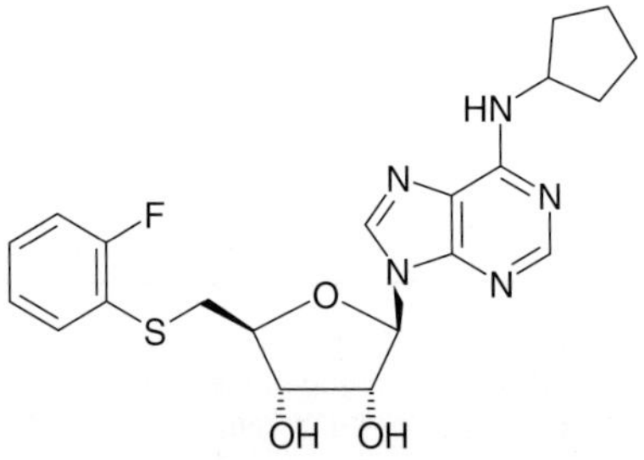

(54) CVT-3619

N^6-Heterocyclic adenosine derivatives

N-[3-(R)-tetrahydrofuranyl]-6-aminopurine riboside (Tecadenoson, CVT-510 (4)) [134, 135] is one of the most prominent compounds within the class of N^6-heterocyclyl analogues of adenosine. Tecadenoson, from CV Therapeutics, is currently in Phase III clinical trials for atrial arrythmias [136]. It is a selective A_1 receptor agonist with 356-fold higher affinity for the A_1 receptor ($K_i = 6.5$ nM) compared to the A_2 receptor ($K_i = 2,315$ nM). In this experiment the compound was tested at the porcine receptor in competition with the radioligand CPX and A_{2A} radioligand CGS21680.

As already mentioned, receptor agonists can be partial agonists or full agonists. A partial agonist produces a sub-maximal response even when present in a sufficiently high concentration to saturate all of relevant receptors in a given tissue. Aiming for partial agonism might help avoid unwanted side effects of adenosine agonists. In the case of atrial fibrillation, the primary goal is to avoid third degree A-V block.

Determination of the degree of partial agonism by evaluation of the agonist-induced increase of [^{35}S]-GTPγS binding [137] revealed Tecadenoson as being a full agonist. The CV-Therapeutics team therefore evaluated a series of 5′-carbamate and 5′-thiocarbamate derivatives of Tecadenoson (Table 4.5). Many of the synthesized compounds within these two classes

Table 4.5 AFFINITIES FOR THE A_1 ADENOSINE RECEPTOR AND ACTIVITY IN GTPΓS BINDING ASSAY AND DEGREE OF AGONISM

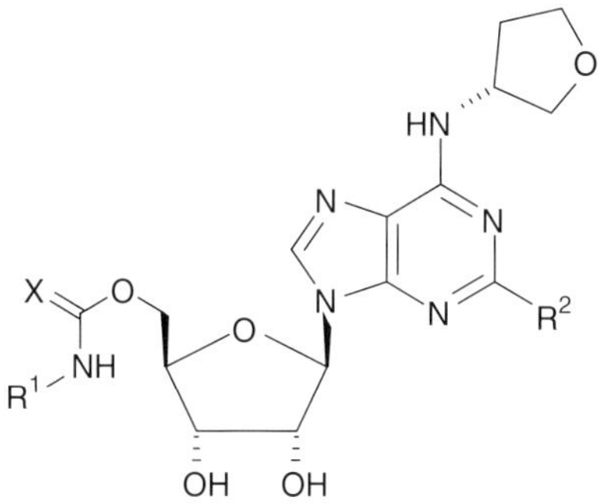

Compound	R^1	R^2	X	K_i (nM) A_1	GTPγS %CPA	Partial or full
(42)	Me	H	O	140 ± 33	76	Partial
(43)	Me	Cl	O	96 ± 34	77	Partial
(44)	Me	H	S	45 ± 11	77	Partial
(45)	Me	Cl	S	71 ± 27	77	Partial
(46)	Et	H	S	33 ± 5	77	Full
(47)	Et	Cl	S	55 ± 31	84	Partial
(48)	Pr	H	S	50 ± 21	89	Partial
(49)	c-Pentyl	H	O	321	82	Partial
(50)	c-Pentyl	Cl	O	226 ± 146	69	Partial
(51)	c-Pentyl	H	S	72	95	Full
(4)	5′-OH	–	–	3 ± 0.4	93	Full

Note: Binding affinity determined using DDT_1 cell membranes (hamster vas deferens smooth muscle cell line) with [^{3}H]-CCPA as radioligand; measurement of stimulation of binding of [^{35}S] GTPγS to G protein-coupled A_1 adenosine receptor in DDT_1 cell membranes of test compound (10 μM), as % of CPA (1 μ).

had a lower intrinsic efficacy, possibly resulting from missing hydrogen bond donor abilities, due to the removal of the 5′-OH group. However, those compounds exhibiting partial agonism (the goal of the study), were either metabolized to the full agonist Tecadenoson or exhibited low *in vitro* metabolic stability.

N^6-Aryl- and arylalkyl-substituted adenosine derivatives

All compounds described above possess an aliphatic cycloalkyl or heterocyclic substituent at the N^6-position. However, this is not a prerequisite for high binding affinity for the A_1 receptor and selectivity over the other adenosine receptors. GW-493838 (59), a compound that originated from the Glaxo Wellcome labs, is a 4-chloro-2-fluorophenyl N^6-substituted adenosine analogue with both high affinity and high selectivity for the A_2 receptor ($EC_{50} = 4\,nM$). (Bayer HealthCare AG, internal experimental data; EC_{50} values have been determined on recombinant human A_1, A_{2A}, A_{2B} and A_3 receptors expressed in Chinese hamster ovarian (CHO) cells. Activation of receptors has been measured by a cAMP modulated luciferase expression read-out system. A_1 and A_3 agonism has been characterized on forskolin-prestimulated cells.) The compound was developed for both oral treatment of neuropathic pain and lipoprotein disorders. However, no further development has been reported recently. In a study evaluating the treatment of peripheral neuropathic pain in humans, no significant changes were seen compared to the placebo group when treated with GW-493838 [138].

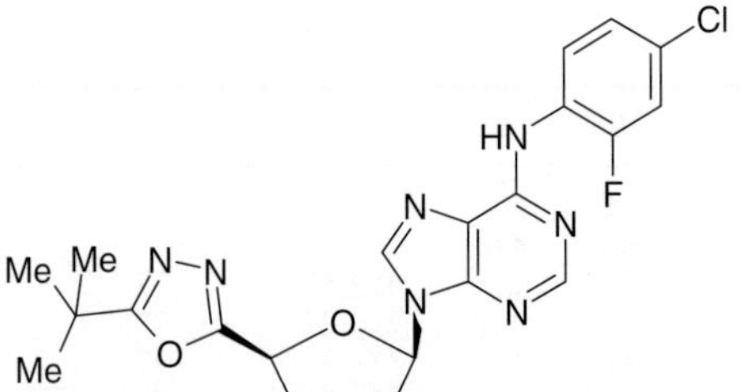

(59) GW-493838

VCP102 (61), another partial adenosine A_1 agonist [139], is characterized by an interesting feature: an N-oxide moiety in the N^6-substitution side chain. This known antioxidant moiety was incorporated into the design of new adenosine A_1 agonists with the view of providing antioxidant activity thereby producing cardioprotective effects at the same time as enhancing the

Table 4.6 BINDING AFFINITIES AND SELECTIVITIES OF VCP102 AND ANALOGUES

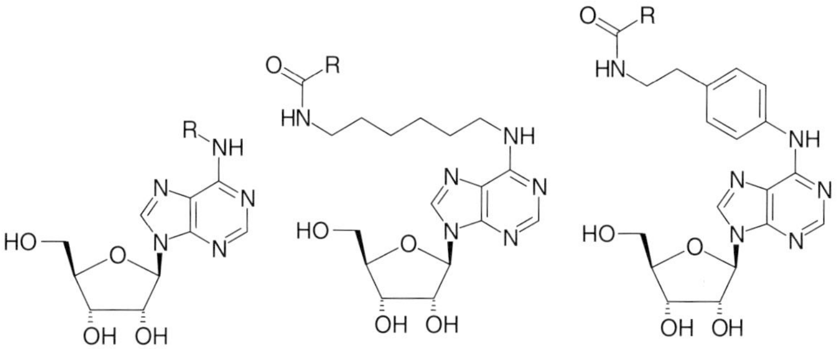

Compound	R	K_i (μM)			
		A_1	A_{2A}	A_{2B}	A_3
(62)	A	0.10 ± 0.03	>10	>10	0.79 ± 0.56
(63)	B	0.05 ± 0.01	>10	>10	8.56 ± 3.20
(64)	A	0.015 ± 0.0003	>10	>10	0.089 ± 0.03
(65)	C	0.040 ± 0.0003	>10	>10	0.014 ± 0.05
(61)	A	0.007 ± 0.001	>10	1.45 ± 0.08	0.023 ± 0.01
(66)	C	0.032 ± 0.006	>10	8.58 ± 0.90	0.084 ± 0.001

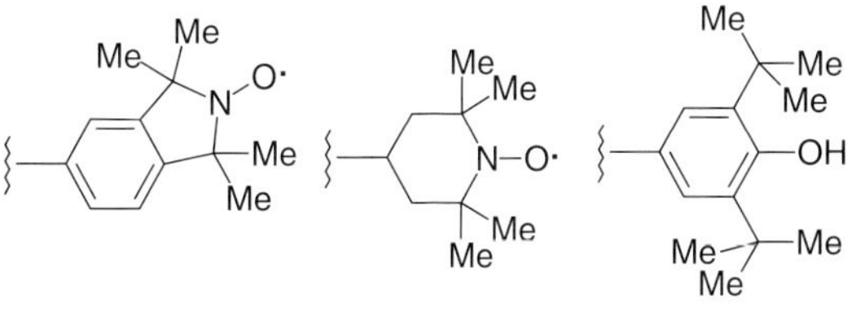

receptor affinity and selectivity [140]. In initial studies, the Monash University group evaluated the cardioprotective effects of VCP102 and analogues (Table 4.6) and found that adding an adenosine A_1 receptor antagonist such as DPCPX (1,3-dipropyl-8-cyclopentyl xanthine) reduced the protective effects. However, these effects were not completely abolished, raising the possibility that there was some benefit derived from the antioxidant moiety attached to the molecules.

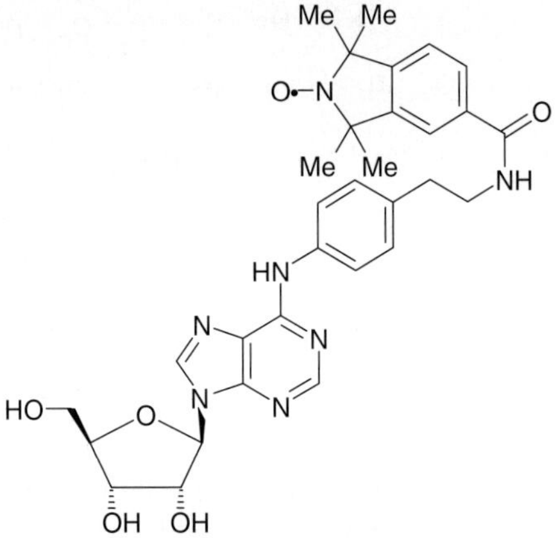

(61) VCP102

C^2-substituted adenosine derivatives

Researchers from CV Therapeutics [141] examined the C^2-substitution on compounds derived from Tecadenoson (Table 4.7). Introduction of a pyrazolyl group into the 2-position of the purine moiety, and further substitution at the 4-position of the pyrazole ring system, yielded compounds with a high to moderate affinity for the A_1 receptor. In general, selectivity versus the A_{2A} and A_{2B} receptors was high; however, high selectivity over the A_3 receptor was only achieved by replacing the N^6-tetrahydrofuran moiety with either cyclopentyl (70) or norbornyl (77, 78). It was found that within this class of compounds steric factors at the 4-position of the pyrazole ring played a crucial role in determining the binding affinity and selectivity for the A_1 and the A_3 receptors. Smaller substituents seemed to be conducive to high A_1 receptor binding affinity, whereas larger substituents led to higher affinity for the A_3 receptor. One of the examples fitting into this hypothesis is compound (73), which displayed a sub-nanomolar binding affinity for the A_1 receptor with a high selectivity of more than 600 over the A_3 and A_{2A} adenosine receptors. Further modification of the 2-position of the ribose moiety did not lead to compounds with higher affinity; however depending on the substitution, the team was able to identify compounds that were selective for the A_3 receptor over the A_1 receptor (83, 85). In addition, some of the compounds with high affinity contained a carboxylic acid group that would be charged at physiological pH and therefore it was speculated that these compounds should

Table 4.7 BINDING AFFINITIES AND SELECTIVITIES

Compound	R^1	R^2	R^3	K_i (nM) A_1	A_3	A_{2A}	A_{2B}
(67)	(R)-THF	COOEt	CH_2OH	3	41	2,410	>6,000
(68)	(R)-THF	COOH	CH_2OH	24	3,210	>4,000	>6,000
(69)	(R)-THF	CONHBn	CH_2OH	2	54	>5,000	>6,000
(70)	Cyclopentyl	CONHMe	CH_2OH	2	1,420	>5,000	>6,000
(71)	Cyclopentyl	CONHEt	CH_2OH	1	20	>5,000	>6,000
(72)	Cyclopentyl	COOH	CH_2OH	9	4,120	>5,000	>6,000
(73)	Cyclopentyl	$CONH_2$	CH_2OH	0.6	380	>5,000	>6,000
(74)	Cyclopentyl	COOEt	CH_2OH	0.9	19	>5,000	>6,000
(75)	Cyclopentyl	H	CH_2OH	34	105	>5,000	>6,000
(76)	Norbornyl	COOEt	CH_2OH	3	250	>5,000	>6,000
(77)	Norbornyl	COOH	CH_2OH	1	>5,000	>5,000	>6,000
(78)	Norbornyl	H	CH_2OH	0.4	1,270	>5,000	>6,000
(79)	Cyclohexyl	COOEt	CH_2OH	30	30	>5,000	>6,000
(80)	Cyclohexyl	CONHMe	CH_2OH	8	158	>5,000	>6,000
(81)	(R)-THF	Ph-4-Me	CH_2OH	74	94	2,320	>6,000
(82)	(R)-THF	2-Pyridyl	CH_2OH	36	3	1,180	>6,000
(83)	(R)-THF	4-Pyridyl	CH_2OH	134	4	1,480	>6,000
(84)	(R)-THF	2-Pyrazinyl	CH_2OH	20	36	1,170	>6,000
(85)	(R)-THF	Ph-4-Me	CONHEt	121	0.7	>5,000	>6,000
(86)	(R)-THF	4-Pyridyl	CONHEt	1,060	NT	NT	>6,000

Note: Binding affinity for A_1 determined using DDT membranes with [^{3}H]-CPX as radioligand; A_3 receptor: determined using CHO-A_3 cells with [^{125}I]-AB-MECA as radioligand; A_{2A} receptor: determined using HEK-A_{2A} cells with [^{3}H]-ZM241385 as radioligand; A_{2B} receptor: determined using HEK-A_{2B} cells with [^{3}H]-ZM241385 as radioligand.

diminish the diffusion across the blood-brain barrier. This might be important as some groups have interpreted the locomotor depression seen with some peripherally administered adenosine A_1 receptor agonists as CNS effects.

A group from the University of Leiden [142] evaluated 3-substituted aminocarbonyltriazene-1-yl groups at the C^2-position. This study underlines that bulky substituents are tolerated in the C^2-position (Table 4.8).

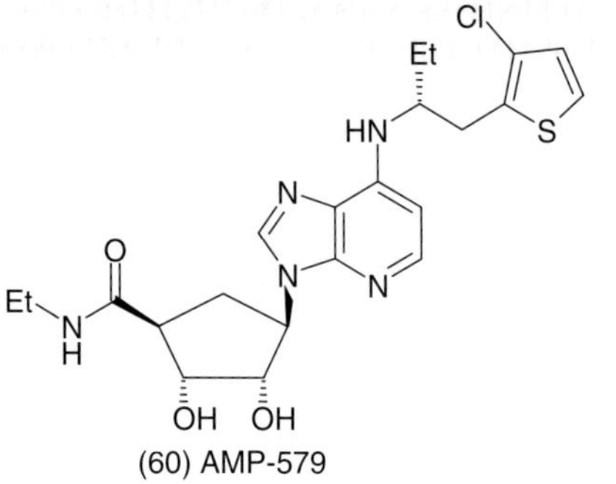

(60) AMP-579

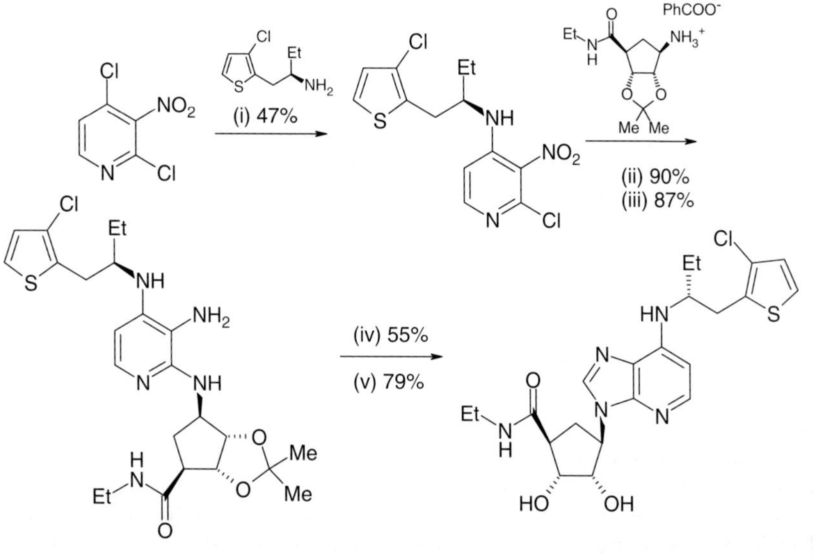

Scheme 4.2 Synthesis of AMP-579 (i) NEt_3, iPrOH; (ii) NEt_3, BuOH; (iii) H_2, Pt; (iv) formamidine acetate, BuOAc and (v) aqueous HCl, THF.

Table 4.10 BINDING AFFINITIES AND SELECTIVITIES OF AMP-579

Compound	Rat brain	Rat adipocyte	Rat brain	Human (recombinant)	Human (recombinant)
	A_1	A_1	A_{2A}	A_{2B}	A_3
(60)	1.7±0.3	4.5±1.0	56±17	>10,000	1,070±150
(15)	0.5±0.02	0.7±0.2	700±44	NT	NT

Table 4.11 BINDING AFFINITIES AND SELECTIVITIES OF RING-CONSTRAINED (N)-METHANOCARBA NUCLEOSIDES

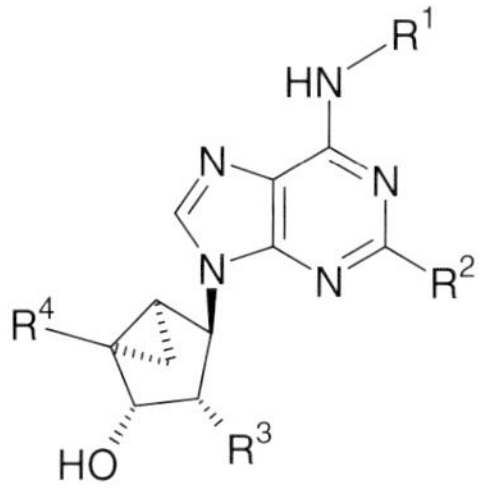

Compound	R^1	R^2	R^3	R^4	K_i (nM)		
					A_1	A_{2A}	A_3
(97)	H	H	OH	CH_2OH	1,680 ± 80	22,500 ± 100	404 ± 70
(98)	H	Cl	OH	CH_2OH	273 ± 36	1,910 ± 240	84.7 ± 19
(99)	Cp	H	OH	CH_2OH	5.06 ± 0.51	6,800 ± 1,800	170 ± 51
(100)	Cp	H	H	CH_2OH	5,110 ± 790	15%[a]	2,880 ± 910
(101)	Cp	Cl	OH	CH_2OH	8.76 ± 0.81	3,390 ± 520	466 ± 58
(102)	Cp	Cl	H	CH_2OH	3,600 ± 780	45 ± 5%[a]	1,090 ± 190
(103)	IB	H	OH	CH_2OH	69.2 ± 9.8	601 ± 236	4.13 ± 1.76
(104)	IB	H	OH	CONHMe	52.7 ± 5.2	548 ± 115	2.39 ± 0.54
(105)	IB	Cl	OH	CH_2OH	141 ± 22	732 ± 207	2.24 ± .45
(106)	IB	Cl	OH	CONHMe	83.9 ± 10.3	1,660 ± 260	1.51 ± 0.23
(107)	IB	Cl	H	CH_2OH	8,730 ± 370	25,400 ± 3,800	912 ± 29

Notes: Displacement of specific [^{3}H]R-PIA binding to A_1 receptors in rat brain membranes; displacement of specific [^{3}H]CGS21680 binding to A_{2A} receptors in rat striatal membranes, and at A_{2B} receptors expressed in HEK-293 cells versus [^{3}H]ZM241,385; displacement of specific [^{125}I]AB-MECA binding to human A_3 receptors expressed in CHO cells, in membranes. Cp, cyclopentyl; IB, 3-iodobenzyl.
[a]% displacement at 100 μM.

Few additional examples of carbocycle replacement have been discussed in the literature. One example, by researchers of the NIH and NCI in Maryland [149], leading to compounds with up to single digit nanomolar affinity for the A_1 receptor and high selectivity over the A_{2A} and A_3 receptors (Table 4.11), does not depend on the bulk of the N^6-substituent for affinity for the A_3 receptor. The researchers included changes at the ribose moiety replacing one hydroxyl functionality with a hydrogen atom.

Returning to the N^6-cycloalkyl-substituted adenosine analogues, SDZ-WAG-994 (108) is an early representative that bears a 2'-O-methyl group at the ribose unit [150]. SDZ-WAG-994 is a full agonist with a K_i of 23 nM and high selectivity over A_{2A} and A_{2B} ($K_i = 25,000$ and $10,000$, respectively) [151]. The compound showed effects in animal models such as spontaneously hypertensive rats (SHR) (dose-related decrease in blood pressure and heart rate) [152], and in hypertensive models in rhesus monkeys (sustained bradycardia, suppression of plasma renin activity, and the plasma free fatty acid and triglyceride concentrations). Beyond examination of cardiovascular effects [153], SDZ-WAG-994 was also evaluated for the treatment of pain [154]. The development of the compound was terminated after Phase I trials.

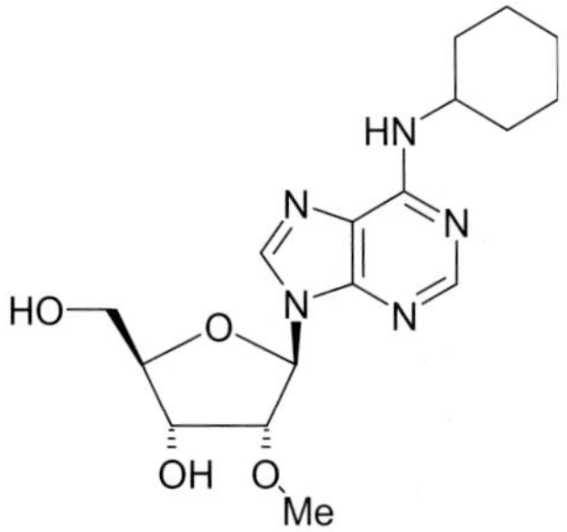

(108)

INO-8875 (formerly known as PJ-875) is an Inotek compound currently in clinical Phase I trials against atrial fibrillation. The chemical structure of INO-8875 has only been published in the Investigational Drug database [155] as (109), being the major example of the corresponding patent [156]. INO-8875 shows a binding affinity of $K_i < 1$ nM for the A_1 receptor with a selectivity about 10,000 over the A_{2A} receptor.

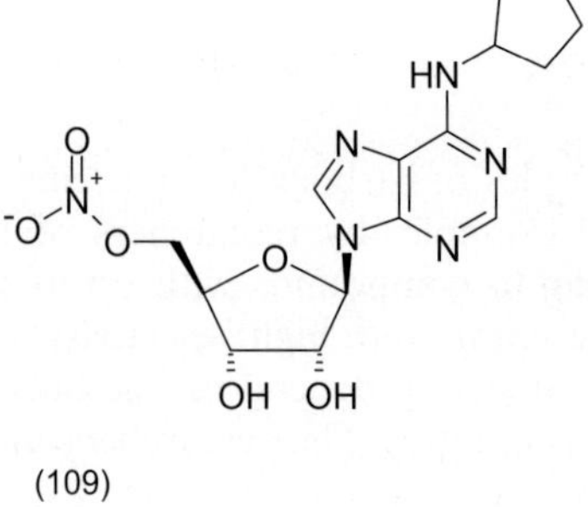

(109)

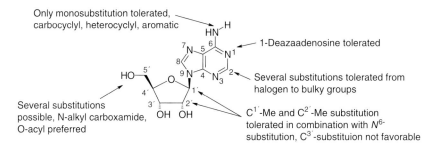

Fig. 4.2 Structure activity relationship of adenosine-derived A_1 receptor agonists.

In rats, INO-8875 exhibited increased dromotropic activity. It did not reduce the arterial blood pressure up to the maximal tested dose (50 μg/kg) [157].

Summary SAR of adenosine derivatives

A large number of ligands for the adenosine receptors have been synthesized and evaluated for binding affinity to the A_1 receptor. The most frequently varied positions are the N^6-position, the 2-position of the purine ring system and the 5′-position of the ribose unit. Mono-substitution at the exocyclic N^6-position is required for high affinity for the A_1 receptor. Several substituents, including cycloalkyl, aryl and alkylaryl are tolerated. The purine ring system itself requires nitrogen atoms at the 3- and 7-positions. The nitrogen at the 1-position can be replaced by a carbon atom (as in AMP-579). Substituents at the C^2-position are well tolerated, ranging from small halogen to bulky groups. In general, the ribose unit is crucial to achieve affinity for the A_1 receptor; exceptions are found with compounds where the ribose moiety has been replaced by carbocycles. Many variations have been explored at the 5′-position of the ribose unit, adjusting selectivity and affinity for the A_1 receptor. These findings are summarized in Figure 4.2.

AGONISTS WITH NON-ADENOSINE-DERIVED STRUCTURES

All compounds described so far originated from the endogenous ligand adenosine, bearing the characteristic ribose moiety and a purine scaffold.

Non-purinergic natural products

In 2002, a research group from the University of Bonn published the isolation of lignans from a hydrophilic extract of the medical plant valerian exhibiting affinity for the adenosine A_1 receptor [158, 159]. Compound (110) exhibited only a moderate affinity ($K_i = 5.28\,\mu M$) for the A_1 receptor, with almost no selectivity over the A_{2A} receptor. It was shown to be a partial agonist.

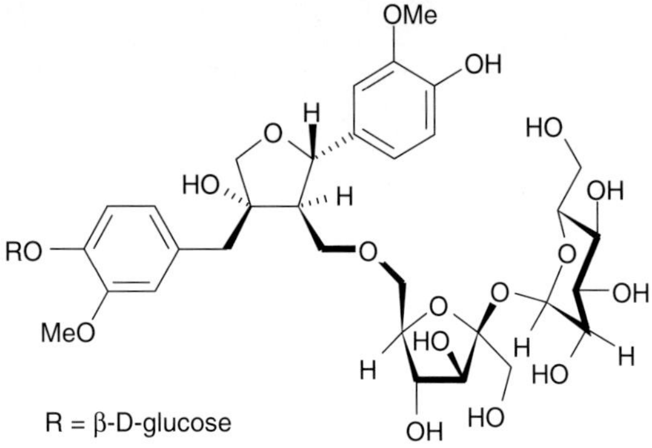

(110)

R = β-D-glucose

Synthetic non-purinergic adenosine A_1 agonists

In 2001 the first patent of a series was published by Bayer AG claiming-substituted 2-thio-3,5-dicyano-4-aryl-6-aminopyridines as adenosine A_1 receptor agonists [160]. The synthesis of the core 3,5-dicyanopyridine structure is based on work by chemists from the Russian Academy of Science [161]. The aromatic aldehyde is reacted with two equivalents of cyanothioacetamide (Scheme 4.3). The resulting thiopyridine can be alkylated using known reaction conditions. This reaction has been the point of interest of some recent publications [162, 163].

The Bayer program started off with a lead compound (112), derived from high throughput screening of CHO cell lines expressing human adenosine receptors and luciferase or aequorin reporter genes. This lead was the starting point for an extensive optimisation program. The liabilities of the lead compound included the low selectivity over the A_{2A} receptor, a marked hepatic metabolism and insufficient pharmacokinetic parameters for p.o. application. First, the poor selectivity for the A_1 receptor was addressed resulting in compound (113) (Table 4.12). The hydroxyethyl derivative exhibited a selectivity of approximately 100 over the A_{2B} receptor and of more than 3,000 over the A_{2A} and A_3 receptors.

Based on this hydroxyethyl compound, a structure activity relationship was established. At the northern site (Table 4.13) promising results were

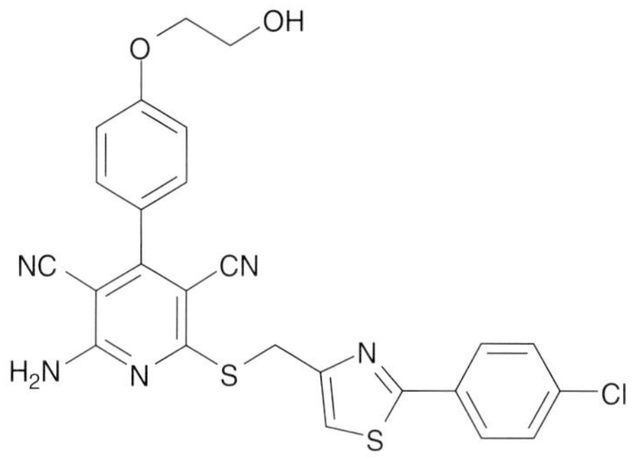

(111)

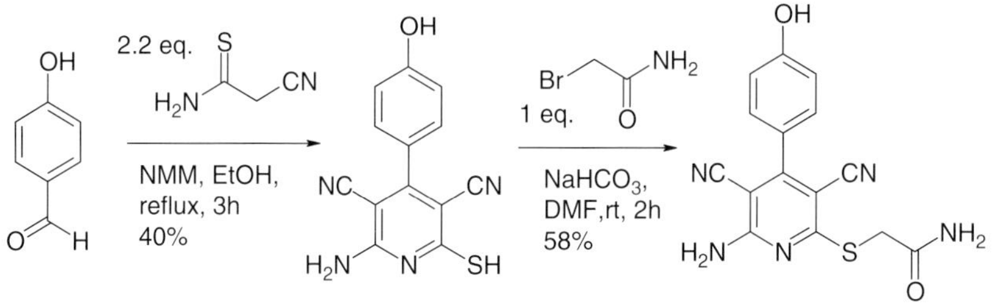

Scheme 4.3   Chemical synthesis of a non-purinergic adenosine A_1 agonist.

achieved with hydroxyl- and alkoxy-substituted aromatic ring systems (114–119). In 2005, a group from the Amsterdam Center for Drug Research [164] re-synthesized explicit examples of the Bayer patents including compound (115) and confirmed the high affinity and selectivity for the A_1 receptor.

The Bayer team explored the substitution of the southeastern site (Table 4.14) and found that substitution with [2-aryl-1,3-thiazol-4-yl]methyl (127) and [2-morpholinyl-1,3-thiazol-4-yl]methyl (126) groups gave compounds with a very good selectivity profile.

Substitution at the amino group in the southwest region was explored to some extent. Examples are given in Table 4.15. The core variation of exchanging the cyanopyrdine by a pyrimidine, or the replacement of one of the cyano groups, resulted in less active compounds; however, some of the resulting compounds exhibited antagonistic activity for the adenosine receptors.

Table 4.12 AGONISTIC ACTIVITY AND SELECTIVITIES OF NON-PURINERGIC ADENOSINE RECEPTOR AGONISTS

(112)

(113)

Compound	*EC$_{50}$ (nM)*			
	A$_1$	*A$_{2A}$*	*A$_{2B}$*	*A$_3$*
(112)	6.3	1,800	16	>3,000
(113)	0.7	>3,000	670	>3,000

Table 4.13 AGONISTIC ACTIVITY AND SELECTIVITIES OF NON-PURINERGIC ADENOSINE RECEPTOR AGONISTS

(114) (115) (116) (117) (118) (119)

Compound	*EC$_{50}$ (nM)*			
	A$_1$	*A$_{2A}$*	*A$_{2B}$*	*A$_3$*
(114)	0.5	>3,000	248	
(115)	0.4	1,510	543	
(116)	2.7	>3,000	>3,000	>3,000
(117)	0.2	>3,000	,2,000	
(118)	8.0	>3,000	>3,000	
(119)	77	>3,000	>3,000	

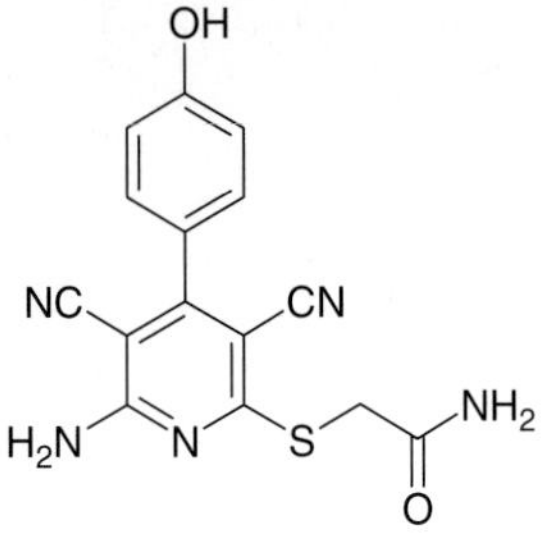

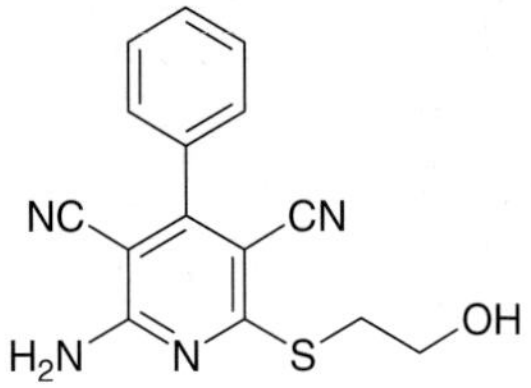

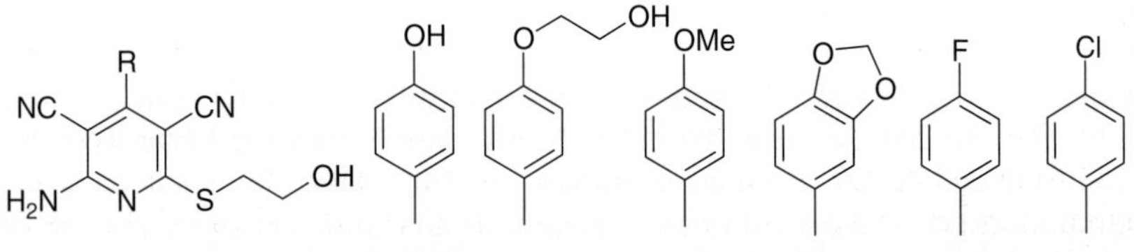

Table 4.14 AGONISTIC ACTIVITY AND SELECTIVITIES OF NON-PURINERGIC ADENOSINE RECEPTOR AGONISTS

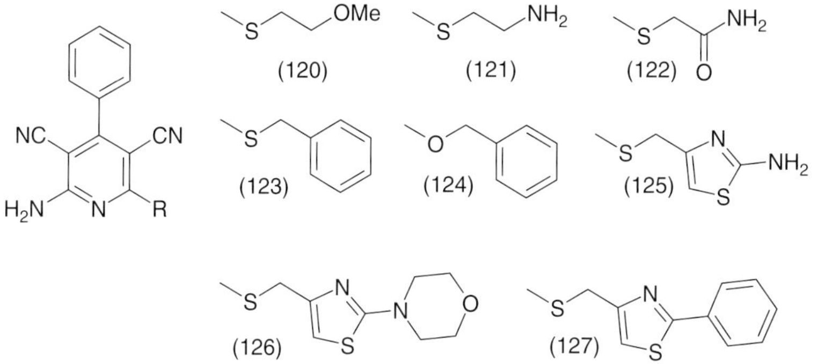

Compound	EC$_{50}$ (nM)			
	A_1	A_{2A}	A_{2B}	A_3
(120)	1.5	>3,000	>3,000	>3,000
(121)	41			
(122)	6.0	2,450	99	
(123)	0.2			
(124)	2.3	>3,000	>3,000	
(125)	0.5			
(126)	0.1			
(127)	0.3			

During the course of the optimisation programme, adenosine receptor agonists with various selectivity patterns were identified (Table 4.16) ranging from selectivity for the A$_1$ receptor (131), for the A$_{2B}$ receptor (132) [165], to mixed selectivity for the A$_1$ and A$_{2B}$ receptors (133). Most of the compounds evaluated were partial agonists.

Further optimisation of the compound class, including various parameters of lead optimisation required for oral administration, led to BAY 68-4986 (Capadenoson) (111) [166], a compound with high affinity for the A$_1$ receptor and high selectivity over the other adenosine receptors. Capadenoson shows a favourable pharmacokinetic profile with a long half-life and high bioavailability. In spontaneously hypertensive rats (SHR) oral dosing of Capadenoson resulted in a long lasting dose-dependent reduction in heart rate and blood pressure. The compound is currently in clinical Phase II studies.

Table 4.15 AGONISTIC ACTIVITY AND SELECTIVITIES OF NON-PURINERGIC ADENOSINE RECEPTOR AGONISTS

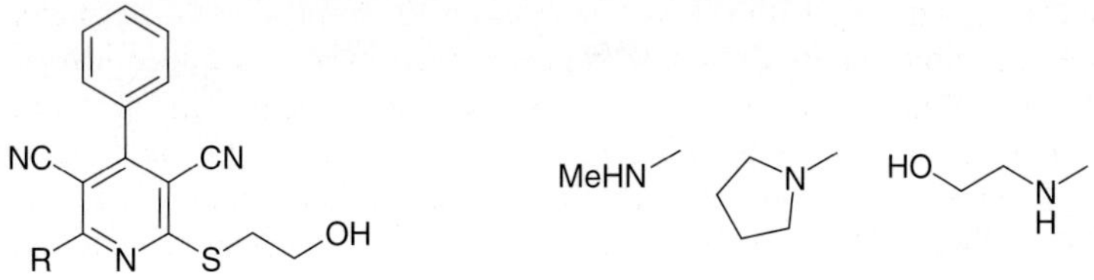

Compound	EC$_{50}$ (nM)			
	A$_1$	A$_{2A}$	A$_{2B}$	A$_3$
(128)	1.8	1,660	>3,000	
(129)	5.0			
(130)	20.0			

Table 4.16 AGONISTIC ACTIVITY AND SELECTIVITIES OF NON-PURINERGIC ADENOSINE RECEPTOR AGONISTS

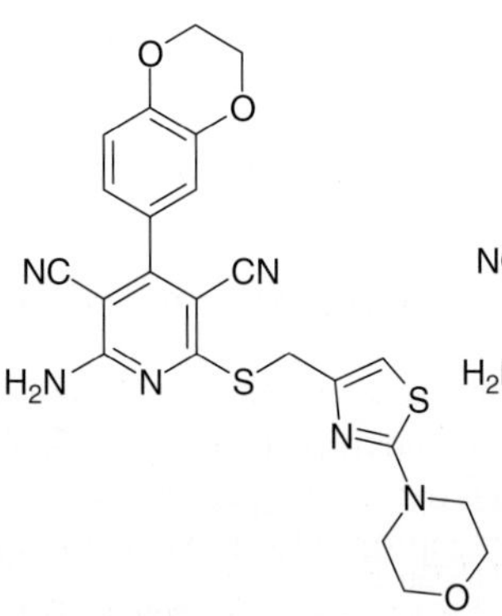

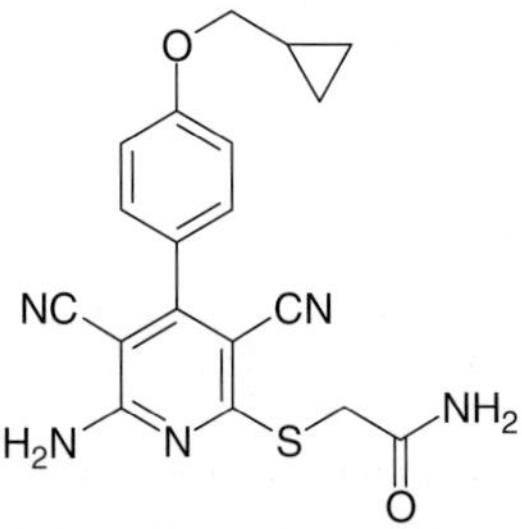

Compound	EC$_{50}$ (nM)			
	A$_1$	A$_{2A}$	A$_{2B}$	A$_3$
(131)	0.5	>3,000	>3,000	740.0
(132)	>3,000	>3,000	10.0	>3,000
(133)	0.9	160	0.5	>3,000

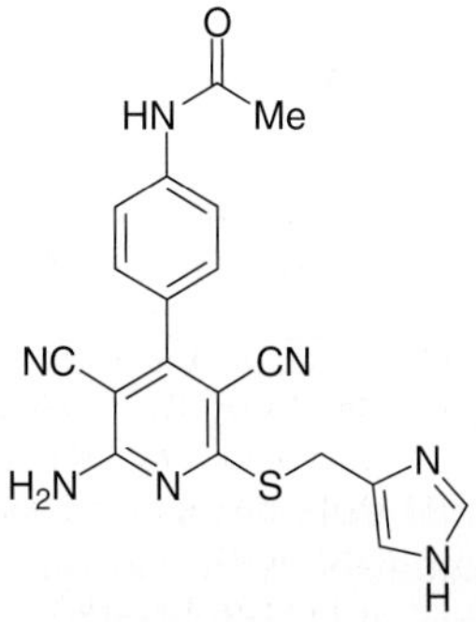

CONCLUSIONS

There is a broad spectrum of therapeutic opportunities for A_1 agonists. This includes cardiovascular diseases, neuroprotection and modulation of lipid metabolism. Potent and selective agonists for the A_1 adenosine receptor have been identified and have been evaluated in clinical studies. The development of A_1-selective agonists for non-cardiovascular indications such as pain and metabolic diseases has not been successful to date. Currently, adenosine A_1 agonists are only in development for the treatment of cardiac arrhythmias such as atrial fibrillation and PSVT. So far, none of these compounds has been registered, which might be due to a problem with the non-partial agonist character of these compounds. The major class of A_1 agonists is derived from adenosine and the structural requirements of these agonists for potency and selectivity is understood. Partial A_1 agonists may be able to overcome the critical issues in development of full A_1 agonists (i.e., side effects). Recently, promising non-purinergic adenosine receptor agonists have emerged.

REFERENCES

[1] Drury, A.N. and Szent-Györgyi, A. (1929) *J. Physiol.* **68**, 213–237.
[2] Pearson, J.D., Hellewell, P.G. and Gordon, J.L. (1983) In "Regulatory Function of Adenosine". Berne, R.M., Rall, T.W. and Rubio, R. (eds), pp. 333–357. Martinius Nijhoff, Boston.
[3] Achterberg, P.W., de Tombe, P.P., Harmsen, E. and de Jong, J.W. (1985) *Biochem. Biophys. Acta* **840**, 393–400.
[4] Jarvis, S.M. (1986) *Mol. Pharmacol.* **30**, 659–665.
[5] Jacobson, K.A., Trivedi, B.K., Churchill, P.C. and Williams, M. (1991) *Biochem. Pharmacol.* **41**, 1399–1410.
[6] Belardinelli, L., Shryock, J.C., Song, Y., Wang, D. and Srinivas, M. (1995) *FASEB* **9**, 359–365.
[7] DiMarco, J.P. (1985) *J. Am. Coll. Cardiol.* **6**, 417–425.
[8] Sollevi, A. (1986) *Prog. Neurobiol.* **27**, 319–349.
[9] Verani, M.S. and Zoghbi, W.A. (1992) *Am. J. Card. Imaging* **6**, 71–80.
[10] Burnstock, G. (1978) In "Cell Membrane Receptors for Drugs and Hormones". Bolis, L. and Straub, R.W. (eds), pp. 107–118. Raven, New York.
[11] Van Calker, D., Müller, M. and Hamprecht, B. (1978) *Nature (London)* **276**, 839–841.
[12] Van Calker, D., Müller, M. and Hamprecht, B. (1979) *J. Neurochem.* **33**, 999–1005.
[13] Londos, C., Cooper, D.M.F. and Wolff, J. (1980) *Proc. Natl. Acad. Sci. USA* **77**, 2551–2554.
[14] Marquardt, D.L., Walker, L.L. and Heinemann, S.J. (1994) *J. Immunol.* **152**, 4508–4515.
[15] Libert, F., Parmentier, M., Lefort, A., Dinsart, C., van-Sande, J., Maenhaut, C., Simons, M.-J., Dumont, J.E. and Vas-sart, G. (1989) *Science (Washington DC)* **244**, 569–572.

[16] Libert, F., Schiffmann, S.N., Lefort, A., Parmentier, M., Gerard, C., Dumont, J.E., Vanderhaeghen, J.-J. and Vassart, G. (1991) *EMBO J.* **10**, 1677–1682.
[17] Mahan, L.C., McVittie, L.D., Smyk-Randall, E.M., Nakata, H., Monsma, F.J., Gerfen, C.R. and Sibley, D.R. (1991) *Mol. Pharmacol.* **40**, 1–7.
[18] Reppert, S.M., Weaver, D.R., Stehle, J.H. and Rivkees, S. (1991) *A. Mol. Endocrinol.* **5**, 1037–1048.
[19] Libert, F., van Sande, J., Lefort, A., Czernilofsky, A., Dumont, J.E., Vassart, G., Ensinger, H.A. and Mendla, K.D. (1992) *Biochem. Biophys. Res. Commun.* **187**, 919–926.
[20] Townsend-Nicholson, A. and Shine, J. (1992) *Mol. Brain Res.* **16**, 365–370.
[21] Ren, H. and Stiles, G.L. (1994) *J. Biol. Chem.* **269**, 3104–3110.
[22] Olah, M.E., Ren, H., Ostrowski, J., Jacobson, K.A. and Stiles, G.L. (1992) *J. Biol. Chem.* **267**, 10764–10770.
[23] Tucker, A.L., Linden, J., Robeva, A.S., D'Angelo, D.D. and Lynch, K.R. (1992) *FEBS Lett.* **297**, 107–111.
[24] Bhattacharya, S., Dewitt, D.L., Burnatowska-Hledin, M., Smith, W.L. and Spielman, W.S. (1993) *Gene* **128**, 285–288.
[25] Weaver, D.R. and Reppert, S.M. (1992) *Am. J. Physiol.* **263**, F991–F995.
[26] Belardinelli, L., Linden, J. and Berne, R.M. (1989) *Prog. Cardiovasc. Dis.* **32**, 73–97.
[27] Hori, M. and Kitakaze, M. (1991) *Hypertension* **18**, 565–574.
[28] Vitzhum, H., Weiss, B., Bachleitner, W., Krämer, B.K. and Kurtz, A. (2004) *Kidney Int.* **65**, 1180–1190.
[29] Daly, J.W., Butts-Lamb, P. and Padgett, W. (1983) *Cell. Mol. Neurobiol.* **3**, 69–80.
[30] Bruns, R.F., Lu, G.H. and Pugsley, T.A. (1986) *Mol. Pharmacol.* **29**, 331–346.
[31] Schulte, G. and Fredholm, B.B. (2003) *Cell. Signal.* **15**, 813–827.
[32] Spicuzza, L., Di Maria, G. and Polosa, R. (2006) *Eur. J. Pharmacol.* **533**, 77–88.
[33] Peart, J.N. and Headrick, J.P. (2007) *Pharmacol. Therapeu.* **114**, 208–221.
[34] Chen, J.-F., Sonsalla, P.K., Pedata, F., Melani, A., Domenici, M.R., Popoli, P., Geiger, J., Lopes, L.V. and de Mendonc, A. (2007) *Prog. Neurobiol.* **83**, 310–331.
[35] Maenhaut, C., van Sande, J., Libert, F., Abramowicz, M., Parmentier, M., Vanderhaeghen, J.-J., Dumont, J.E., Vassart, G. and Schiffmann, S. (1990) *Biochem. Biophys. Res. Commun.* **173**, 1169–1178.
[36] Fink, J.S., Weaver, D.R., Rivkees, S.A., Peterfreund, R.A., Pollack, A.E., Adler, E.M. and Reppert, S.M. (1992) *Mol. Brain Res.* **14**, 186–195.
[37] Chern, Y., Kling, K., Lai, H.-L. and Lai, H.-T. (1992) *Biochem. Biophys. Res. Cummun.* **185**, 304–309.
[38] Furlong, T.J., Pierce, K.D., Selbie, L.A. and Shine, J. (1992) *Mol. Brain Res.* **15**, 62–66.
[39] Meng, F., Xie, G., Chalmers, D., Morgan, C., Watson, S.J. and Akil, H. (1994) *Neurochem. Res.* **19**, 613–621.
[40] Stehle, J.H., Rivkees, S.A., Lee, J.J., Weaver, D.R., Deeds, J.D. and Reppert, S.M. (1992) *Mol. Endocrinol.* **6**, 384–393.
[41] Rivkees, S.A. and Reppert, S.M. (1992) *Mol. Endokrinol.* **6**, 1598–1604.
[42] Pierce, K.D., Furlong, T.J., Selbie, L.A. and Shine, J. (1992) *Biochem. Biophys. Res. Cummun.* **187**, 86–93.
[43] Meyerhof, W., Müller-Brechlin, R. and Richter, D. (1991) *FEBS Lett.* **284**, 155–160.
[44] Zhou, Q.-Y., Li, C., Olah, M.-E., Johnson, R.A., Stiles, G.L. and Civelli, O. (1992) *Proc. Natl. Acad. Sci. USA* **89**, 7432–7436.
[45] Van Muijlwijk-Koenzen, J.E., Timmerman, H. and Ijzerman, A.P. (2001) *Prog. Med. Chem.* **38**, 61–113.

[46] Headrick, J.P. and Peart, J. (2005) *Vasc. Pharmacol.* **42**, 271–279.
[47] Bar-Yehuda, S., Silverman, M.H., Kerns, W.D., Ochaion, A., Cohen, S. and Fishman, P. (2007) *Expert Opin. Investig. Drugs* **16**, 1601–1613.
[48] Van Galen, P.J., Stiles, G.L., Michaels, G. and Jacobson, K.A. (1992) *Med. Res. Rev.* **12**, 423–471.
[49] Dixon, A.K., Gubitz, A.K., Sirinathsinghji, D.S.J., Richardson, P.J. and Freemann, T.C. (1996) *Br. J. Pharmacol.* **118**, 1461–1468.
[50] Hussain, T. and Mustafa, S.J. (1995) *Circ. Res.* **77**, 194–198.
[51] Musser, B., Morgan, M.E., Leid, M., Murray, T.F., Linden, J. and Vestal, R.E. (1993) *Eur. J. Pharmacol.* **246**, 105–111.
[52] Freissmuth, M., Selzer, E. and Schutz, W. (1991) *Biochem. J.* **276**, 651–656.
[53] Akbar, M., Okajima, F., Tomura, H., Shimegi, S. and Kondo, Y. (1994) *Mol. Pharmacol.* **45**, 1036–1042.
[54] Jockers, R., Linden, J., Hohenegger, M., Nanoff, C., Bertin, B. and Strosberg, A.D. (1994) *J. Biol. Chem.* **269**, 32077–32084.
[55] Dobson, J.G., Jr. (1983) *Circ. Res.* **52**, 151–160.
[56] Dobson, J.G., Jr., Ordway, R.W. and Fenton, R.A. (1986) *Am. J. Physiol.* **251**, H455–H462.
[57] Romano, F.D. and Dobson, J.G., Jr.. (1990) *J. Mol. Cell. Cardiol.* **22**, 1359–1370.
[58] Schütte, F., Burgdorf, C., Richardt, G. and Kurz, T. (2006) *Can. J. Physiol. Pharmacol.* **84**, 573–577.
[59] Yuan, K., Cao, C., Han, J.H., Kim, S.Z. and Kim, S.H. (2005) *Hypertension* **46**, 1381–1387.
[60] Dickenson, J.M. and Hill, S.J. (1998) *Eur. J. Pharmacol.* **355**, 85–93.
[61] Munshi, R., Pang, I.H., Sternweis, P.C. and Linden, J. (1991) *J. Biol. Chem.* **266**, 22285–22299.
[62] Freund, S., Ungerer, M. and Lohse, M.J. (1994) *Naunyn-Schmiederbergs's Arch. Pharmacol.* **350**, 49–56.
[63] Belardinelli, L. and Isenberg, G. (1983) *Am. J. Physiol.* **244**, H734–H737.
[64] Trusell, L.O. and Jackson, M.B. (1985) *Proc. Natl. Acad. Sci. USA* **82**, 4857–4861.
[65] Dolphin, A.C., Forda, S.R. and Scott, R.H. (1986) *J. Physiol. (London)* **373**, 47–61.
[66] Scholz, K.P. and Miller, R.J. (1991) *J. Physiol. (London)* **435**, 373–393.
[67] Mogul, D.J., Adams, M.E. and Fox, A.P. (1993) *Neuron* **10**, 327–334.
[68] Wang, D. and Belardinelli, L (1994) *Am. J. Physiol.* **267**, H2420–H2429.
[69] Martynuyk, A.E., Kane, K.A., Cobbe, S.M. and Rankin, A.C. (1996) *Plügers Arch. Eur. J. Physiol.* **431**, 452–457.
[70] Liang, B.T. and Donovan, L.A. (1990) *Circ. Res.* **67**, 406–414.
[71] Green, A. (1987) *J. Biol. Chem.* **262**, 15702–15707.
[72] Lee, H.T., Thompson, C.I., Hernandez, A., Lewy, J.L. and Belloni, F.L. (1993) *Am. J. Physiol.* **265**, H1916–H1927.
[73] Roman, V., Keijser, J.N., Luiten, P.G. and Meerlo, P. (2008) *Brain Res.* **29**, 69–74.
[74] Parsons, W.J. and Stiles, G.L. (1987) *J. Biol. Chem.* **262**, 841–847.
[75] Longabaugh, J.P., Didsbury, J., Spiegel, A. and Stiles, G.L. (1989) *Mol. Pharmacol.* **36**, 681–688.
[76] Bhattacharya, S. and Linden, J. (1996) *Mol. Pharmacol.* **50**, 104–111.
[77] Linden, J., Tucker, A.L., Robeva, A.S., Graber, S.G. and Munshi, R. (1993) *Drug Dev. Res.* **28**, 232–236.
[78] Poulson, S.-A. and Quinn, R.J. (1998) *Bioorg. Med. Chem.* **6**, 619–641.

[79] Fredholm, B.B., Izerman, A.P., Jacobson, K.A., Klotz, K.N. and Linden, J. (2001) *Pharmacol. Rev.* **264**, 527–552.
[80] Böhm, M., Pieske, B., Ungerer, M. and Erdmann, E. (1989) *Circ. Res.* **65**, 1201–1211.
[81] Froldi, G. and Belardinelli, L. (1990) *Circ. Res.* **67**, 960–978.
[82] Tucker, A.L., Robeva, A.S., Taylor, H.E., Holeton, D., Bockner, M., Lynch, K.R. and Linden, J. (1994) *J. Biol.Chem.* **269**, 27900–27906.
[83] Poole-Wilson, P.A. (1983) *Postgrad. Med. J.* **59**(Suppl. 3), 11–21.
[84] Tendera, M., Sosnowski, M., Gaszewska-Zurek, E., Parma, Z., Neuser, D. and Seman, L. (2007) *J. Am. Coll. Cardiol.* **49**(Suppl. A), 232A.
[85] Tsang, T.S. and Gersh, B. (2002) *Am. J. Med.* **1**, 127–139.
[86] Blaauw, Y. and Crijns, H.J. (2007) *J. Intern. Med.* **262**, 593–614.
[87] Wyse, D.G., Waldo, A.L., DiMarco, J.P., Domanski, M.J., Rosenberg, Y., Schron, E.B., et al. (2002) *N. Engl. J. Med.* **347**, 1825–1833.
[88] Wang, D., Shryock, J.C. and Belardinelli, L. (1996) *Circ. Res.* **78**, 697–706.
[89] Isomoto, S., Kondo, C. and Kurachi, Y. (2007) *Jpn. J. Physiol.* **47**, 11–39.
[90] Endoh, M., Maruyama, M. and Taira, N.J. (1983) *Cardiovasc. Pharmacol.* **5**, 131–142.
[91] Bois, P., Guinamard, R., Chemaly, A.E., Faivre, J.F. and Bescond, J. (2007) *Curr. Pharm. Des.* **13**, 2338–2349.
[92] De Haan, S., Verheule, S., Kuiper, M., van Hunnik, A., Dhalla, A., Belardinelli, L., Allessie, M. and Schotten, U. (2007) *JACC* **49**(Suppl. 1), 25A.
[93] McLaurin, B.T., San, W.G., Chen, C., Kosinski, E.J., Dillon, P.M. and O'Dell, S.W. (2003). Abstract AHA congress.
[94] Wu, L., Belardinelli, L., Zablocki, J.A., Palle, V. and Shryock, J.C. (2001) *Am. J. Physiol.* **280**, H334–H343.
[95] Zablocki, J.A., Wu, L., Shryock, J.C. and Belardinelli, L. (2004) *Curr. Top. Med. Chem.* **4**, 839–854.
[96] Bolli, R., Becker, L., Gross, G., Mentzer, R., Jr.., Balshaw, D. and Lathrop, D.A. (2004) *Circ. Res.* **95**, 125–134.
[97] Otani, H. (2008) *Antioxid. Redox Signal.* **10**, 207–247.
[98] Sommerschild, H.T. and Kirkebøen, K.A. (2002) *Acta Anaesthesiol. Scand.* **46**, 123–137.
[99] Baxter, G.F. (2002) *Cardiovasc. Res.* **55**, 483–494.
[100] Baxter, G.F., Hale, S.L., Miki, T., Kloner, R.A., Cohen, M.V., Downey, J.M. and Yellon, D.M. (2000) *Cardiovasc. Drugs Therap.* **14**, 607–614.
[101] Thornton, J.D., Liu, G.S., Olsson, R.A. and Downey, J.M. (1992) *Circulation* **85**, 659–665.
[102] Liu, G.S., Thornton, J.D., Van Winkle, D.M., Stanley, A.W., Olsson, R.A. and Downey, J.M. (1991) *Circulation* **84**, 350–356.
[103] Hausenloy, D.J. and Yellon, D.M. (2006) *Cardiovasc. Res.* **70**, 240–253.
[104] Dana, A., Baxter, G.F., Walker, J.M. and Yellon, D.M. (1998) *J. Am. Coll. Cardiol.* **31**, 1142–1149.
[105] Murray, T.F., Franklin, P.H., Zhang, G. and Tripp, E. (1992) *Epilepsy Res. Suppl.* **8**, 255–261.
[106] Zhang, G., Franklin, P.H. and Murray, T.F. (1994) *Eur. J. Pharmacol.* **255**, 239–243.
[107] Pan, H.L., Xu, Z., Leung, E. and Eisenach, J.C. (2001) *Anesthesiology* **95**, 416–420.
[108] Zahn, P.K., Straub, H., Wenk, M. and Pogatzki-Zahn, E.M. (2007) *Anesthesiology* **107**, 797–806.
[109] Heurteaux, C., Lauritzen, I., Widmann, C. and Lazdunski, M. (1995) *Proc. Natl. Acad. Sci. USA* **92**, 4666–4670.

[110] Lassmann, H., Bruck, W. and Lucchinetti, C. (2001) *Trends Mol. Med.* **7**, 115–121.
[111] Tsutsui, S., Schnermann, J., Noorbakhsh, F., Henry, S., Yong, V.W., Winston, B.W., Warren, K. and Power, C. (2004) *J. Neurosci.* **24**, 1521–1529.
[112] Ferre, S., Ciruela, F., Borycz, J., Solinas, M., Quarta, D., Antoniou, K., et al. (2008) *Front. Biosci.* **13**, 2391–2399.
[113] Franco, R., Lluis, C., Canela, E.I., Mallo, l.J., Agnati, L., Casadó, V., Ciruela, F., Ferré, S. and Fuxe, K. (2007) *J. Neural. Transm.* **114**, 94–104.
[114] Dhalla, A.K., Shryock, J.C., Shreeniwas, R. and Belardinelli, L. (2003) *Curr. Top. Med. Chem.* **3**, 369–385.
[115] Fain, J.N. (1979) *Biochem. Biophys. Acta* **573**, 510–520.
[116] Ferrannini, E., Barrett, E.J., Bevilacqua, S. and DeFronzo, R.A. (1983) *J. Clin. Invest.* **72**, 1737–1747.
[117] DeFronzo, R.A. (1988) *Diabetes* **37**, 667–687.
[118] Dong, Q., Ginsberg, H.N. and Erlanger, B.F. (2001) *Diabetes Obes. Metab.* **3**, 360–366.
[119] Shah, B., Rohatagi, S., Natarajan, C., Kirkesseli, S., Baybutt, R. and Jensen, B.K. (2004) *Am. J. Ther.* **11**, 175–189.
[120] Ashton, T.D., Aumann, K.M., Baker, S.P., Schiesser, C.H. and Scammels, P.J. (2007) *Bioorg. Med. Chem. Lett.* **17**, 6779–6784.
[121] Kato, K., Hayakawa, H., Tanaka, H., Kumamoto, H. and Miyasaka, T. (1995) *Tetrahedron Lett.* **36**, 6507–6510.
[122] Trivedi, B.K., Bridges, A.J., Patt, W.C., Priebe, S.R. and Bruns, R.F. (1989) *J. Med. Chem.* **32**, 8–11.
[123] Moos, W.H., Szotek, D.S. and Bruns, R.F. (1985) *J. Med. Chem.* **28**, 1383–1384.
[124] Bruns, R.F., Daly, J.W. and Snyder, S.H. (1980) *Proc. Natl. Acad. Sci. USA* **77**, 5547–5551.
[125] Knutsen, L.J.S., Lau, J., Petersen, H., Thomsen, C., Weis, J.U., Shalmi, M., Judge, M.E., Hansen, A.J. and Sheardown, M.J. (1999) *J. Med. Chem.* **42**, 3463–3477.
[126] Goadsby, P.J., Hoskin, K.L., Storer, R.J., Edvinsson, R.J. and Connor, H.E. (2002) *Brain* **125**, 1392–1401.
[127] Ashton, T.D. and Scammells, P.J. (2006) *Bioorg. Med. Chem. Lett.* **16**, 4564–4566.
[128] Hutchinson, S.A., Baker, S.P., Linden, J. and Scammells, P.J. (2004) *Bioorg. Med. Chem.* **12**, 4877–4884.
[129] Ashton, T.D., Baker, S.P., Hutchinson, S.A. and Scammells, P.J. (2008) *Bioorg. Med. Chem.* **16**, 1861–1873.
[130] Merkel, L.A., Rivera, L.M., Colussi, D.J., Perrone, M.H., Smits, G.J. and Cox, B.F. (1993) *J. Pharmacol. Exp. Ther.* **265**, 699–706.
[131] Fatholai, M., Xiang, Y., Wu, Y., Li, Y., Wu, L., Dhalla, A.K., Belardinelli, L. and Shryock, J.C. (2006) *J. Pharmacol. Exp. Ther.* **317**, 676–684.
[132] Dhalla, A.K., Santikul, M., Smith, M., Wong, M-Y., Shryock, J.C. and Belardinelli, L. (2007) *J. Pharmacol. Exp. Ther.* **321**, 327–333.
[133] Drug Data Report. (2007) 29, p. 822.
[134] Cheung, J.W. and Lerman, B.B. (2003) *Cardiovasc. Drug Rev.* **21**, 277–292.
[135] Ellenbogen, K.A., O'Neill, G., Prystowsky, E.N., Camm, J.A., Meng, L., Lieu, H.D., Jerling, M., Shreeniwas, R., Belardinelli, L. and Wolff, A.A. (2005) *Circulation* **111**, 3202–3208.
[136] CV Therapeutics. Available at www.cvt.com (April, 2008).
[137] Palle, V.P., Varkhedkar, V., Ibrahim, P., Ahmed, H., Li, Z., Gao, Z., et al. (2004) *Bioorg. Med. Chem. Lett.* **14**, 535–539.

[138] 11th World Congress Pain, August 21–26, Sydney, 2005, Abstr. 637-P243.
[139] Butcher, A., Gregg, A., Scammells, P.J. and Meyer, R.B.R. (2007) *Drug Dev. Res.* **68**, 529–537.
[140] Gregg, A., Bottle, S.E., Devine, S.M., Figler, H., Linden, J., White, P., Pouton, C.W., Urmaliya, V. and Scammells, P.J. (2007) *Bioorg. Med. Chem. Lett.* **17**, 5437–5441.
[141] Elzein, E., Kalla, R., Li, X., Perry, T., Marquart, T., Micklatcher, M., Li, Y., Wu, Y., Zeng, D. and Zablocki, J. (2007) *Bioorg. Med. Chem. Lett.* **17**, 161–166.
[142] Beukers, M.W., Wanner, M.J., Van Frijtag Drabbe Künzel, J.K., Klaasse, E.C., Ijzerman, A.P. and Koomen, G.-J. (2003) *J. Med. Chem.* **46**, 1492–1503.
[143] Cappellacci, L., Barboni, G., Palmieri, M., Pasqualini, M., Grifantini, M., Costa, B., Martini, C. and Franchetti, P. (2002) *J. Med. Chem.* **45**, 1196–1202.
[144] Franchetti, P., Cappellacci, L., Marchetti, S., Trincavelli, L., Martini, C., Mazzoni, M.R., Lucacchini, A. and Grifantini, M. (1998) *J. Med. Chem.* **41**, 1708–1715.
[145] Cappellacci, L., Franchetti, P., Pasqualini, M., Petrelli, R., Vita, P., Lavecchia, A., et al. (2005) *J. Med. Chem.* **48**, 1550–1562.
[146] Sledeski, A.W., Kubiak, G.G., O'Brien, M.K., Powers, M.R., Powner, T.H. and Truesdale, L.K. (2000) *J. Org. Chem.* **65**, 8114–8118.
[147] Clark, K.L., Merkel, L., Zannnikos, P., Kelly, M.F., Boutouyrie, B. and Perrone, M.H. (2000) *Cardiovasc. Drug Rev.* **18**, 183–210.
[148] Jähne, G. (2005). GDCh Meeting: Frontiers in Medicinal Chemistry, Leipzig.
[149] Lee, K., Ravi, G., Ji, X.-d., Marquez, V.E. and Jacobson, K.A. (2001) *Bioorg. Med. Chem. Lett.* **11**, 1333–1337.
[150] Belardinelli, L., Lu, J., Dennis, D., Martens, J. and Shryock, J.C. (1994) *J. Pharm. Exp. Ther.* **271**, 1371–1382.
[151] Wagner, H., Milavec-Krizman, M., Gadient, F., Menninger, K., Schoeffter, P., Tapparelli, C., Pfannkuche, H.-J. and Fozard, J.R. (1995) *Drug Disc. Res.* **34**, 276–288.
[152] Ishikawa, J., Mitani, H., Bandoh, T., Kimura, M., Totsuka, T. and Hayashi, S. (1998) *Diabetes Res. Clin. Pract.* **39**, 3–9.
[153] Bertolet, B.D., Anand, I.S., Bryg, R.J., Mohanty, P.K., Chatterjee, K., Cohn, J.N., Khurmi, N.S. and Pepine, C.J. (1996) *Circulation* **94**, 1212–1215.
[154] Seymour, R.A., Hawkesford, J.E., Hill, C.M., Frame, J. and Andrews, C. (1999) *Br. J. Clin. Pharm.* **47**, 675–680.
[155] Thomson Scientific. Available at www.iddb.com
[156] Jagtap, P., Abo, C. and Salzman, A.L. (2005) *PCT Int. Appl.*, WO 2005 117910, CAN 144:51838.
[157] Etzion, Y., Shalev, A., Mor, M., Moran, A. and Katz, A. (2008) *J. Am. Coll. Cardiol.* **51**(10), Abstract 1001–101.
[158] Schumacher, A., Scholle, S., Hölzl, J., Khudeir, N., Hess, S. and Müller, C.E. (2002) *J. Nat. Prod.* **65**, 1479–1485.
[159] Müller, C.E., Schumacher, B., Brattström, A., Abourashed, E.A. and Koetter, U. (2002) *Life Sci.* **71**, 1939–1949.
[160] Rosentreter, U., Henning, R., Bauser, M., Krämer, T., Vaupel, A., Hübsch, W., et al. (2001). *PCT Int. Appl.*, WO 01 25210, CAN 134:295744.
[161] Dyachenko, V.D. and Litvinov, V.P. (1998) *Chem. Heterocyl. Comp.* **34**, 188–194.
[162] Evdokimov, N.M., Magedov, I.V., Kireev, A.S. and Kornienko, A. (2006) *Org. Lett.* **8**, 899–902.
[163] Guo, K., Mutter, R., Heal, W., Reddy, T.R.K., Cope, A., Pratt, S., Thompson, M.J. and Chen, B. (2008) *Eur. J. Med. Chem.* **43**, 93–106.

[164] Chang, L.C.W., von Frijtag Drabbe Künzel, J.K., Mulder-Krierger, T., Spanjersberg, R.F., Roerink, S.F., van den Hout, G., Beukers, M.W., Brussee, J. and Ijzerman, A.P. (2005) *J. Med. Chem.* **48**, 2045–2053.
[165] Eckle, T., Krahn, T., Grenz, A., Köhler, D., Mittelbronn, M., Ledent, C., et al. (2007) *Circulation* **115**, 1581–1590.
[166] Rosentreter, U., Krämer, T., Shimada, M., Hübsch, W., Diedrichs, N., Krahn, T., Henninger, K., Stasch, J.-P. and Wischnat, R. (2003) *PCT Int. Appl.*, WO 03 053441 CAN 139:36452.

5 Endothelin Receptor Antagonists: Status and Learning 20 Years On

MICHAEL J. PALMER

Sandwich Discovery Chemistry, Pfizer Global Research and Development, Sandwich Laboratories, Ramsgate Road, Sandwich, Kent CT13 9NJ, UK

INTRODUCTION	203
Aims of Chapter	203
The Discovery and Biology of Endothelin	204
Structural Biology	207
Clinical Status	207
Safety	209
MEDICINAL CHEMISTRY	210
General Overview	210
Peptide Antagonists	211
Sulfonamide Antagonists	214
Carboxylic Acid Antagonists	224
N-Acylsulfonamide Antagonists	230
Miscellaneous Antagonist Series	232
CONCLUSIONS	232
REFERENCES	234

INTRODUCTION

AIMS OF CHAPTER

This chapter will summarise the current biology and clinical status of endothelin receptor antagonists and then review the array of different antagonist series that have been discovered. Given that initial research progress has been documented in previous reviews, focus will be on recent developments and the key medicinal chemistry aspects that have led to successful clinical candidates.

THE DISCOVERY AND BIOLOGY OF ENDOTHELIN

Since the proposal of an endothelial vasoconstricting factor by Hickey and co-workers in 1985 [1], the field of endothelin and associated endothelin receptor antagonists has been an area of intense scientific research with some 27,000 publications to date. There is as yet no significant sign of interest waning with approximately 1,500 new papers published in 2007. Yanagisawa and colleagues characterised endothelin-1 (ET-1) in 1988 [2] as a 21 amino acid peptide and this was followed by the identification of two further human forms, endothelin-2 (ET-2) and endothelin-3 (ET-3) [3]. Detailed study followed and the three human endothelins quickly became established as a family (Figure 5.1) with important biological roles. The full endothelin system, in terms of biosynthesis, properties and function has subsequently been well researched and documented [4] and is outlined for endothelin-1 in Figure 5.2.

Each endothelin is encoded by a different gene and has a distinct tissue distribution. Endothelins are derived from precursor Big endothelin as a result of cleavage controlled by endothelin converting enzymes (ECE). ET-1 is the major isoform in the human cardiovascular system and is a highly potent constrictor of human vessels. ET-1 functions as a locally released paracrine mediator from vascular endothelial cells and contributes to maintaining vascular tone [5]. ET-1 concentrations in plasma and other tissues are comparatively low, consistent with local release and a paracrine mode of action. In addition to regulating vascular tone, ET-1 is implicated in a range of additional processes including cell proliferation, endothelial dysfunction, inflammation and fibrosis [6]. ET-2 differs by only two amino acid residues from ET-1 and is as potent a vasoconstrictor. ET-2 may also contribute to maintaining vascular tone *via* local release, as ET-2 peptide and messenger RNA are found in cardiovascular tissue [4, 7], and Big ET-2 and ET-2 messenger RNA have both been found in the cytoplasm of endothelial cells [4, 8]. However, the precise role of the ET-2 isoform is not fully understood. ET-3 is not synthesised by endothelial cells but together with the Big ET-3 precursor has been detected in plasma and a range of tissues including heart and brain. ET-3 is the only peptide family member that has a substantially different binding affinity for the two types of endothelin receptor (see later), leading to speculation of a beneficial role in regulating unwanted vasoconstriction [4].

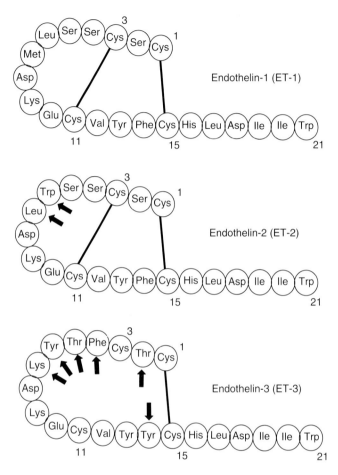

Fig. 5.1 Primary amino acid structure of the human endothelin family. Highlighted residues for ET-2 and ET-3 are those residues not conserved in ET-1.

The biological effects of the three endothelins are mediated *via* two receptors, the ET_A receptor and the ET_B receptor (Figure 5.2) [9, 10]. The two receptor sub-types both belong to class 1 of the G-protein-coupled receptor (GPCR) family. Both receptors are expressed in a wide variety of tissues and cell types, and locations include cardiovascular tissue, lung, kidney and brain. The ET_A receptor binds ET-1 and ET-2 with equal affinity, whereas ET-3 has little or no affinity at physiological concentration. In contrast, the ET_B receptor binds all three endothelin isoforms with

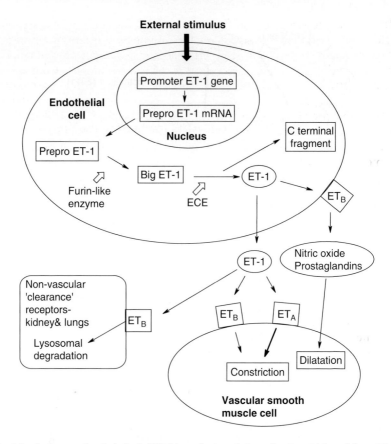

Fig. 5.2 Pathways of endothelin-1 (ET-1) synthesis and sites of action. (Adapted from Dhaun et al. [23], Copyright (2007), with permission from Elsevier.)

similar affinity [4]. Whilst both receptors are found on vascular smooth muscle cells for mediation of vasoconstrictor effects [11, 12], the ET_A receptor dominates and represents some 85% of the endothelin receptor population. However, the ET_B receptor is also found within vascular endothelial cells wherein activation results in vasodilator effectors mediated through release of nitric oxide (NO) and prostacyclin [13] (Figure 5.2). Additionally, the unique role of ET-3 in only activating the ET_B receptor has raised theories of a role for ET-3 in effecting the release of NO and prostacyclin. The ET_B receptor is the predominant sub-type in kidney and appears to have an ET-1 clearance role that helps to regulate cardiovascular tone [14].

Increasing knowledge of the functional role of the endothelin system led to the conclusion that specific antagonists might have utility in a range of indications such as hypertension. This energised the search for antagonists and the development of the necessary screening paradigms. Competitive receptor binding assays using membranes derived from appropriate endothelin receptor expressing cells have been developed. These allow ready profiling of target compounds, with radiolabelled ET-1 ligand or a suitable agonist/antagonist [4, 15–17]. Secondary functional assays have used appropriate tissue strips, such as rat carotid artery rings for ET_A receptors. Tool compounds quickly emerged that formed the basis of research programmes and allowed the potential of endothelin antagonists to be realised in animal models. Initial focus centred on cardiovascular-based indications such as essential hypertension, heart failure and atherosclerosis [18]; the field then quickly expanded to actively involve most pharmaceutical research organisations.

STRUCTURAL BIOLOGY

Structural biology has played a role in the discovery of a number of the potent endothelin receptor antagonist series that have emerged. The crystal structure has been solved for ET-1 [19], as have various NMR structures [20]. The NMR structures have differed in terms of the C-terminus conformation, and arguably the crystal structure has found greater use in SAR assessment and pharmacophore model construction [21, 22]. Whilst no ET receptor or receptor-ligand crystal structures have been obtained for docked antagonists or agonists, 3D working models for the ET_A receptor–ligand complex have subsequently been proposed and used for antagonist design [15, 22].

CLINICAL STATUS

To date, three endothelin receptor antagonists have received regulatory approval and been launched, with a fourth currently in pre-registration and a further six in Phase III clinical development (Table 5.1). Following the launch of the dual ET_A/ET_B receptor antagonist bosentan (Tracleer^{TM}) for pulmonary arterial hypertension (PAH), two ET_A selective antagonists sitaxsentan (Thelin^{TM}) and ambrisentan (Letairis^{TM}, Volibris^{TM}) have subsequently been launched for the same indication. Bosentan has also been approved for indications in the field of digital ulcer disease. Atrasentan, a selective ET_A antagonist, is in pre-registration targeting the treatment of cancer.

Table 5.1 ENDOTHELIN RECEPTOR ANTAGONISTS THAT HAVE REACHED MARKET, OR ARE CURRENTLY IN PRE-REGISTRATION OR PHASE III

Name (Company)	Status	Selectivity	Indication; route
Bosentan (Roche)	Launched	Dual ET_A/ET_B	PAH, digital ulcer disease; PO
Sitaxsentan (Encysive)	Launched	ET_A selective	PAH; PO
Ambrisentan (Gilead)	Launched	ET_A selective	PAH; PO
Atrasentan (Abbott)	Pre-registration	ET_A selective	Cancer (prostate, solid tumour); PO
Tezosentan (Actelion)	Phase III	Dual ET_A/ET_B	Right ventricular failure; IV
Actelion-1 (Actelion)	Phase III	Dual ET_A/ET_B	PAH, Cardiovascular disease; PO
Zibotentan (AstraZeneca)	Phase III	ET_A selective	Cancer (prostate, solid tumour); PO
Darusentan (Gilead)	Phase III	ET_A selective	Hypertension; PO
Avosentan (Speedel)	Phase III	ET_A selective	Diabetic nephropathy; PO
Clazosentan (Actelion)	Phase III	ET_A selective	Brain haemorrhage; IV

The question as to whether selective ET_A receptor antagonism, as opposed to dual ET_A/ET_B antagonism, is preferable clinically remains a subject of debate and scientific interest with no clear definitive outcome yet reached [23]. In the field of PAH, endothelin receptor antagonism has emerged as a beneficial treatment option with both dual and ET_A selective receptor antagonists providing clinical benefit [24, 25]. Although the ET_A receptor population dominates in the pulmonary vasculature, and blockade clearly appears to be beneficial, the case for ET_B receptor antagonism in PAH remains unproven. A recent STRIDE-2 trial for PAH compared a dual agent (bosentan) with an ET_A selective agent (sitaxsentan) and a similar benefit was seen with both drugs [26]. However, the comparatively low dose of bosentan (125 mg) and the open arm nature of the study made assessment of the degree of ET_B receptor blockade difficult. Additionally, support for a protective role for the ET_B receptor is provided by transgenic animal studies [27].

ET-1 levels are increased in the vascular smooth muscle cells of hypertensive patients and the ET_A selective and dual endothelin receptor antagonists, darusentan and bosentan have both shown blood pressure lowering capacity in hypertensive humans [23, 28]. Although no clear verdict has yet emerged as to whether selective ET_A or dual receptor antagonism is preferable, a rise in mean arterial pressure in rats has been observed upon dosing with an ET_B selective agent [29]. In other cardiovascular indications, data suggest potential benefit of ET receptor blockade in atherosclerosis and chronic kidney disease with a possible leaning towards selective ET_A blockade. Long-term patient studies for chronic heart failure with endothelin antagonists have shown no clear benefit to date [23]. The ET_A selective agent clazosentan has been shown to prevent cerebral vasospasm following subarachnoid haemorrhage [30].

The endothelin peptides impart regulatory control over cellular processes important for normal cell functioning such as growth and survival. ET-1 acting primarily through the ET_A receptor has been implicated in the neoplastic growth of multiple cell types [31]. Clinical testing with the ET_A selective agent atrasentan has demonstrated clear potential for benefit in cancer therapy, notably in studies for prostate and lung cancer. Further studies with atrasentan, together with other agents, such as ZD-4054 (25), will more fully define the cancer therapeutic potential for this class of agent [31, 32]. In terms of other indications, the full potential of endothelin antagonists within the cardiovascular field continues to be explored [33], together with opportunities in other therapeutic areas such as pain [34].

SAFETY

The safety profiles of the endothelin receptor antagonists taken into late stage development thus far have generally been acceptable [35]. Side effects are, however, relatively common and clinical symptoms commonly observed include headache, peripheral oedema, nausea and nasal congestion. Reversible liver toxicity has been seen at high doses both with dual ET_A/ET_B antagonists and with ET_A selective agents, manifesting as an asymptomatic increase in liver enzymes. This has led to a requirement for liver function tests in patients at onset of treatment and at monthly intervals. Whilst these side effects have proven manageable, the upper safe clinical dosing limit has been restricted for some of the lead agents. A number of endothelin receptor antagonists are substrates and inducers of enzymes in the cytochrome P450 system. For instance, some sulfonamide-based antagonists can inhibit or induce CYP2C9, indicating the potential for drug–drug interactions with agents that are principally metabolised by this enzyme, such as warfarin [35]. Similar effects have been seen across sulfonamide and carboxylic acid antagonist classes with CYP3A4, and drug–drug interactions have been described with the strong CYP3A4 inhibitor ketoconazole [36]. Additionally, the endothelin system is implicated in foetal development, appearing to play a crucial role in craniofacial and cardiovascular development, based on studies with ET-1 or ECE-1 deficient mice. Hence all endothelin receptor antagonists are teratogenic and contraindicated in pregnancy.

MEDICINAL CHEMISTRY

GENERAL OVERVIEW

The endothelin ET_A and ET_B receptors fall into the peptidic GPCR family, and recent analyses support the hypothesis that this receptor family generally represents a challenging drug target [37, 38]. A study of documented endothelin antagonists tends to support this view with only three launched agents to date from some 500 preclinical compounds documented as targeting the ET_A receptor (Table 5.2). (These data originate from the following databases: Prous Integrity, IDDB, the CEREP Bioprint database, Inpharmatica's Drugstore and Drugbank.)

Where a structure has been disclosed for compounds documented as targeting either the ET_A or ET_B receptor (These data originate from the following databases: Prous Integrity, IDDB, the CEREP Bioprint database, Inpharmatica's Drugstore and Drugbank), further analysis of physical properties indicates that whilst the discovery of preclinical agents with desirable drug-like properties is achievable, the majority of these compounds do not have ideal physical properties, so for instance, approximately 68% have at least one "rule of five" violation [39]. Given that endothelin antagonists routinely contain at least one acidic centre, physical properties influencing permeability are also likely to carry a high tariff. For instance, $c\log P$ and MWt are generally on the high side of recent averages for marketed drugs [40], and both H-bond acceptor and polar surface area values are also often high (Figure 5.3). Overall, the retention of good absorption/metabolism (ADME) properties during lead optimisation has proven challenging and appears to be a key reason for the low success rate of clinical candidates. As this section will highlight, the successful agents have generally derived from efficient small molecule starting points that in turn were mostly discovered from file screening of compound library collections. Antagonists will be discussed in terms of ligand efficiency (LE), where the binding free energy is divided by the number of heavy atoms [41]

Table 5.2 DOCUMENTED ENDOTHELIN ET_A RECEPTOR ANTAGONISTS AND STATUS REACHED

Receptor	Discovery	Phase I	Phase II	Phase III	Registration	Launched
Endothelin-1 receptor (ET_A)	478	8	4	6	1	3

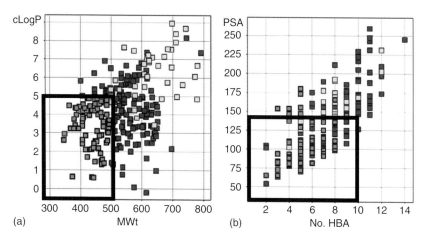

Fig. 5.3 Documented ET_A and ETB antagonists with disclosed structures. (a) $c\log P$ vs MWt, only 32% meet rule of five criteria, (b) PSA vs number of H-bond acceptor atoms, 60% meet property criteria when polar surface area greater than 140 taken as the cut-off point.

to indicate efficiency of binding relative to mass. Binding free energy was calculated from $-RT \ln K_{eq}$ using ET_A K_i or, where not available, ET_A IC_{50} as the surrogate for dissociation constant, and RT equal to 1.4 [42]. Ligand-lipophilicity efficiency (LLE) is another important concept that will be discussed, and reflects the minimally acceptable lipophilicity per unit of *in vitro* potency, giving an indication of specific hydrophobic binding requirement [43]. LLE was calculated as ($-\ln K_{eq}$ minus $c \log P$) with the dissociation constant defined as for LE. Structural biology application to antagonist design strategy, in terms of ET-1 mimicking or pharmacophore models, will also be highlighted in the various chemical classes of antagonist that have emerged.

PEPTIDE ANTAGONISTS

The first reported endothelin receptor antagonists were peptide based and discovered either from compound library screening and subsequent optimisation or from rational design techniques based on the natural ET-1 agonist. Notable examples of the former are BQ-123 (1) and FR 139317 (2). BQ-123 (1) arose initially from a series of D-Glu-Ala-based cyclic pentapeptides isolated from *Streptomyces* sp. 7338. Structure-activity relationship (SAR) studies followed and in particular, replacement of the

D-Glu subunit by D-Asp and substitution of Pro for the Ala unit both gave potency enhancement. Combination of these SAR ploys gave rise to BQ-123 (1), a compound with low nanomolar potency for the ET_A receptor [44] and with efficacy in animal models of hypertension and acute renal failure [45]. The screening of natural products also gave rise to FR 139317, a potent linear tripeptide that is also ET_A selective [46]. The dual ET_A/ET_B hexapeptide PD-145065 (3), on the other hand, arose from antagonist design based on the C-terminal unit of ET-1 [47]. The cyclic hexapeptide based TAK-044 (4) is another potent ET_A selective agent that was isolated from a culture broth of Streptomyces misakiensis and identified in subsequent SAR studies [48]. These compounds have mainly found use as research tools for the early exploration of the pharmacology of the endothelin system. TAK-044 (4) progressed to clinical evaluation for hypertension and improving ventricular function [49, 50], but development was hampered by low oral bioavailability and appears to have been discontinued, whilst BQ-123 is in Phase II for myocardial infarction as CGS-26343. (These data originate from the following databases: Prous Integrity, IDDB, the CEREP Bioprint database, Inpharmatica's Drugstore and Drugbank.) In physical property terms, these compounds all display rule of five violations that point to development challenges. In particular, high molecular weight together with high H-bond donor and acceptor values have generally made for low oral bioavailability and restricted use to parenteral administration (Table 5.3). In terms of LE, these agents generally score on the low side with FR-139317 (2) setting the benchmark at 0.30. The calculation of LLE for the above peptide agents reveals some reasonable benchmark scores in the 5–6 region when polar amino acid sub-units were incorporated, (FR-139317, 2; TAK-044, 4) however, polar surface area (PSA) is high for the latter compound.

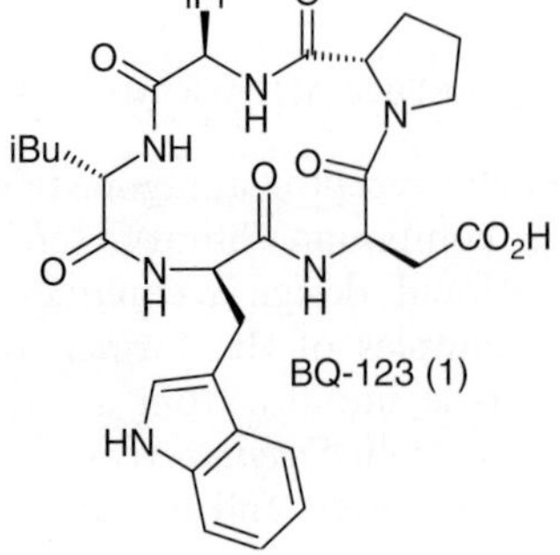

BQ-123 (1)

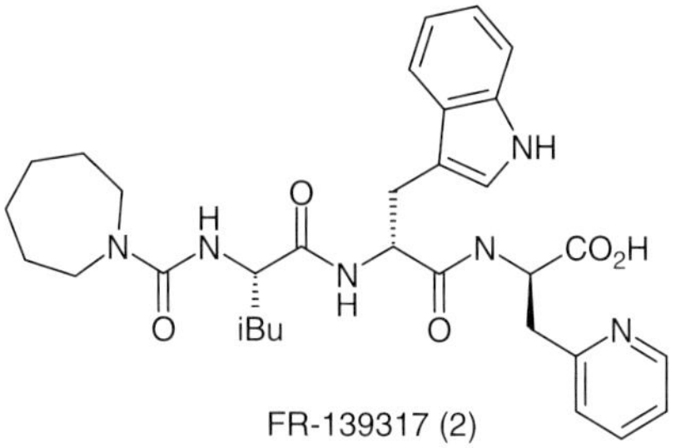

FR-139317 (2)

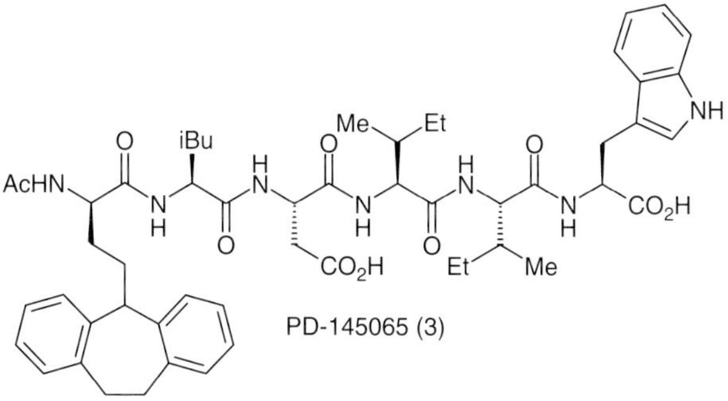

PD-145065 (3)

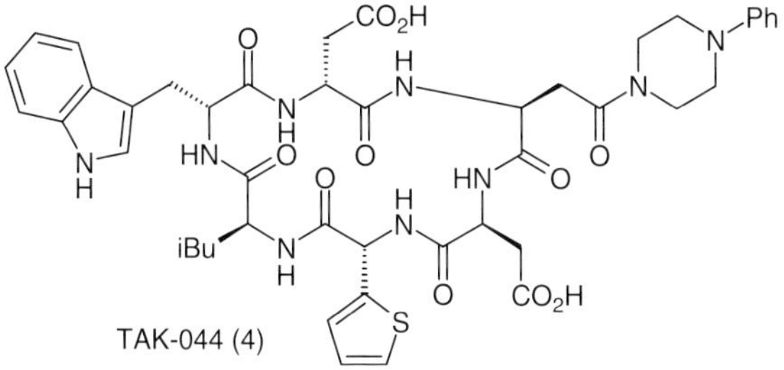

TAK-044 (4)

Table 5.3 PROPERTIES OF REPRESENTATIVE PEPTIDE-BASED ENDOTHELIN RECEPTOR ANTAGONISTS

ID	ET_A IC_{50} (nM)	ET_B IC_{50} (nM)	ET_A Sel.	MW	c log P	LE ET_A	LLE ET_A	HBA	HBD	PSA	Rof5 fails
BQ-123 (1)	22	18,000	818	611	4.1	0.24	3.56	13	6	190	3
FR-139317 (2)	0.53	4050	7,641	591	3.72	0.30	5.56	11	5	157	2
PD-145065 (3)	4	15	3.75	978	8.07	0.17	0.33	17	9	265	4
TAK-044 (4)	0.08	120	1,463	928	4.16	0.21	5.93	20	9	289	3

SULFONAMIDE ANTAGONISTS

Pyrimidine sulfonamides

Bosentan (Tracleer™) (5) was the first endothelin antagonist to achieve regulatory approval and is currently indicated for the treatment of PAH and digital ulcer disease. The programme originated from partial screening of the Roche chemical compound library and identification of the pyrimidine sulfonamide (6) as a micromolar level ET_A receptor hit. SAR studies on (6) indicated that an electron rich 5-position aryl substituent was important for potency and that introduction of a 2-substituent gave a further potency increase that was compatible with good functional activity [51].

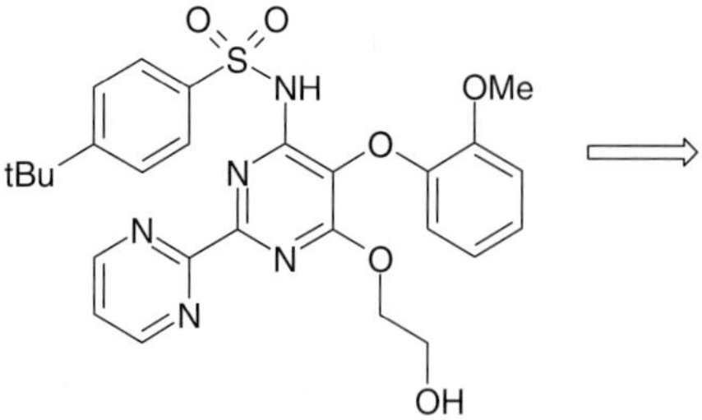

Bosentan (5)
ET_A K_i (Hu) 4.7nM
LE 0.30
pA_2 (rat) 7.4
ET_B K_i (Hu) 95nM

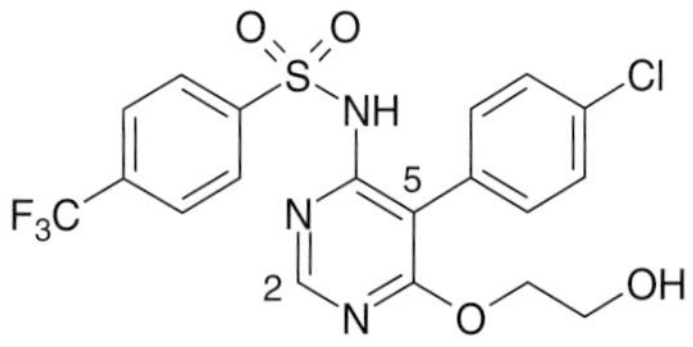

(6)
ET_A IC_{50} (Sf9) 18µM
LE 0.21
pA_2 (rat) 5.8

Bosentan (5) displays a mixed endothelin antagonist binding profile against human receptors, and has a functional pA_2 of 7.4 on rat aorta. Following activity in animal models, such as hypertensive rats, from both oral and i.v. dosing, bosentan (5) proceeded into development and represented a step forward in terms of physicochemical properties relative to the early peptide based antagonists. H-bond donor and acceptor counts (Table 5.4) are significantly diminished. These properties translate to bosentan having an oral bioavailability of 50% in man and a half-life of approximately 5 h, allowing twice-daily dosing. Pharmacokinetics (PK) are dose proportional up to 500 mg/kg/day (multiple doses). Clearance for a 250 mg human dose is 2 ml/min/kg, and steady state volume is 0.25 L/Kg in keeping with an acidic molecule confined to plasma [52]. Metabolism occurs in the liver *via* cytochrome P450 enzymes, notably CYP3A4 and CYP2C9. The compound is also an inducer of these enzymes, raising the possibility of drug–drug interactions, but to date most of the effects

Table 5.4 PROPERTIES OF REPRESENTATIVE PYRIMIDINE SULPHONAMIDE-BASED ENDOTHELIN RECEPTOR ANTAGONISTS

ID	$ET_A IC_{50}$ (nM)	$ET_B IC_{50}$ (nM)	ET_A Sel.	MW	c log P	LE ET_A	LLE ET_A	HBA	HBD	PSA	Ro f5 fails
Bosentan (5)	4.7[a]	95[a]	20	552	4.17	0.30	4.12	11	2	146	2
(6)	18000			474	4.03	0.21	0.71	7	2	101	0
Tezosentan (7)	0.3[a]	10–21[a]	33–70	606	3.34	0.31	6.18	15	3	200	2
Clazosentan (8)	1.9[a]	1930[a]	1016	578	2.41	0.30	6.31	15	3	200	2
Avosentan (9)	<50			480	3.58	>0.3	>3.7	10	1	125	0
TA-0201 (10)	0.015[a]	41[a]	2733	613	6.51	0.39	4.31	8	2	111	2
YM-598 (11)	3.1	1200	387	492	3.7	0.34	4.81	10	1	125	0
Shionogi (12)	140	0.9	155 (ET_B)	667	7.42	0.26 (ET_B)	1.63 (ET_B)	10	2	129	2
K-8794 (13)				683	5.76			12	2	155	3

[a] K_i rather than IC_{50}.

described have been modest. Exceptions are interactions with glyburide and cyclosporine, and there is a need for caution with CYP2C9 and CYP3A4 inhibitors [53]. Bosentan (5) also causes reversible elevation of liver transaminase levels in up to 11% of patients. Overall, bosentan (5) has fulfilled an invaluable pathfinding role in the endothelin field. In PAH, an indication that is very difficult to treat, bosentan (5) has proven effective and durable with favourable effects on exercise capacity, functional class and hemodynamics. Bosentan (5) is also approved for the reduction of new digital ulcers in patients with systemic sclerosis and ongoing digital ulcer disease [54], and development is going ahead to extend the indication to both pulmonary fibrosis and Raynaud's disease secondary to scleroderma [55].

A number of other significant antagonists have arisen from the bosentan programme. The dual endothelin receptor agent tezosentan (7) progressed to Phase III development as an i.v. agent for right ventricular failure relating to bypass surgery [56], but recent trial results have not supported continuation. Clazosentan (8), an ET_A selective agent achieved as a result of cutting back the sulfonamide group, is also in Phase III as an i.v. agent for subarachnoid hemorrhage [57]. Subsequent efforts looked to attain optimal physicochemistry with lower mass structures. Compounds such as the ET_A selective agent avosentan (SPP-301, 9) are representative of the advances that were made [51]. Avosentan is in Phase III for diabetic nephropathy, and appears generally well tolerated with peripheral oedema the main adverse effect reported. Whilst no definitive activity data are available, the minimised structure and improved physicochemistry have resulted in linear oral human pharmacokinetics up to a 100 mg dose [58].

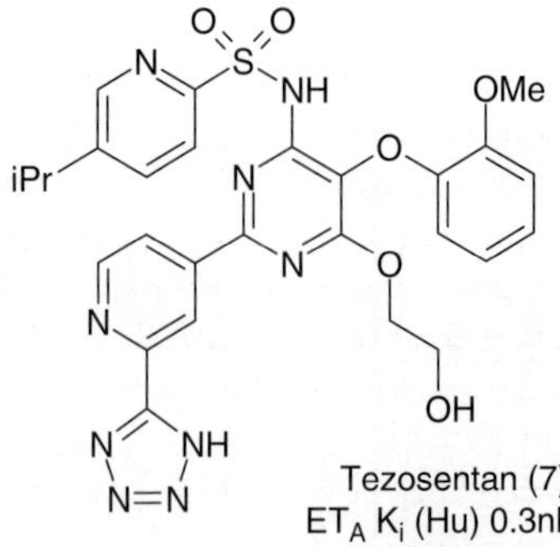

Tezosentan (7)
ET_A K_i (Hu) 0.3nM
LE 0.31
pA_2 (rat) 9.5
ET_B K_i (Hu) 10-21nM

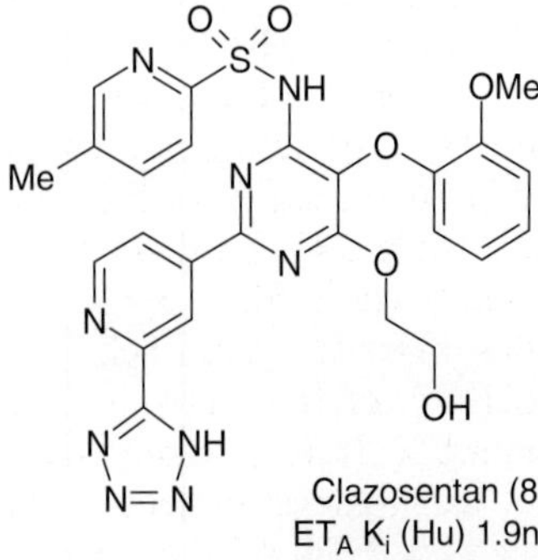

Clazosentan (8)
ET_A K_i (Hu) 1.9nM
LE 0.3
pA_2 (rat) 9.4
ET_B K_i (Hu) 1930nM

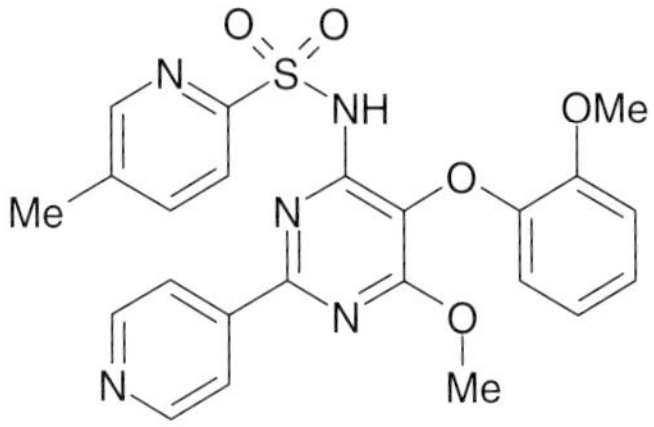

Avosentan (9)
ET_A IC_{50} (Hu) <50nM
pA_2 (rat) >8

Several other prominent endothelin antagonists have subsequently emerged from this class. The ET_A selective Tanabe compound TA-0201 (10) has picomolar level activity and features an extended 6-position alkoxy group and no 2-substituent [51, 59]. Tanabe's rationale was that the transmembrane-2 region was important for ET_A selectivity and targeted this with the extended 6-substituent. A hydroxyl group was also introduced onto the sulfonamide to give aqueous solubility. TA-0201 (10) shows bioavailability of 83% in rat and a plasma half-life of 0.9 h. TA-0201 (10) has shown activity *in vivo* in canine models for CHF and PAH but no development has been reported. Long duration of action has been achieved *in vivo*, and this may be due to an active metabolite with an extended half-life [60]. The Yamanouchi compound YM-598 (nebentan, 11) is an ET_A selective agent that is structurally similar to bosentan but features a minimised ether substituent and modified sulfonamide group to realise a ligand efficient ET_A agent [61]. The styryl sulfonamide motif appears isosteric with the bosentan arylsulfonamide. YM-598 (11) appears to have good oral bioavailability based on achieving inhibition of the increase in blood pressure affected by Big ET-1 in rat. YM-598 (11) progressed into Phase II for prostate cancer and associated pain but development has been discontinued. TA-0201 (10), YM-598 (11) and avosentan (9) all represent advances in terms of LE for this pyrimidine class, with the latter two molecules also meeting rule of five criteria (Table 5.4). LLE score is generally above five for the pyrimidine sulfonamide class. Shionogi scientists have disclosed ET_B selective agents from this class. The crucial change to achieve the selectivity switch has been the extension of the 6-position ether substituent with amidic substituents. Compound (12) is a representative example with picomolar level ET_B activity, but no development has been reported [51]. K-8974 (13) is a related orally active ET_B selective antagonist reported by Kowa. (These data originate from the following databases: Prous Integrity, IDDB, the CEREP Bioprint database, Inpharmatica's Drugstore and Drugbank.)

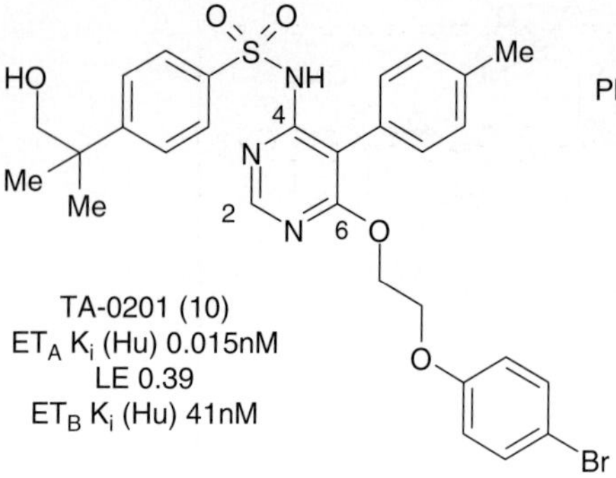

TA-0201 (10)
ET_A K_i (Hu) 0.015nM
LE 0.39
ET_B K_i (Hu) 41nM

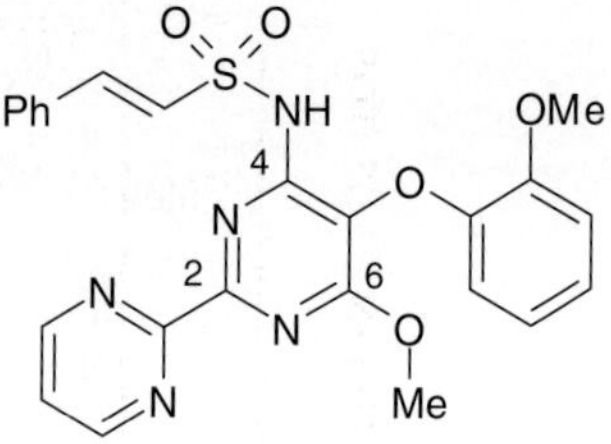

YM-598 (11)
ET_A IC_{50} (Hu) 3.1nM
LE 0.34
ET_B IC_{50} (Hu) 1200nM

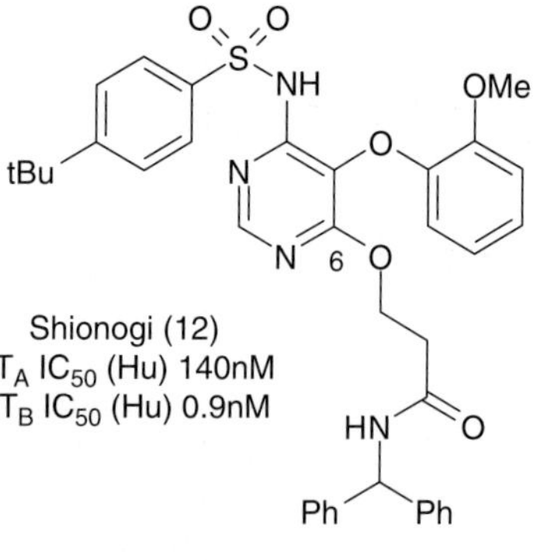

Shionogi (12)
ET_A IC_{50} (Hu) 140nM
ET_B IC_{50} (Hu) 0.9nM

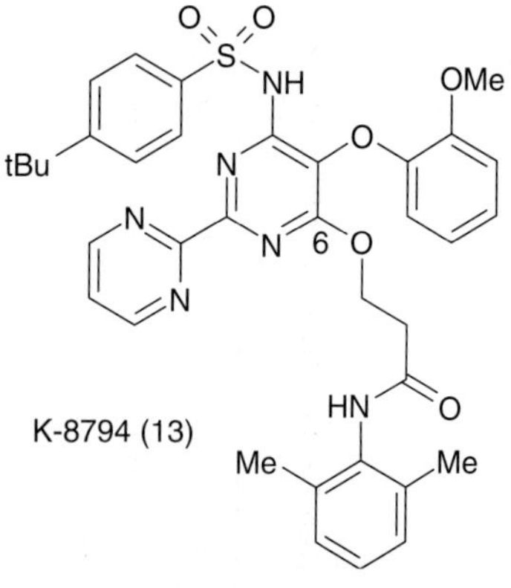

K-8794 (13)

Heteroaryl sulfonamides and related sulfamides

The heteroaryl sulfonamide class has also been a successful chemical area for endothelin antagonists realising one launched agent, sitaxsentan (Thelin™, 21) and several others that remain in development.

ET_A selective antagonists from Bristol-Myers Squibb originated from file screening of the in-house compound collection and yielded the small molecule isoxazole sulfonamide (14) as a ligand efficient starting point. Optimisation resulted in the identification of the clinical candidates BMS-182874 (15), BMS-193884 (16) and BMS-207940 (17), all of which represent highly ligand efficient molecules (Table 5.5). Key SAR aspects that gave rise

Table 5.5 PROPERTIES OF REPRESENTATIVE HETEROARYL SULPHONAMIDE-BASED ENDOTHELIN RECEPTOR ANTAGONISTS

ID	$ET_A\ IC_{50}$ (nM)	$ET_B\ IC_{50}$ (nM)	ET_A Sel.	MW	clogP	LE	LLE	HBA	HBD	PSA	Ro5 fails
(14)	780			267	0.22	0.48	5.89	6	3	98.2	0
BMS-182874 (15)	55[a]	$>2 \times 10^{5a}$	>3,636	345	2.56	0.42	4.66	6	1	75.4	0
BMS-193884 (16)	1.4[a]	1,900[a]	1,357	395	2.77	0.44	6.08	7	1	98.2	0
BMS-207940 (17)	0.01[a]	810[a]	81,000	537	3.14	0.41	7.86	9	1	119	1
PS-433540 (18)	9.3[a]			593	6.0	0.27	2.03	9	1	114	2
(19)	838			366	0.51	0.45	5.57	7	3	115	0
(20)	15			412	2.9	0.42	4.92	7	2	101	0
Sitaxsentan (21)	1.4	9,800	7,000	455	3.44	0.43	5.41	8	1	108	0
TBC-3711 (22)	0.08	35,300	$>4 \times 10^5$	448	2.58	0.47	7.52	8	2	118	0
(23)				393	4.16			7	1	84.4	0
ZD-1611 (24)				457	4.01			9	2	131	0
ZD-4054 (25)				424	2.09			10	1	133	0
UK-419106 (26)	38[a]	1,626[a]	43	416	2.87	0.36	4.55	8	1	101	0
UK-434643 (27)	8[a]	6,856[a]	857	488	1.41	0.33	6.89	10	1	122	0

[a] K_i rather than IC_{50}.

to the candidates were optimisation of the aryl group, initially to a naphthalene-based motif that improved potency more than 10 fold, as in BMS-182874 (15). Subsequent SAR studies led to biphenylsulfonamide derivatives with improved low nanomolar level binding affinity. Key to this process was identification of a 4′-oxazole motif that further enhanced potency together with improving metabolic stability, and which gave rise to BMS-193884 (16) [62]. Further SAR studies identified that a 3-amino-isoxazole displays significantly improved metabolic stability relative to the 5-regioisomer. Additionally, incorporation of a 2′-group into the biphenyl motif of (16), further improved potency, and after optimisation of the 2′-group for permeability and metabolic stability, BMS-207940 (17) was identified [63]. BMS-182874 (15) displayed both oral and i.v. activity in rats and attenuated blood pressure elevation, however, development was discontinued. In addition to being more potent, BMS-193884 (16) had an ADME profile (rat PK: half-life $(t_{1/2})$ 2 h, clearance (Cl) 2.6 ml/min/kg, steady state volume (V_{ss}) 0.08 L/kg, bioavailability 43% and human plasma protein binding (PPB) 99.3%) that encouraged progression to man. In Phase I clinical trials, BMS-193884 displayed 95% oral bioavailability and a half-life of 2 h and was well tolerated with headache the major adverse event. The compound progressed to Phase II for heart failure but no clear advantage over current therapies was displayed and development was subsequently discontinued [64]. For BMS-207940 (17), the neopentyl amide motif represented optimisation of the 2′-substituent for ADME properties (rat PK: $t_{1/2}$ 3.4 h, Cl 5.4 ml/min/kg, V_{ss} 1.0 L/kg, bioavailability 100%). In Phase I clinical trials BMS-207940 (17) had 80% oral bioavailability and an improved plasma elimination half-life of 15–20 h relative to (16) [64]. BMS-207940 (17) was positioned for the treatment of cardiac failure but no further development has been reported. In terms of their properties, this heteroaryl sulfonamide series set new standards in the endothelin antagonist field for LE, LLE and physical properties generally.

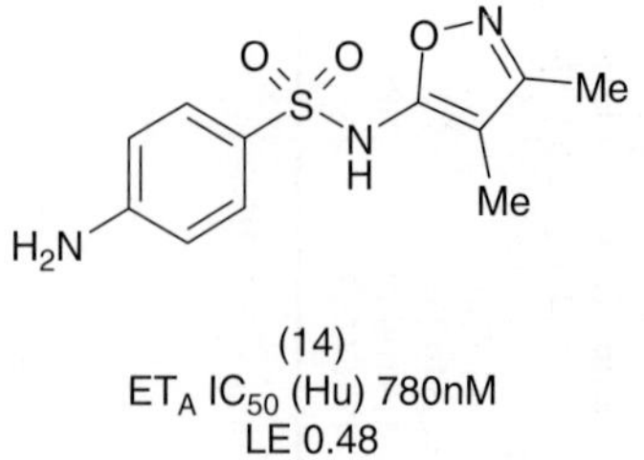

(14)
ET_A IC_{50} (Hu) 780nM
LE 0.48

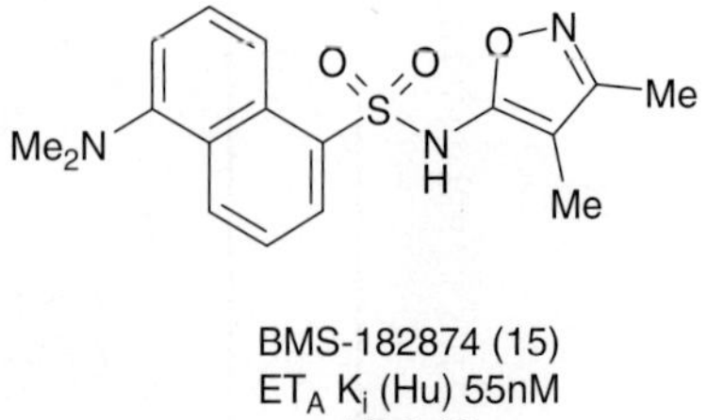

BMS-182874 (15)
ET_A K_i (Hu) 55nM
LE 0.42
ET_B K_i (rat) >200μM

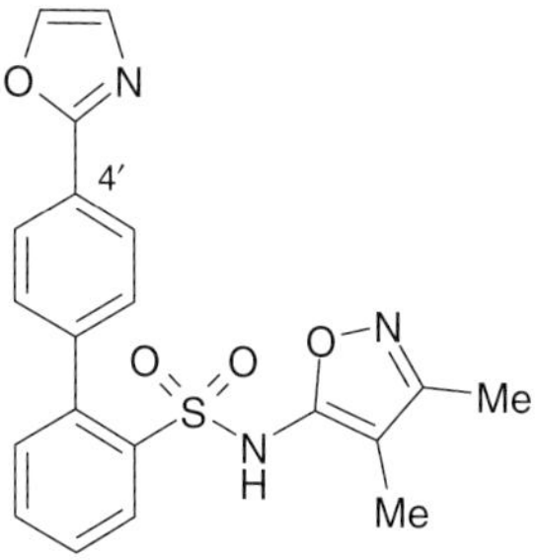

BMS-193884 (16)
ET$_A$ K$_i$ (Hu) 1.4nM
LE 0.44
ET$_B$ K$_i$ (Hu) 1.9μM

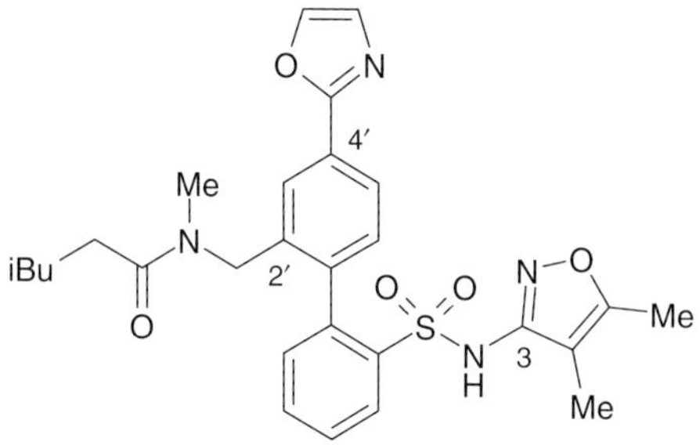

BMS-207940 (17)
ET$_A$ K$_i$ (Hu) 0.01nM
LE 0.41
ET$_B$ K$_i$ (Hu) 810nM

A further evolution of the BMS series was the discovery of dual antagonists of angiotensin II and endothelin A receptors. BMS-346567 (18) represents the optimisation of a research programme based on BMS-193884 (16) and the angiotensin antagonist losartan. The acidic sulfonamide function characteristic of the ET$_A$ selective agents served as the acidic group required for activity at both receptors, whilst tuning of the upper phenyl ring and attached substituents was able to satisfy both pharmacophore requirements [65, 66]. Initial leads suffered from CYP3A4 metabolism but optimisation gave BMS-346567 (18), which had good oral bioavailability. BMS-346567 (18) was subsequently licensed by Pharmacopeia, becoming PS-433540 and is currently in Phase II for the potential treatment of hypertension and diabetic neuropathy. ET$_A$ LE was compromised relative to related ET$_A$ selective agents to achieve the required dual profile (Table 5.5).

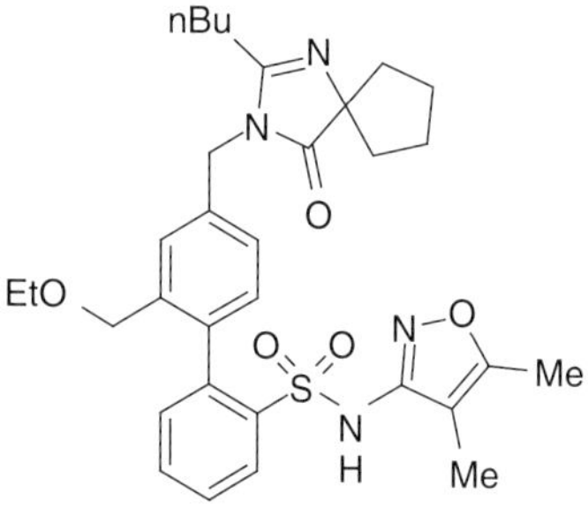

BMS-346567
PS-433540 (18)
ET$_A$ K$_i$ (Hu) 9.3nM
LE 0.27
AT$_1$ K$_i$ (Hu) 0.8nM

A structurally related series of ET_A selective isoxazole sulfonamides was also discovered by Immuno Pharmaceuticals/Texas Biotechnology. An initial thiazole-based lead (19) was developed into the 4-methyl benzamide (20), with the small para-substituent found to be important for potency. Metabolism of the amide group restricted *in vivo* activity and whilst ortho substitution of the phenyl ring adjacent to the amide motif improved metabolic stability, ultimately replacement of the amide by a keto-methylene was necessary to give a suitable *in vivo* profile. These enhancements led to sitaxsentan (21), which displayed both good *in vitro* and *in vivo* potency [67]. Sitaxsentan (21) has good pharmacokinetics (rat PK: $t_{1/2}$ 6.7 h, oral bioavailability 60%), and displays a long plasma half-life and high oral bioavailability in man (human PK for single 100 mg oral dose: $t_{1/2}$ 9 h, Cl 1.21 ml/min/kg, V_d 0.93 L/kg, human PPB 98.6%) [68]. Sitaxsentan (21) is safe and well tolerated, offers once-daily dosing [69] and has been marketed for the treatment of PAH in Europe and Australia by Encysive Pharmaceuticals. TBC-3711 (22) is a related derivative in which the issue of amide metabolism was addressed with 2,6-ortho substitution of the adjacent benzene and is in Phase II clinical development for CHF and hypertension. TBC-3711 has a $t_{1/2}$ in man of 6–7 h and oral bioavailability of >80% [70]. This series again displays high LE and good physical properties (Table 5.5).

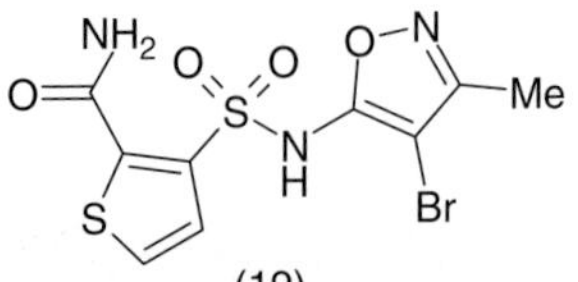

(19)
ET_A IC_{50} (Hu) 838nM
LE 0.45

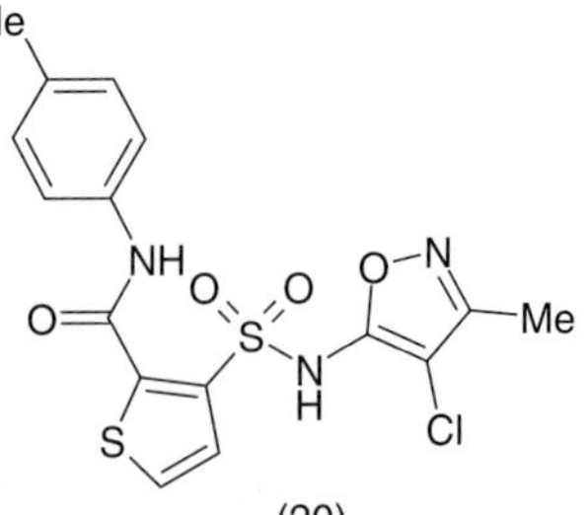

(20)
ET_A IC_{50} (Hu) 15nM
LE 0.42

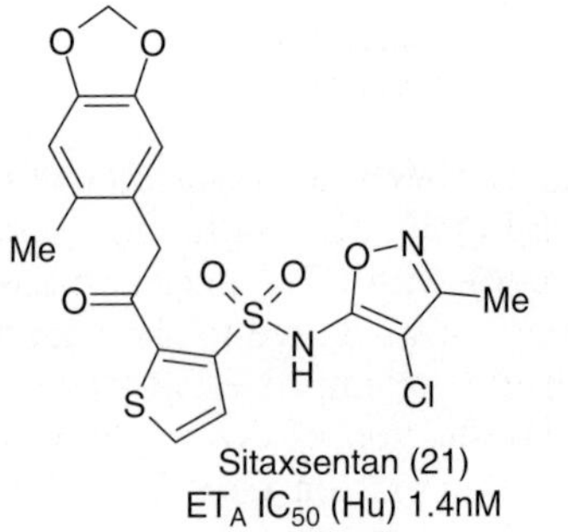

Sitaxsentan (21)
ET_A IC_{50} (Hu) 1.4nM
LE 0.43
ET_B IC_{50} (Hu) 9.8μM

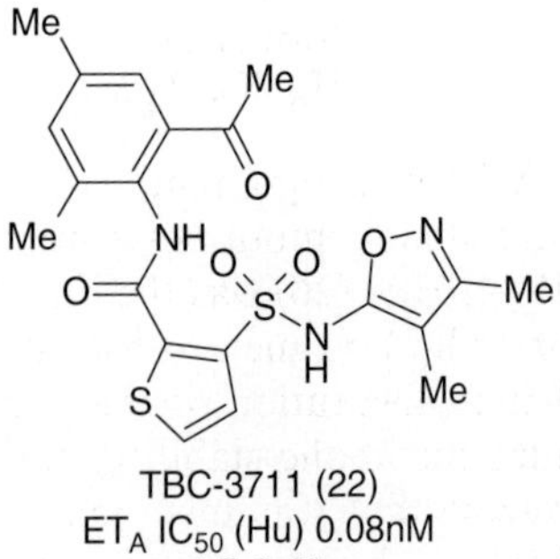

TBC-3711 (22)
ET_A IC_{50} (Hu) 0.08nM
LE 0.47
ET_B IC_{50} (Hu) 35.3μM

AstraZeneca has developed a related ET_A selective heteroaryl series of significance. A study of alternative heterocyclic replacements for the isoxazole of BMS-182874 gave rise to the pyrazine (23) that displayed oral activity *in vivo* as an initial lead. Optimisation saw evolution to biaryl systems ultimately leading to the 4′-carboxylic acid derivative ZD-1611 (24) [51]. ZD-1611 (24) was taken forward into development but was subsequently replaced by ZD-4054 (Zibotentan 25). This latter compound is currently in Phase III and is being targeted towards hormone-resistant prostate cancer. The compound is reported to be orally bioavailable, and Phase II clinical trial data indicates that a 15 mg/kg daily dose is well tolerated and that overall survival rates in metastatic prostate cancer patients were increased. (These data originate from the following databases: Prous Integrity, IDDB, the CEREP Bioprint database, Inpharmatica's Drugstore and Drugbank.) Based on the limited data available, good properties are again retained in this series.

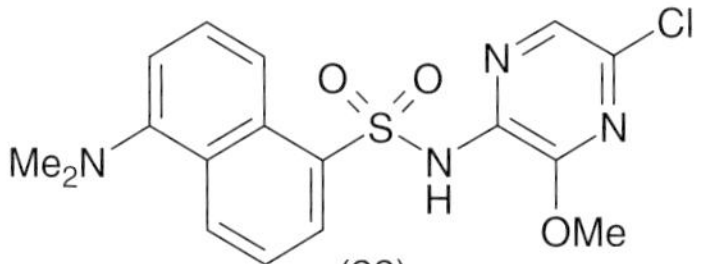

(23) ET_A pIC_{50} 8.0

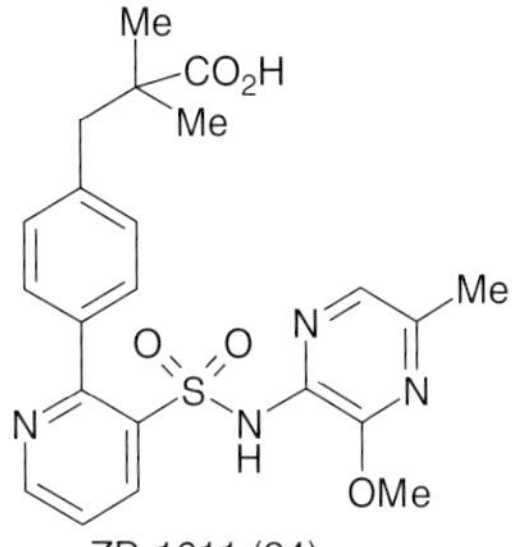

ZD-1611 (24) ET_A pIC_{50} 8.6

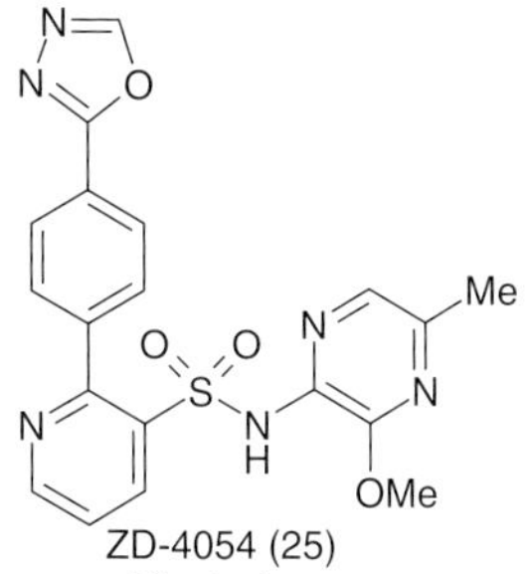

ZD-4054 (25) Zibotentan

A Pfizer programme studying low molecular weight acid isosteres of the sulfonamide motif, discovered a related ET_A selective sulfamide series. UK-419106 (26) was the initial lead, and overlay with ET-1 suggested scope for ortho and meta substitution of the phenyl ring relative to the oxazole. Ortho substitution proved most successful and both improved potency and aided metabolic stability, possibly by steric protection of the neighbouring oxazole that was metabolically vulnerable. Optimisation gave UK-434643 (27), but the compound still displayed high clearance in rat despite *in vitro* metabolic stability (human liver microsomal $t_{1/2} > 120$ min; rat PK: $t_{1/2}$ 0.4 h,

Cl 19 ml/min/kg, V_d 0.6 L/Kg) [71]. Compounds typically displayed a 40-fold activity fall-off between binding K_i and assessment of functional activity on human umbilical vein and no development has been reported from this series. In terms of properties, the series has lower LE. The acidic sulfamide motif carries additional atoms relative to sulfonamide, and the differing cyclopentyl/aryl groups relative to BMS 193884 (16) do not deliver sufficient potency enhancements to sustain a high (>0.4) LE.

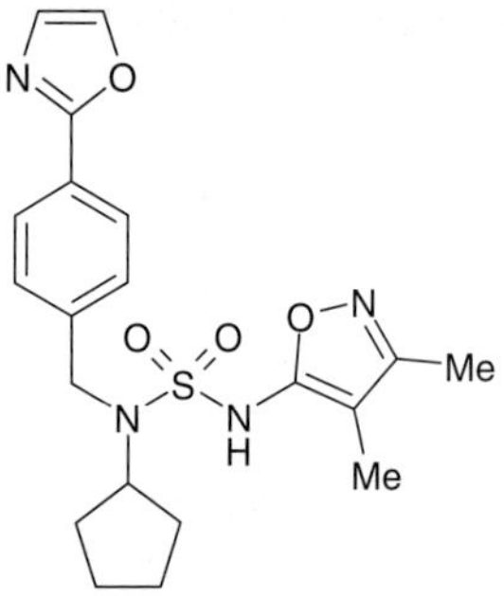

UK-419106 (26)
ET_A K_i (hu) 38nM
LE 0.36
ET_B K_i (hu) 1626nM

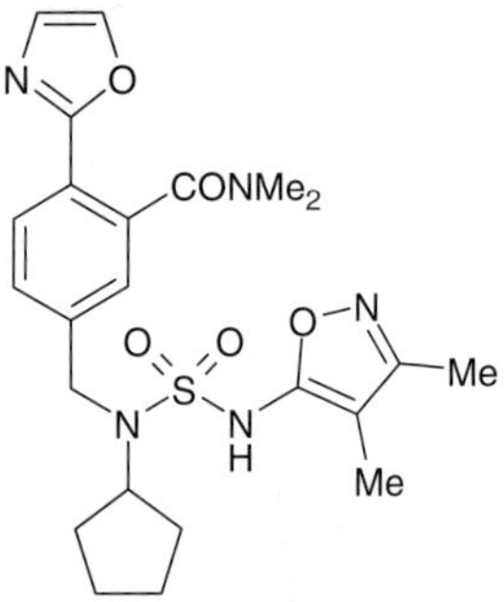

UK-434643 (27)
ET_A K_i (hu) 8nM
LE 0.33
ET_B K_i (hu) 6856nM

CARBOXYLIC ACID ANTAGONISTS

Diarylpropionic acid derivatives

Diarylpropionic acids represent another ligand efficient class of antagonist that has yielded clinical candidates and a marketed agent. In a programme initiated by BASF/Knoll, file screening of the BASF compound library gave rise to nanomolar range ET_A selective hits such as (28). Optimisation of these early leads revealed an SAR that indicated meta substitution of the phenyl rings was tolerated, whilst ortho- and para-groups diminished activity. Additionally, symmetrical analogues with directly attached phenyl rings retained activity and allowed structural simplification through loss of a chiral centre. Resolution of the remaining chiral centre, and establishment that (S) stereochemistry was preferable for ET_A activity, subsequently gave rise to the discovery of darusentan (29) [72]. Close-in modification of the beta-methoxy group (nPr, Ph, H) diminished activity, whilst addition of methylene spacers leading to an aryl group cap gave larger, more balanced antagonists (30) [73]. This ploy did not yield viable candidates, however, and further success was gained by reverting to the simpler darusentan-like structure. The SAR around the pyrimidine group was fairly restrictive. However, akin to the AstraZeneca series, methyl substitution was favourable and this ploy led to a second ET_A selective agent, ambrisentan (31). Gilead Sciences have developed and launched ambrisentan (31) for the treatment of PAH [74], and darusentan (29) is in Phase III development for hypertension [28]. Both agents are reported to be highly bioavailable and to have a long duration of action. These good PK properties are in keeping with their high LE values and good physicochemistry akin to the heteroaryl sulfonamides (Table 5.6).

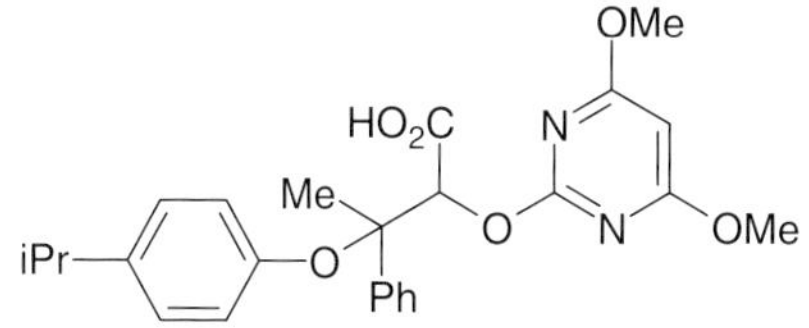

(28)
ET_A K_i 250nM
LE 0.28
ET_B K_i 4700nM

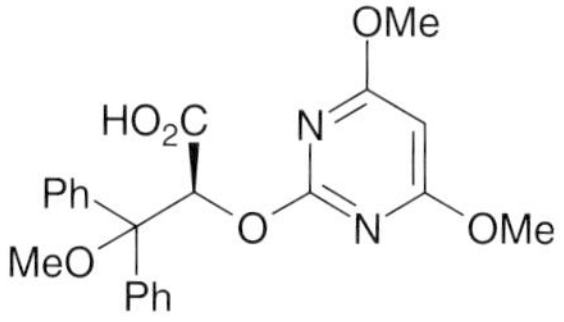

Darusentan (29)
ET_A K_i (Hu) 1.5nM
LE 0.41
Approx 160x Sel vs ET_B

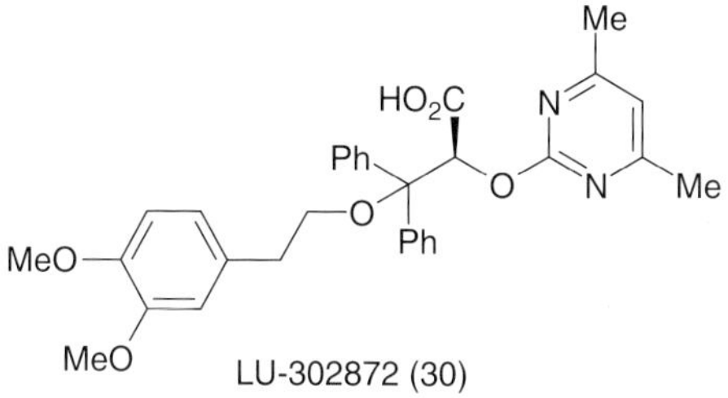

LU-302872 (30)
ET_A K_i 2.15nM
LE 0.31
ET_B K_i 4.75nM

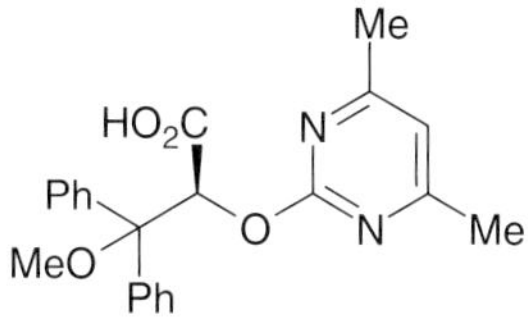

Ambrisentan (31)
ET_A K_i (Hu) 0.63nM
LE 0.46
ET_B K_i (Hu) 48.7nM

Indane-2-carboxylic acids

Indane-2-carboxylic acids represent another chemical area that has yielded clinical candidates. Researchers at SmithKline Beecham initially discovered an unstable indene from file screening (32). Optimisation of this structure led to SB-209670 (33) which reached Phase II for acute renal failure, but was subsequently discontinued. Evolution of the i.v. agent SB-209670 (33) gave rise to the closely related enrasentan (34) [51, 75, 76]. Enrasentan is orally

Table 5.6 PROPERTIES OF REPRESENTATIVE CARBOXYLIC ACID-BASED ENDOTHELIN RECEPTOR ANTAGONISTS

ID	$ET_A\ IC_{50}\ (nM)$	$ET_B\ IC_{50}\ (nM)$	ET_A Sel.	MW	clogP	LE	LLE	HBA	HBD	PSA	Ro5 fails
(28)	250[a]	4,700[a]	19	452	6.31	0.28	0.29	8	1	100	1
Darusentan (29)	1.5[a]		160	410	4.22	0.41	4.60	8	1	100	0
LU-302872 (30)	2.15[a]	4.75[a]	2	529	5.37	0.31	3.30	8		100	2
Ambrisentan (31)	0.63[a]	48.7[a]	77	378	3.75	0.46	5.45	6	1	81.5	0
(32)	3,700[a]			312	5.24	0.32	0.19	2		37.3	1
SB-209670 (33)	0.43[a]	14.7[a]	34	521	4.74	0.35	4.63	9	2	121	1
Enrasentan (34)	1.1[a]	111[a]	101	507	4.58	0.34	4.38	8	2	104	1
J-104132 (35)	0.034[a]	0.10[a]	3	532	5.4	0.38	5.07	8	2	115	2
PD-012527 (36)	430	27,000	63	421	5.09	0.30	1.28	5		65	1
PD-156707 (37)	0.31	417	1,345	507	4.24	0.36	4.27	9	1	110	1
PD-180988 (38)	0.46	2,200	4,783	517	5.44	0.36	3.90	7	1	93.1	2
Atrasentan (39)	0.0034[a]	63[a]	18,529	511	3.66	0.43	7.81	8	1	88.5	1
(40)	2.0[a]	60[a]	30	547	3.32	0.32	5.38	9		106	1
A-192621 (41)		4.5[a]		559	3.39	0.29 (ET_B)	3.96	8	2	97.3	1
ABT-546 (42)	0.46[a]	13,000[a]	28,261	533	5.61	0.34	3.73	8	1	88.5	2

[a] K_i, rather than IC_{50}.

active but has a short human half-life (rat PK: $t_{1/2}$ 3.3 h, bioavailability 66%). Development proceeded to Phase II for cardiac failure and PAH but has subsequently been discontinued. A structurally similar Banyu compound J-104132 (35) used a pyridine-cyclopentane core, and oral activity was seen (rat PK: 1 mg/kg: Cl 18 ml/min/kg, V_d 1.0 L/Kg, oral bioavailability 44%) [77]. The compound was taken into Phase IIa trials for heart failure, but no development has subsequently been reported. As a class, these agents are less ligand efficient than the successful diaryl carboxylic acids and contain rule of five violations. (Table 5.6)

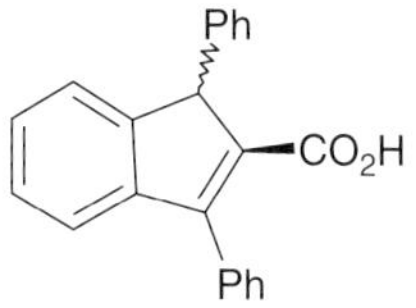

(32)
ET_A K_i 3.7μM
LE 0.32

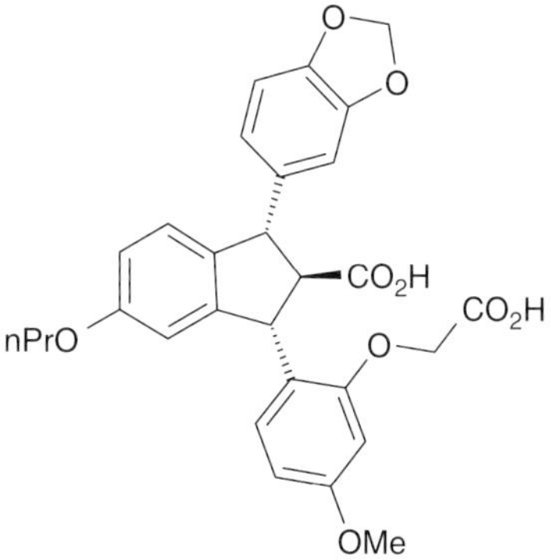

SB-209670 (33)
ET_A K_i (Hu) 0.43nM
LE 0.35
ET_B K_i (Hu) 14.7nM

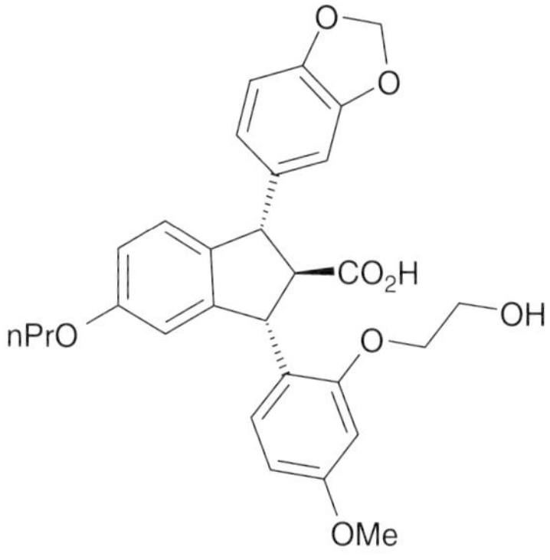

Enrasentan (34)
ET_A (Hu) K_i 1.1nM
LE 0.34
ET_B (Hu) K_i 111nM

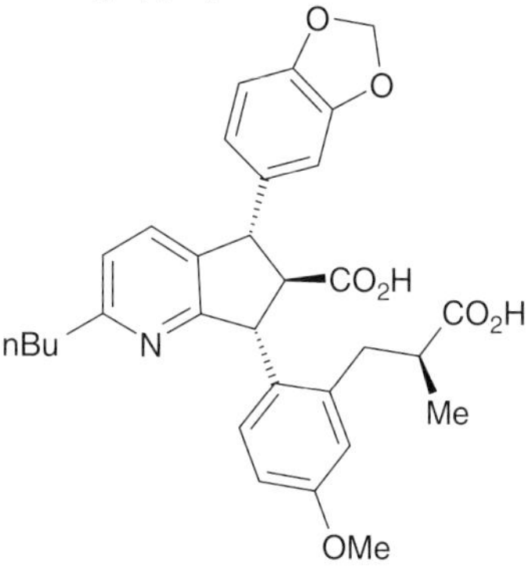

J-104132 (35)
ET_A K_i (Hu) 0.034nM
LE 0.38
ET_B K_i (Hu) 0.104nM

Butenolides and benzothiazine-based acids

A Parke-Davis compound library-screening programme identified PD-012527 (36) as an initial hit. Optimisation using a Topliss "decision tree" approach led to the potent PD-156707 (37) [78]. The open chain salt form of the butenolide motif is obtained by addition of base and this form was generally used for testing. PD-156707 displayed good oral bioavailability but had a short half-life (rat PK: $t_{1/2}$ 1 h, oral bioavailability 41%) and development was not pursued. Subsequently PD-180988 (CI-1034, 38) was identified from compound library screening as an alternative benzothiazine-based series with improved pharmacokinetics (rat PK: $t_{1/2}$ 2.2 h, oral bioavailbility 100%) [79] and progressed into Phase I, but no further development was reported. The LE and properties of this series were comparable to the indane carboxylic acids.

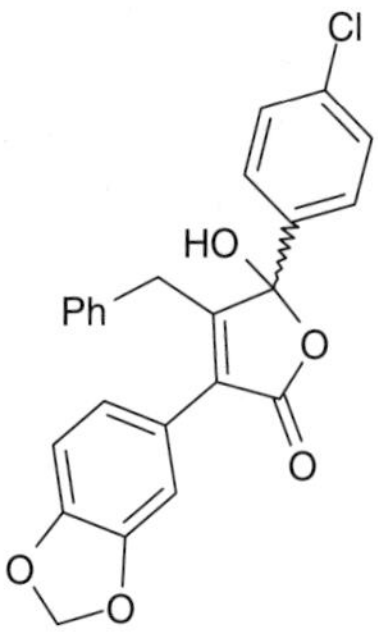

PD-012527 (36)
ET_A IC_{50} (Hu) 430 nM
LE 0.30
ET_B IC_{50} (Hu) 27μM

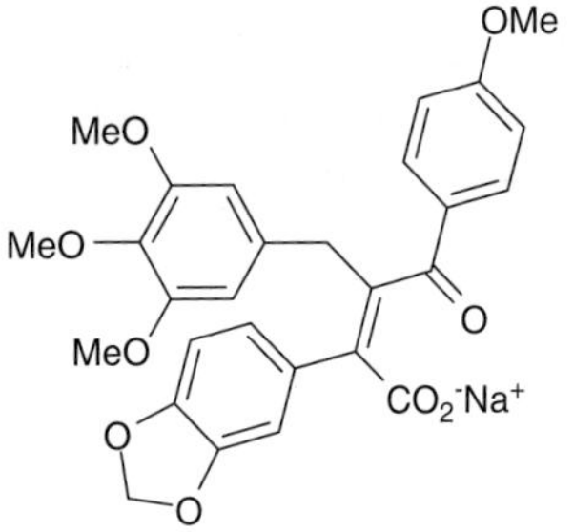

PD-156707 (37)
ET_A IC_{50} (Hu) 0.31nM
LE 0.36
ET_B IC_{50} (Hu) 417nM

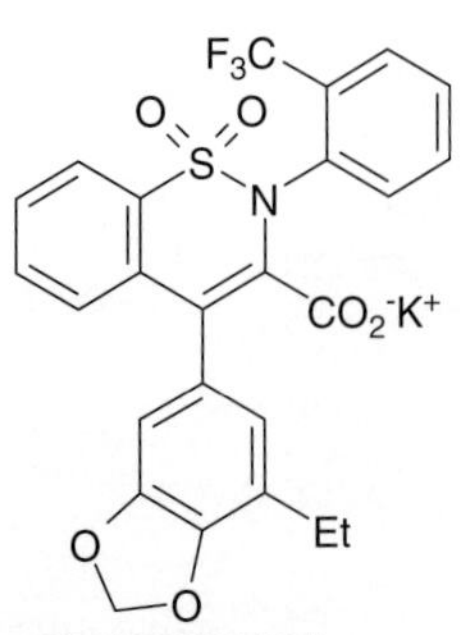

PD-180988 (38)
ET_A IC_{50} (Hu) 0.46nM
LE 0.36
ET_B IC_{50} (Hu) 2.2μM

Pyrrolidine-3-carboxylic acids

An Abbott programme used a pyrrolidine-3-carboxylic acid as its key element. An initial aza-indole lead that bore resemblance to SB-209670 (33) was evolved using a pharmacophore-based approach to yield the versatile pyrrolidine motif [80]. Optimisation gave rise to a series of potent endothelin antagonists that were either ET_A selective as in atrasentan (39), dual antagonists as in (40) through use of a sulfonamide side chain, or ET_B selective through conformational restriction of the amidic side chain (A-192621) (41) [29, 81]. Atrasentan shows good pharmacokinetics (human PK: $t_{1/2}$ 20–25 h, Cl 5–6.4 ml/min/kg, V_d 6 L/Kg [82]) and is currently in pre-registration for the treatment of cancer, whilst (41) was used as a research tool and produced a 25 mmHg rise in mean arterial blood pressure in rats, a potential pointer to the benefit of ET_A dominated agents. An ET_A selective backup to atrasentan (39), ABT-546 (42) has also been reported [83]. This agent features an additional methoxy group on the benzodioxolane ring that adds hydrophilicity, but no development has been documented. The successful agent from this class, atrasentan (39), is more ligand efficient than the other leads from this and related fields, and its exquisite potency, as reflected in a high LLE value, is probably a key factor in enabling progression.

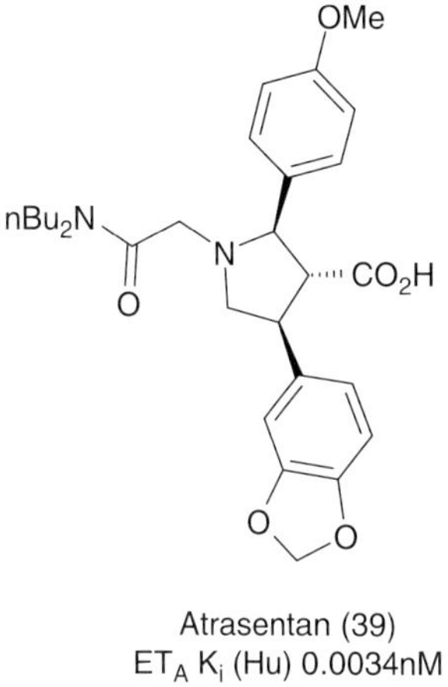

Atrasentan (39)
ET_A K_i (Hu) 0.0034nM
LE 0.43
ET_B K_i (Hu) 63nM

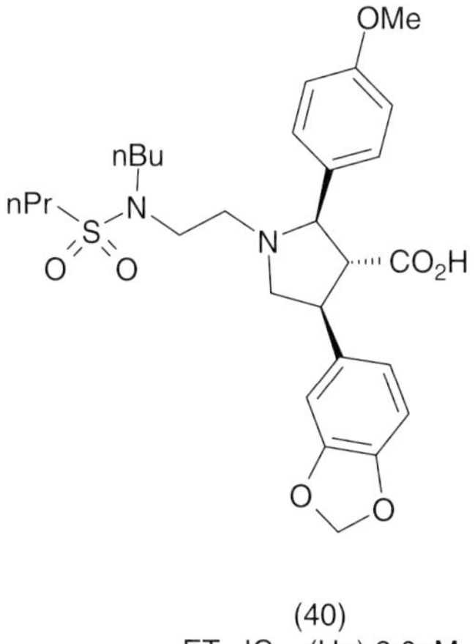

(40)
ET_A IC_{50} (Hu) 2.0nM
LE 0.32
ET_B IC_{50} (Hu) 60nM

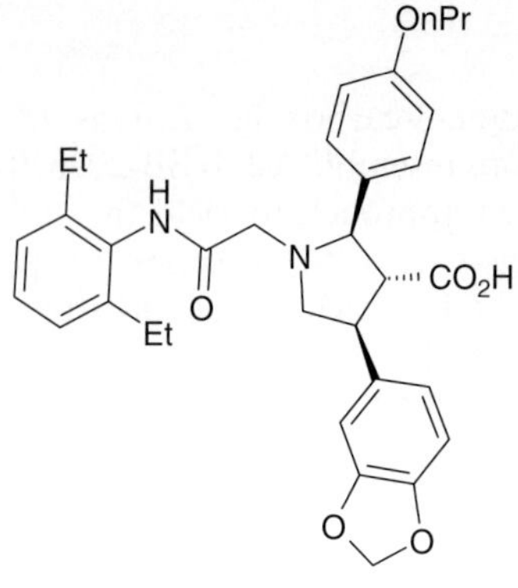

A-192621 (41)
ET$_B$ K$_i$ (Hu) 4.5nM
LE 0.29

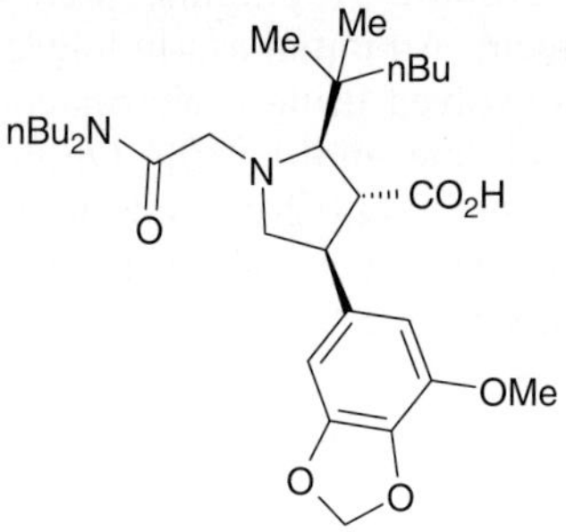

ABT-546 (42)
ET$_A$ K$_i$ (Hu) 0.46nM
LE 0.34
ET$_B$ K$_i$ (Hu) 13μM

N-ACYLSULFONAMIDE ANTAGONISTS

Screening of a selected library of compounds from their angiotensin II antagonist programme led Merck to discover glycolic acid based systems such as (43). Generally such systems appear to have suffered from metabolic vulnerability and more success has been achieved with related N-acylsulfonamide derivatives, notably L-754142 (44) [51]. However, no development has been reported. Applying principles of conformational restraint led Pfizer to discover related indole-containing systems (45) [84]. However, these indole 6-carboxamides displayed very poor absorption. Modification to a closely related hydroxymethyl derivative (46) gave an orally active agent which was then metabolised *in vivo* to the corresponding acid (47). The metabolite (47), whilst poorly absorbed in its own right, displayed the profile of a potent ET$_A$ selective agent and was long lived *in vivo*. Both compounds displayed good functional activity and (46) was targeted for hypertension but no development has been reported [85]. As a chemical class, these agents display high-end LLE, mid-range LE and physical properties relative to previously discussed series (Table 5.7).

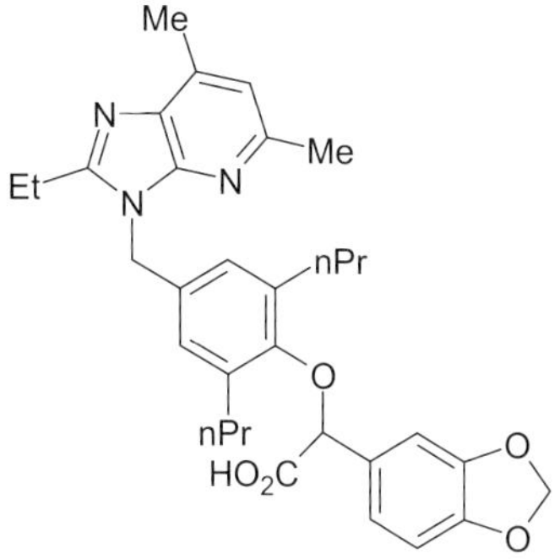

(43)

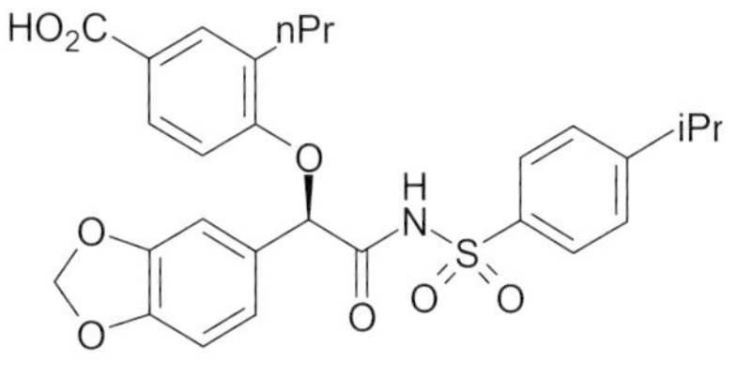

L-754142 (44)
ET$_A$ K$_i$ (Hu) 0.062nM
LE 0.38
ET$_B$ K$_i$ (Hu) 2.25nM

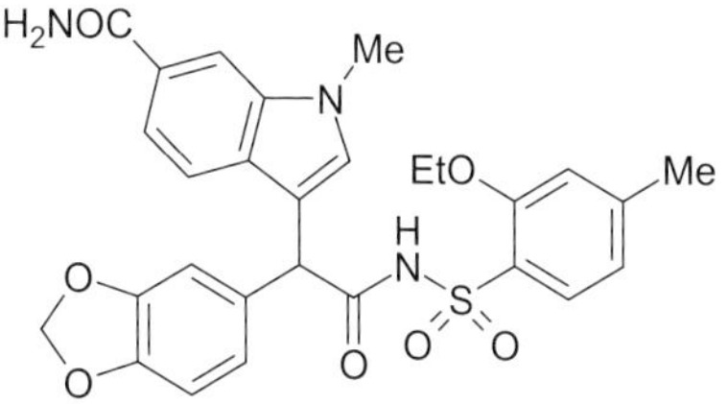

(45)
ET$_A$ IC$_{50}$ (Hu) 0.55nM
LE 0.33
ET$_B$ IC$_{50}$ (Hu) 397nM

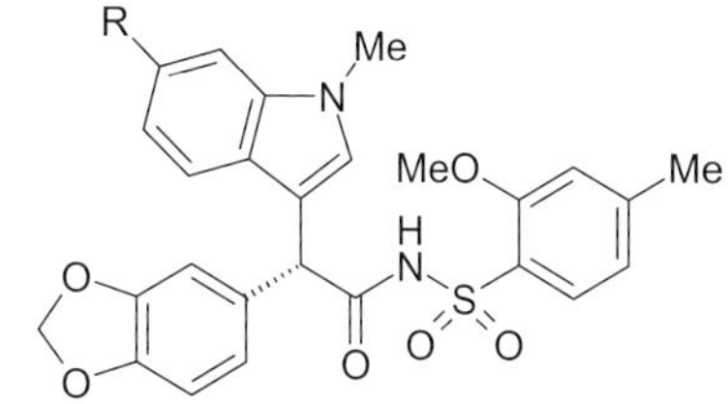

(46)
R = CH$_2$OH
ET$_A$ IC$_{50}$ (Hu) 2.2nM
LE 0.33
ET$_B$ IC$_{50}$ (Hu) 683nM
(47) R = CO$_2$H
ET$_A$ IC$_{50}$ (Hu) 4.2nM
LE 0.31
ET$_B$ IC$_{50}$ (Hu) 347nM

Table 5.7 PROPERTIES OF REPRESENTATIVE *N*-ACYLSULPHONAMIDE-BASED AND MISCELLANEOUS ENDOTHELIN RECEPTOR ANTAGONISTS

ID	ET$_A$ IC$_{50}$ (nM)	ET$_B$ IC$_{50}$ (nM)	ET$_A$ Sel.	MW	c*log* P	LE	LLE	HBA	HBD	PSA	Rof5 fails
(43)				544	7.57			8	1	95.7	2
L-754142 (44)	0.062[a]	2.25[a]	36	540	6.42	0.38	3.79	9	2	128	2
(45)	0.55	387	704	550	2.98	0.33	6.23	10	3	139	1
(46)	2.2	683	310	523	2.71	0.33	5.95	9	2	116	1
(47)	4.2	347	83	537	3.62	0.31	4.76	10	2	133	1
(48)	2.5	83.6	33	563	2.83	0.29	5.77	8	2	105	1
(49)	1[a]	670[a]	670	744	8.17	0.23	0.83	10	3	156	2

[a]K_i rather than IC$_{50}$.

MISCELLANEOUS ANTAGONIST SERIES

Actelion scientists designed a new series of benzo[1,4]diazepin-2-ones based on ambrisentan using superimposition and active site model studies [15]. This programme led to (48). Compound (48) displayed superior efficacy to ambrisentan in both *in vitro* and animal model studies but no development has been reported. LE and properties are mid-range and comparable to the acylsulfonamides. Shionogi have S-0139 (49), a myriceric acid A derivative in Phase II trials for the potential treatment of hemorrhagic and ischemic stroke as an injectable agent [86].

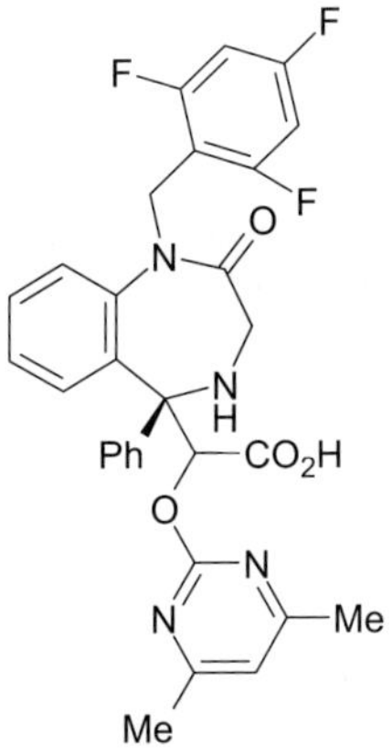

(48)
ET_A IC_{50} (Hu) 2.5nM
LE 0.29
ET_B IC_{50} (Hu) 83.6nM

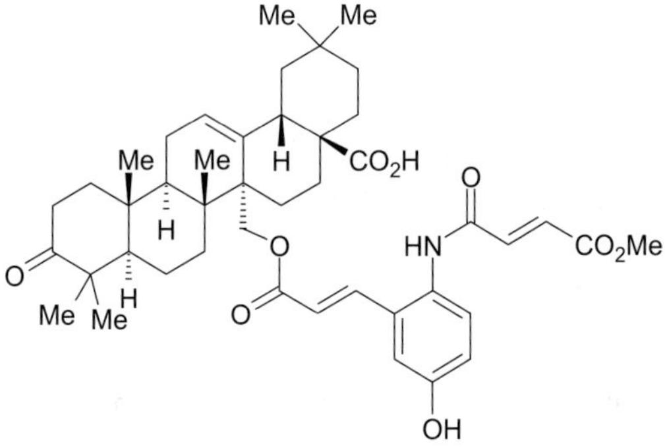

S-0139 (49)
ET_A K_i (Hu) 1nM
LE 0.23
ET_B K_i (Hu) 670nM

CONCLUSIONS

The field of endothelin receptor antagonists represents one of the most researched areas in medicinal chemistry. The vast array of important biological effects that are influenced by the endothelin family of peptides and the plethora of different chemical classes that have subsequently arisen provide a great opportunity to study the targeting of a difficult peptidic GPCR. In an area where structural biology has proven challenging, compound library collections have been very effective in providing initial leads, together with adaptation from related arenas such as angiotensin II.

In general it is the highly ligand efficient, good property based starting entities that have progressed to provide the successful clinical candidates, as with the heteroaryl sulfonamides. Alternatively, a modestly efficient lead has been initially transformed in a manner that raises LE, as with the diaryl carboxylic acids. Identification of an initial or second phase ligand efficient, good property based lead is no guarantee of success however, rather merely a starting point with potential. In a challenging target area, improving lead potency and tuning selectivity has proven relatively easy. However, optimisation of initial leads to provide ligand efficient candidates that also possess the necessary physicochemistry to allow oral delivery, and provide a suitable pharmacokinetic/pharmacodynamic (PK/PD) profile has proven challenging. Here again the heteroaryl sulfonamide and diaryl carboxylic acid chemical classes have proven successful in maintaining a high LE (>0.4) through to candidate and achieving market approval. In the endothelin field, meeting the rule of five is generally important to success. Where there are exceptions, as with bosentan (5) or atrasentan (39), one can hypothesise that there are special circumstances. Bosentan (5) manages to maintain a sufficient property balance and the sulfonamide motif perhaps does not carry a full H-bond acceptor tariff for each heteroatom. Atrasentan (39) is exquisitely potent and probably highly bound to plasma protein. Given that acidic molecules generally have a blood/plasma constrained low volume of distribution, controlling clearance is key to pharmacokinetics, and atrasentan's (39) potency and plasma protein binding are likely to help in this respect and enable a PK/PD profile that makes for a suitable dose size. Additionally, where a group has some metabolic liability, as with the oxazole motif that featured in several series such as BMS-193884 (16) and UK-419106 (26), the ET_A receptor target is sufficiently challenging that fixing the liability *via* addition of extra mass (to provide Log P tuning or steric blocking groups) generally tips the LE and PK/PD balance unfavourably.

Clinically, the accumulating body of evidence appears to point towards ET_A selective antagonism being preferable, although modest ET_B antagonism certainly appears to do no harm in some fields, such as PAH. Managing safety has proven challenging and generally the resulting dose restrictions appear to have hampered efforts to show compelling clinical efficacy in the wider arena beyond PAH. Ultimately, the challenge of achieving a suitable therapeutic index may prove a key factor in defining the potential of endothelin receptor antagonists. Some very good clinical agents have been discovered, there appears no sign of an agent that significantly betters the current pathfinders (Table 5.1), and hence these agents and their application will probably define this arena.

REFERENCES

[1] Hickey, K.A., Rubanyi, G., Paul, R.J. and Highsmith, R.F. (1985) *Am. J. Physiol.* **248**, C550–C556.
[2] Yanagisawa, M., Kurihara, H., Kimura, S., Tomobe, Y., Kobayashi, M., Mitsui, Y., Yazaki, Y., Goto, K. and Masaki, T. (1988) *Nature (London)* **332**, 411–415.
[3] Inoue, A., Yanagisawa, M., Kimura, S., Kasuya, Y., Miyauchi, T., Goto, K. and Masaki, T. (1989) *Proc. Natl. Acad. Sci. USA* **86**, 2863–2867.
[4] Davenport, A.P. and Maguire, J.J. (2006) *Handb. Exp. Pharmacol.* **176**(1), 295–329.
[5] Haynes, W.G. and Webb, D.J. (1994) *Lancet* **344**, 852–854.
[6] Dhaun, N., Goddard, J. and Webb, D.J. (2006) *J. Am. Soc. Nephrol.* **17**, 943–955.
[7] O'Reilly, G., Charnock-Jones, D.S., Morrison, J.J., Cameron, I.T., Davenport, A.P. and Smith, S.K. (1993) *Biochem. Biophys. Res. Commun.* **193**, 834–840.
[8] Howard, P.G., Plumpton, C. and Davenport, A.P. (1992) *J. Hypertens.* **10**, 1379–1386.
[9] Arai, H., Hori, S., Aramori, I., Ohkubo, H. and Nakanishi, S. (1990) *Nature (London)* **348**, 730–732.
[10] Sakarai, T., Yanagisawa, M., Takuwa, Y., Miyazaki, H., Kimura, S., Goto, K. and Masaki, T. (1990) *Nature (London)* **348**, 732–735.
[11] Maguire, J.J. and Davenport, A.P. (1995) *Br. J. Pharmacol.* **115**, 191–197.
[12] Schiffrin, E.L. (2001) *Am. J. Hypertens.* **14**, S83–S89.
[13] De Nucci, G., Thomas, R., D'Orleans-Juste, P., Antunes, E., Walder, C., Warner, T.D. and Vane, J.R. (1988) *Proc. Natl. Acad. Sci. USA* **85**, 9797–9800.
[14] Attina, T., Camidge, R., Newby, D.E. and Webb, D.J. (2005) *Heart* **91**, 825–831.
[15] Bolli, M.H., Marfurt, J., Grisostomi, C., Boss, C., Binkert, C., Hess, P., et al. (2004) *J. Med. Chem.* **47**, 2776–2795.
[16] Murugesan, N., Gu, Z., Stein, P.D., Bisaha, S., Spergel, S., Gorotra, R., et al. (1998) *J. Med. Chem.* **41**, 5198–5218.
[17] Heilker, R., Zemanova, L., Valler, M.J. and Nienhaus, G.U. (2005) *Curr. Med. Chem.* **12**, 2551–2559.
[18] Benigni, A., Perico, N. and Remuzzi, G. (2004) *Expert Opin. Investig. Drugs* **13**, 1419–1435.
[19] Janes, R.W., Peapus, D.H. and Wallace, B.A. (1994) *Nature Struct. Biol.* **1**, 311–319.
[20] Krystek, S.R., Bassolino, D.A., Novotny, J., Chen, C., Marschner, T.M. and Andersen, N.H. (1991) *FEBS Lett.* **281**, 212–218.
[21] Huggins, J.P., Pelton, J.T. and Miller, R.C. (1993) *Pharmacol. Ther.* **59**, 55–123.
[22] Orry, A.J.W. and Wallace, B.A. (2000) *Biophys. J.* **79**, 3083–3094.
[23] Dhaun, N., Pollock, D.M., Goddard, J. and Webb, D.J. (2007) *Trends Pharmacol. Sci.* **28**, 573–579.
[24] Rubin, L.J., Badesch, D.B., Barst, R.J., Galie, N., Black, C.M., Keogh, A., Pulido, T., Frost, A., Roux, S., Leconte, L., et al. (2002) *New Engl. J. Med.* **346**, 896–903.
[25] Barst, R.J., Langleben, D., Frost, A., Horn, E.M., Oudiz, R., Shapiro, S., et al. (2004) *Am. J. Respir. Crit. Care Med.* **169**, 441–447.
[26] Barst, R.J., Langleben, D., Badesch, D.B., Frost, A., Lawrence, E.C., Shapiro, S., Naeije, R. and Galie, N. (2006) *J. Am. Coll. Cardiol.* **47**, 2049–2056.
[27] Nishida, M., Okada, Y., Akiyoshi, K., Eshiro, K., Takaoka, M., Gariepy, C.E., Yanagisawa, M. and Matsumura, Y. (2004) *Eur. J. Pharmacol.* **496**, 159–165.
[28] Nakov, R., Pfarr, E. and Eberle, S. (2002) *Am. J. Hypertens.* **15**, 583–589.

[29] Von Geldern, T.W., Tasker, A.S., Sorensen, B.K., Winn, M., Szczepankiewicz, B.G., Dixon, D.B., et al. (1999) *J. Med. Chem.* **42**, 3668–3678.
[30] Macdonald, R.L., Pluta, R.M. and Zhang, J.H. (2007) *Nature Clin. Pract. Neurol.* **3**, 256–263.
[31] Lalich, M., McNeel, D.G., Wilding, G. and Liu, G. (2007) *Cancer Invest.* **25**, 785–794.
[32] Chiappori, A.A., Haura, E., Rodriguez, F.A., Boulware, D., Kapoor, R., Neuger, A.M., et al. (2008) *Clin. Cancer Res.* **14**, 1464–1469.
[33] Sudano, I., Hermann, M. and Ruschitzka, F.T. (2007) *Curr. Hypertens. Rep.* **9**, 59–65.
[34] Gulati, A., Bhalia, S. and Matwyshyn, G. (2004) *J. Cardiovasc. Pharmacol.* **44**, S129–S131.
[35] Motte, S., McEntee, K. and Naeije, R. (2006) *Pharmacol. and Ther.* **110**, 386–414.
[36] Van Giersbergen, P.L., Halabi, A. and Dingemanse, J. (2002) *Br. J. Clin. Pharmacol.* **53**, 589–595.
[37] Cheng, A.C., Coleman, R.G., Smyth, K.T., Cao, Q., Soulard, P., Caffrey, D.R., Salzberg, A.C. and Huang, E.S. (2007) *Nat. Biotechnol.* **25**, 71–75.
[38] Hopkins, A.L. and Groom, C.R. (2002) *Nat. Rev.* **1**, 727–730.
[39] Lipinski, C.A. (2004) *Drug Discov. Today Technol.* **1**, 337–341.
[40] Leeson, P.D. and Springthorpe, B. (2007) *Nat. Rev.* **6**, 881–890.
[41] Hopkins, A.L., Groom, C.R. and Alex, A. (2004) *Drug Discov. Today* **9**, 430–431.
[42] Kuntz, I.D., Chen, K., Sharp, K.A. and Kollman, P.A. (1999) *Proc. Nat. Acad. Sci. USA* **96**, 9997–10002.
[43] Leach, A.R., Hann, M.M., Burrows, J.N. and Griffen, E.J. (2006) *Mol. Biosyst.* **2**, 429–446.
[44] Ihara, M., Noguchi, K., Saeki, T., Fukuroda, T., Tsuchida, S., Kimura, S., Fukami, T., Ishikawa, K., Nishikibe, M. and Yano, M. (1992) *Life Sci.* **50**, 247–255.
[45] Bradbury, R.H., Bath, C., Butlin, R.J., Dennis, M., Heys, C., Hunt, S.J., et al. (1997) *J. Med. Chem.* **40**, 996–1004.
[46] Sogabe, K., Nirei, H., Shoubo, M., Nomoto, A., Ao, S., Notsu, Y. and Ono, T. (1993) *J. Pharmacol. Exp. Ther.* **264**, 1040–1046.
[47] Cody, W.L., Doherty, A.M., He, J.X., DePue, P.L., Waite, L.A., Topliss, J.G., et al. (1993) *Med. Chem. Res.* **3**, 154–162.
[48] Watanabe, T. and Fujino, M. (1996) *Cardiovasc. Drug Rev.* **14**, 36–46.
[49] Taddei, S., Virdis, A., Ghiadoni, L., Sudano, I., Notari, M. and Salvetti, A. (1999) *Circulation* **100**, 1680–1683.
[50] Suetsch, G., Fleisch, M., Yan, X.W., Wenzel, R., Binggeli, C., Bianchetti, M.C., Kiowski, W. and Luescher, T. (1997) *J. Am. Coll. Cardiol.* **29**(Suppl. A), 444A.
[51] Boss, C., Bolli, M. and Weller, T. (2002) *Curr. Med. Chem.* **9**, 349–383.
[52] Dingemanse, J. and Van Giersbergen, P.L.M. (2004) *Clin. Pharmacokinet.* **41**, 1089–1115.
[53] Chin, K. and Channick, R. (2004) *Expert Rev. Cardiovasc. Ther.* **2**, 175–182.
[54] Launay, D., Diot, E., Pasquier, E., Mouthon, L., Boullanger, N., Olivier, F., Jego, P., Carpentier, P., Hatron, P.-Y. and Hachulla, E. (2006) *Presse Medicale* **35**, 587–592.
[55] Moore, S.C. and Hermes DeSantis, E.R. (2008) *Am. J. Health Syst. Pharm.* **65**, 315–321.
[56] O'Connor, C.M., Gattis, W.A., Adams, K.F., Hasselblad, V., Chandler, B., Frey, A., et al. (2003) *J. Am. Coll. Cardiol.* **41**, 1452–1457.

[57] Vajkoczy, P., Meyer, B., Weidauer, S., Raabe, A., Thome, C., Ringel, F., Breu, V. and Schmiedek, P. (2005) *J. Neurosurg.* **103**, 9–17.
[58] Dieterle, W., Mann, J. and Kutz, K. (2004) *J. Clin. Pharmacol.* **44**, 59–66.
[59] Takahashi, M., Taniguchi, T., Tanaka, T., Kanamaru, H., Okada, K. and Muramatsu, I. (2003). *Eur. J. Pharmacol.* **467**, 185–189.
[60] Hoshino, T., Yamauchi, R., Kikkawa, K., Yabana, H. and Murata, S. (1998) *J. Pharmacol. Exp. Ther.* **286**, 643–649.
[61] Harada, H., Kazami, J.-I., Watanuki, S., Tsuzuki, R., Sudoh, K., Fujimori, A., Tokunaga, T., Tanaka, A., Tsukamoto, S. and Yanagisawa, I. (2001) *Chem. Pharm. Bull.* **49**, 1593–1603.
[62] Murugesan, N., Gu, Z., Stein, P.D., Spergel, S., Mathur, A., Leith, L., et al. (2000) *J. Med. Chem.* **43**, 3111–3117.
[63] Murugesan, N., Gu, Z., Spergel, S., Young, M., Chen, P., Mathur, A., et al. (2003) *J. Med. Chem.* **46**, 125–137.
[64] Hulpke-Wette, M. and Buchhorn, R. (2002) *Curr. Opin. Inv. Drugs* **3**, 1057–1061.
[65] Murugesan, N., Tellew, J.E., Gu, Z., Kuntz, B.L., Fadnis, L., Cornelius, L.A., et al. (2002) *J. Med. Chem.* **45**, 3829–3835.
[66] Murugesan, N., Gu, Z., Fadnis, L., Tellew, J.E., Baska, R.A.F., Yang, Y., Beyer, S.M., Monshizadegan, H., Dickinson, K.E., Valentine, M.T., Humphreys, W.G., Lan, S-J., Ewing, W.R., Carlson, K.E., Kowala, M.C., Zahler, R. and Macor, J.E. (2005) *J. Med. Chem.* **48**, 171–179.
[67] Wu, C., Chan, M.F., Stavros, F., Raju, B., Okun, I., Mong, S., Keller, K.M., Brock, T., Kogan, T.P. and Dixon, R.A.F. (1997) *J. Med. Chem.* **40**, 1690–1697.
[68] Dhaun, N., Melville, V., Kramer, W., Stavros, F., Coyne, T., Swan, S., Goddard, J. and Webb, D.J. (2007) *Br. J. Clin. Pharmacol.* **64**, 733–737.
[69] Waxman, A.B. (2007) *Vasc. Health and Risk Manag.* **3**, 151–157.
[70] Wu, C., Decker, E.R., Blok, N., Bui, H., You, T.J., Wang, J., et al. (2004) *J. Med. Chem.* **47**, 1969–1986.
[71] Palmer, M.J. (2002) *Trends in Medicinal Chemistry*, SMR Symposium, London, UK.
[72] Riechers, H., Albrecht, H.P., Amberg, W., Baumann, E., Bernard, H., Bohm, H.J., et al. (1996) *J. Med. Chem.* **39**, 2123–2128.
[73] Amberg, W., Hergenroeder, S., Hillen, H., Jansen, R., Kettschaau, R., Kling, A., Klinge, D., Raschack, M., Riechers, H. and Unger, L. (1999) *J. Med. Chem.* **42**, 3026–3032.
[74] Vatter, H. and Seifert, V. (2006) *Cardiovasc. Drug Rev.* **24**, 63–76.
[75] Webb, M.L. and Meek, T.D. (1997) *Med. Res. Rev.* **17**, 17–67.
[76] Cosenzi, A. (2003) *Cardiovasc. Drug Rev.* **21**, 1–16.
[77] Nishikibe, M., Ohta, H., Okada, M., Ishikawa, K., Hayama, T., Fukuroda, T., et al. (1999) *J. Pharmacol. Exp. Ther.* **289**, 1262–1270.
[78] Patt, W.C., Edmunds, J.J., Repine, J.T., Berryman, K.A., Reisdorph, B.R., Lee, C., et al. (1997) *J. Med. Chem.* **40**, 1063–1074.
[79] Bunker, A.M., Cheng, X.-M., Doherty, A.M., Lee, C., Repine, J.T., Skeean, R., Edmunds, J.J. and Kanter, G.D. (1999). *PCT Int. Appl.*, WO 9912916A1.
[80] Winn, M., Von Geldern, T.W., Opgenorth, T.J., Jae, H.S., Tasker, A.S., Boyd, S.A., et al. (1996) *J. Med. Chem.* **39**, 1039–1048.
[81] Jae, H.-S., Winn, M., Dixon, D.B., Marsh, K.C., Nguyen, B., Opgenorth, T.J. and Von Geldern, T.W. (1997) *J. Med. Chem.* **40**, 3217–3227.

[82] Samara, E., Dutta, S., Cao, G., Granneman, G.R., Dordal, M.S. and Padley, R.J. (2001) *J. Clin. Pharmacol.* **41**, 397–403.
[83] Liu, G., Henry, K., Szczepankiewicz, B.G., Winn, M., Kozmina, N.S., Boyd, S.A., et al. (1998) *J. Med. Chem.* **41**, 3261–3275.
[84] Rawson, D.J., Dack, K.N., Dickinson, R.P. and James, K. (2002) *Bioorg. Med. Chem. Lett.* **12**, 125–128.
[85] Rawson, D.J., Dack, K.N., Dickinson, R.P. and James, K. (2004) *Med. Chem. Res.* **13**, 149–157.
[86] Sakurawi, K., Yasuda, F., Tozyo, T., Nakamura, M., Sato, T., Kikuchi, J., et al. (1996) *Chem. Pharm. Bull.* **44**, 343–351.

6 Cytochrome P450 Metabolism and Inhibition: Analysis for Drug Discovery

BARRY C. JONES, DONALD S. MIDDLETON and KURESH YOUDIM

Pfizer Global Research and Development, Ramsgate Road, Sandwich, Kent CT13 9NJ, UK

INTRODUCTION	240
SOURCES OF DRUG METABOLISING ENZYMES	241
Organs and Tissue Slices	242
Isolated Human Hepatocytes	242
Subcellular Fractions	243
IN VITRO STRATEGIES TO ASSESS DRUG METABOLISM	244
Metabolic Stability and Phenotyping	244
REVERSIBLE INHIBITION	247
MECHANISM-BASED INHIBITION	248
IN VITRO–IN VIVO EXTRAPOLATIONS (IVIVE) AND ASSOCIATED PREDICTION OF CLINICAL RISK	249
CYP STRUCTURE–ACTIVITY RELATIONSHIPS	252
CYP1A2	252
CYP2C9	253
CYP2C19	254
CYP2D6	254

CYP3A4	255
OTHER CYP ENZYMES	256
MEDICINAL CHEMISTRY DESIGN CASE STUDIES	256
CONCLUSIONS	260
REFERENCES	260

INTRODUCTION

Drug metabolism studies are routinely performed in laboratory animals, particularly rats and dogs, but they are often not sufficiently accurate to predict the metabolic profile of the drug in humans. In addition, they are often time consuming and labour intensive. As a result, the objective of pharmaceutical companies is to establish reliable and relevant *in vitro* models with strong predictive power for human metabolism, speeding up the selection and limiting the number of late-stage failures of new chemical entities (NCEs) [1].

The "drug metabolising enzymes" (DMEs) are a diverse group of proteins that are responsible for metabolising a vast array of xenobiotic compounds including drugs, environmental pollutants and endogenous compounds such as steroids and prostaglandins. These processes occur through the action of single or multiple specialized enzymatic systems. Thus, the DMEs are noted for their broad substrate specificity; with some members of the cytochrome P450 (CYP) and flavin mono-oxygenase (FMO) families known to metabolise numerous structurally diverse compounds [2, 3]. Hence, an understanding of the structure–activity relationships for the DMEs and their substrates is an important area of research that impacts on pharmacology, toxicology and basic enzymology.

DMEs are often divided into two groups; Phase I and Phase II. Firstly, oxidative enzymes, largely contributing to so-called Phase I metabolism, which include CYPs and FMOs. These enzymes catalyse the introduction of an oxygen atom into substrate molecules, generally resulting in hydroxylation or demethylation. If the metabolites of Phase I reactions are sufficiently polar, they may be readily excreted at this point. However, many Phase I products are not eliminated rapidly and can then undergo subsequent Phase II reactions in which an endogenous substrate combines with the newly incorporated functional group or a pre-existing functional group to form a highly polar conjugate. In this regard, Phase II reactions involve conjugative enzyme families such as UDP-glucuronosyltransferases (UGTs), glutathione transferases (GSTs), sulfotransferases (SULTs) and *N*-acetyltransferases (NATs).

Table 6.1 TYPES OF REACTIONS THAT DRUGS MIGHT UNDERGO WITHIN DIFFERENT SUBCELLULAR FRACTIONS OF THE LIVER

Reaction	Subcellular fraction	Enzyme
Phase I		
Oxidation	Microsomes	Cytochrome P450
	Microsomes	Flavin containing mono-oxygenases
	Microsomes	Monoamine oxidase
	Mitochondria	Xanthine oxidase
	Mitochondria	Monoamine oxidase
	Cytosol	Aldehyde oxidase
	Cytosol	Alcohol dehydrogenase
	Cytosol	Aldehyde dehydrogenase
Reduction	Cytosol	Carbonyl reductase
	Cytosol microsomes	Quinone reductase
	Microflora microsomes	Azo-nitrile reductase
	Cytosol microsomes	Reducto dehalogenation
Hydrolysis	Microsomes cytosol	Carbonyl esterase
	Microsomes cytosol	Epoxide hydrolase
Phase II		
Conjugation	Microsomes	UDP-glucuronosyltransferases (UGTs)
	Cytosol	Sulfotransferases (SULTs)
	Cytosol microsomes	Glutathione transferase (GST)
	Mitochondria	Amino acid conjugation
	Mitochondria cytosol	N-acetyltransferase (NAT)
	Cytosol	Methylation

Table 6.1 lists the major Phase I and Phase II enzyme systems involved in drug metabolism.

Quantitatively, the smooth endoplasmic reticulum of the liver cell is the principal organ of drug metabolism (Table 6.1), although every biological tissue has some ability to metabolise drugs. Factors responsible for the liver's contribution to drug metabolism include its large size, that it is the first organ perfused by chemicals absorbed in the gut and that there are very high concentrations of most DMEs relative to other organs.

SOURCES OF DRUG METABOLISING ENZYMES

A number of different tissue matrices can be employed to study drug metabolism (Table 6.2) as will be briefly alluded to below. These fall into four groups: organs, cells (primary cultures and cell lines), subcellular fractions (S9, cytosol and microsomes) and isolated enzymes (purified and recombinant systems).

Table 6.2 DIFFERENT SOURCES OF DRUG METABOLISING ENZYMES AND THEIR APPLICABILITY TO *IN VITRO* STUDIES

	Metabolite profiling	Metabolic clearance	Inhibitory potential	Phenotyping
Liver slices	Secondary	Secondary	Not used	Not used
Hepatocytes	Primary	Primary	Secondary	Secondary
S9	Secondary	Secondary	Not used	Not used
Cytosol	Secondary	Secondary	Not used	Not used
Microsomes	Primary	Primary	Primary	Primary
Recombinant	Secondary	Primary	Secondary	Primary

ORGANS AND TISSUE SLICES

Use of a whole liver would clearly be most representative of the *in vivo* scenario when investigating drug metabolism. Besides the inherent difficulties in maintaining post-operative viability, this approach is not practical when wishing to screen the metabolic liability of hundreds or thousands of compounds weekly during early discovery. An alternative approach is the use of liver slices. The precision-cut liver slice system has been applied successfully in metabolism and toxicity studies [4] and more recently the induction of Phase I [5] and Phase II enzymes [4]. In these cases, the success of the system was shown to be dependant upon the incubation conditions used to study drug metabolism. The benefit afforded by this system, over use of isolated cells such as hepatocytes, is the retention of the extracellular matrix which provides the necessary architecture and cell–cell interactions, i.e., Kupffer and biliary cells. Moreover, they provide an "immediate" source to study the activity of the DMEs. However, it should be noted that reproducibility is highly dependant on the quality of these slices ($<250\,\mu m$ thick), requiring high-precision tissue slicers (e.g., Krumdieck or Brendal-Vitron). Unfortunately, the significant resources required to prepare these slices, together with the impracticality of automation and cell damage around the perimeter of the slices, precludes their widespread use during early discovery assessment of drug metabolism or biotransformation. Furthermore, this approach does not allow characterisation of the individual isoforms involved in a compound's metabolism.

ISOLATED HUMAN HEPATOCYTES

Primary hepatocytes isolated by collagenase perfusion techniques are a popular *in vitro* model owing to the close resemblance to their *in vivo* counterparts. Despite this, limited supply and issues around long-term storage have contributed to the increased use of cryopreserved human hepatocytes.

Modern cryopreservation techniques allow a high percentage of viable and plateable cells following resuscitation [6]. Once isolated, cells can be used either as suspensions or cultured under defined conditions for long periods of time. Using the suspension approach, cells can exhibit metabolic activity for short periods of time, while fixed monolayers can be maintained for several weeks, although a decrease in enzyme activity is observed [7]. Hence, optimisation of cell culture conditions is important. In this regard, attention has been paid to use of specific media with and without serum, hormones and/or inducers, different matrices such as rat tail collagen or Matrigel, as well as co-culturing with epithelial cells. More recently, the application of a collagen sandwich has been reported, providing a more realistic structural and functional bile canalicular network.

SUBCELLULAR FRACTIONS

These fractions are obtained by a number of successive centrifugations of tissue homogenates. The first fraction is collected after a 9,000 g centrifugation step and hence referred to as S9. The microsomal (pellet) and cytosolic fractions are isolated following subsequent centrifugation at 100,000 g. Hence, compared to the original tissue homogenate these subcellular fractions represent a more concentrated enzyme preparation. The advantages afforded by these fractions are commercial availability of individual donors and pooled donor livers, both of which allow assessment of the diverse inter-individual variability for DMEs. Also important are their amenability to high-throughput automation, a key feature for the pharmaceutical industry during lead development and candidate seeking.

Human liver S9

Liver S9 fractions are subcellular fractions containing DMEs including the CYPs, FMOs and UGTs (Table 6.1). These fractions are a major tool for studying xenobiotic metabolism [8] and offer a complete representation compared to microsomes and cytosol individually. Pooled lots of human S9 have been prepared from several livers, enabling use of this product to evaluate "average human" metabolism of a chosen compound. They provide a similar metabolic capacity to that of hepatocytes without the added influence of a membrane barrier and its associated transporters.

Human liver cytosol

This fraction contains the soluble Phase I and Phase II enzymes. In order to assess the catalytic activity of Phase II enzymes, supplementation with exogenous co-factors is required. The primary use of this system is to guide drug metabolism performed through the action of these soluble enzymes. A more complete system can be achieved through co-incubation with microsomes.

Human liver microsomes

Microsomes are the most popular *in vitro* model currently in use; this is largely due to their affordability and ease of use. Microsomes consist of vesicles of the hepatocyte endoplasmic reticulum, predominantly expressing Phase I enzymes, e.g., CYPs and some Phase II enzymes, e.g., UGTs. This system is preferred over the use of other tissue sources of DMEs, such as hepatocyte, owing to the fact that kinetic measurements are not confused with other metabolic and/or uptake processes.

Recombinant enzymes

The availability of specifically expressed recombinant human DMEs such as the CYPs (rhCYP) allows thorough investigation of the contribution made by individual enzymes to the overall metabolism. The identification of the DMEs responsible for the biotransformations commonly encountered in drug discovery may help to develop metabolic structure-activity relationship (SAR), as well as to predict inter-individual variability and prioritise drug–drug interaction (DDI) studies [9]. This system, together with human liver microsomes as discussed above, will be the focus of the *in vitro* strategies described below.

IN VITRO STRATEGIES TO ASSESS DRUG METABOLISM

Twenty years ago it was estimated that 80% of candidate drugs failed in their development due to poor pharmacokinetics, the major contributors being DDIs and metabolic stability [10]. As a consequence the incorporation of *in vitro* studies to investigate these parameters is now commonplace in drug discovery programmes.

METABOLIC STABILITY AND PHENOTYPING

"Metabolic stability refers to the susceptibility of compounds to biotransformation in the context of selecting and/or designing drugs with favourable pharmacokinetic properties" [11]. Often referred to as intrinsic clearance (CL_{int}) it is now a commonly measured *in vitro* parameter. Once suitably characterised, this value can be extrapolated to predict *in vivo* hepatic clearance (CL_H). Reaction phenotyping on the other hand is the semi-quantitative *in vitro* estimation of the relative contributions of specific DMEs to the metabolism of a test compound. Such data are useful in drug design as well as in the prediction of pharmacokinetic DDIs and inter-patient variability in drug exposure, for example, with respect to genetic polymorphisms such as that seen with CYP2D6. Approximately two-thirds of marketed drugs are metabolised by members of the CYP super-family, in particular the sub-families CYP1, 2 and 3 [2], and therefore, greater emphasis is given to the design of studies aimed at identifying which specific CYP enzymes are involved in the metabolism of a given compound. However, while such approaches are well defined for some specific CYP enzymes (CYP1A2, 2C8, 2C9, 2C19, 2D6, 2E1 and 3A), a similar level of analysis is not yet in place for other CYPs such as CYP1A1, 1B1, 2A6 and 2B6. Current methods used to phenotype CYP contribution employ either microsomes, in the presence of specific CYP inhibitors, and/or rhCYPs. Pooled liver microsomes are still considered to be most physiologically relevant, although rhCYPs can often provide additional data. Complementation of both approaches is considered the most comprehensive approach.

During early discovery, application of *in vitro* CYP reaction phenotyping studies allows estimation of the fraction of the dose metabolised via all CYPs (f_m) and the contribution made to this by each individual CYP ($f_{m\ CYP}$). The sum of each of these clearances can be used to predict the overall CL_{int} as well as guide predictions around the relative clinical risk associated with inhibition of one or more of the metabolising enzymes.

Chemical inhibitors

In these studies, microsomes should be incubated with substrate concentrations lower than or equal to the K_m. Since CYP enzymes are very versatile, they can bind and metabolise a wide range of substrates and inhibitors with diverse chemical structures. Thus, it is unlikely that "absolutely specific inhibitors" can be found. However, well characterised "highly selective inhibitors" with respect to individual human CYP enzymes have been identified (Table 6.3).

The application of chemical inhibitors in studies of this type can provide sound scientific information to make early decisions during drug discovery. However, as the majority of high turnover compounds are screened out, a large proportion of compounds for which phenotyping information is required, are low to intermediate clearance compounds. As such, care needs

Table 6.3 COMMON PROBE SUBSTRATES AND CHEMICAL INHIBITORS USED IN DISCOVERY *IN VITRO* INHIBITION AND PHENOTYPING STUDIES, AND THEIR RECOMMENDED CONCENTRATIONS

CYP form(s)	Inhibitor	Concentrations (μM)	Probe	K_{mapp} (μM)
CYP1A2	Furafylline	10–30[a]	Tacrine	2 [44]
CYP2C9	Sulfaphenazole	10	Diclofenac	5 [18]
CYP2C19	Benzylnirvanol	1	(S)-Mephenytoin	40 [18]
CYP2D6	Quinidine	1	Dextromethorphan	5 [18]
CYP3A4	Ketoconazole	1	Midazolam	2 [18]

[a]Inhibition potency increases with pre-incubation in the presence of NADPH (time-dependent inhibitors). A 15–30 min pre-incubation is recommended. More detailed conditions for additional P450s can be found at http://www.fda.gov/cder/drug/drugInteractions/tableSubstrates.htm#inVitro

to be taken when interpreting data from these more metabolically stable compounds. This is best illustrated for compounds metabolised by more than one CYP, where the inclusion of a specific inhibitor results in no turnover, precluding an exact measurement of a percentage contribution.

Recombinant human P450 enzymes

The use of rhCYPs has grown over recent years, due in part to the increased availability of materials provided by commercial vendors and also because the results from studies utilising these systems can be unambiguously assigned to a particular CYP. However, when choosing a system, it is important to use well characterised sources that report important information such as the CYP protein:reductase:cytochrome b_5 ratio. Differences in this ratio between the recombinant systems can lead to inappropriate study designs [12]. Experiments using rhCYPs follow a similar approach to that of chemical inhibitors. Parent loss ($[S] \leq K_m$) is followed in the presence of the rhCYPs of choice. The individual CL_{int} values ($\mu L/min/pmol$ P450) calculated for each rhCYP require subsequent scaling to the corresponding CL_{int} values for that CYP in human liver microsomes [13]. The sum of the individual CL_{int} values should approach that measured in liver microsomes. This scaling factor, referred to as relative activity factor (RAF), is needed to account for the difference between the CL_{int} per unit amount of CYP (intrinsic activity or turnover number) in liver microsomes and in rhCYP enzymes (Equation (6.1)) [14].

$$RAF_{h,j}(\text{pmol rhCYP/mg HLM}) = \frac{V_{\max} \text{ HLM}_{h,j}(\text{nmol/min/mg HLM})}{V_{\max} \text{ rhCYP}_j(\text{nmol/min/pmol rhCYP})}$$

(6.1)

To overcome inter-individual variability in CYP expression and substrate specificity, an intersystem extrapolation factor (ISEF) is applied which incorporates the correction of CYP abundance (Equation (6.2)). Such corrections can be made using nominal specific contents of individual CYP proteins in liver microsomes or, more appropriately, by employing modelling and simulation software (e.g., SIMCYP™, www.simcyp.com), which takes into account population-based variability in individual CYP content.

$$VISEF_j = \frac{V_{max}\ HLM_j\ (nmol/min/mg\ HLM)}{V_{max}\ rhCYP_j\ (nmol/min/pmol\ rhCYP) \cdot HLM\ CYP_j\ abundance\ (pmol/mg\ HLM)} \quad (6.2)$$

Although corrections for ISEF in Equation (6.2), are based upon substrate V_{max} values, ISEF values based on substrate CL_{int} can also be used. The $f_{m\ CYP}$ for any contributing CYP is then determined from the sum of the individual contributing rhCYP CL_{int} (μL/min/mg protein).

REVERSIBLE INHIBITION

Reversible inhibition can be classified into a number of different mechanisms, namely competitive, uncompetitive, mixed or noncompetitive. These processes have been described in detail elsewhere [15]. Here we will discuss the methodologies that simply characterise the inhibitory potential of NCEs.

During early discovery, speed is very important. As high-throughput strategies are often employed to allow rapid turn around of inhibition data without compromising on quality, the major emphasis is on CYPs 1A2, 2C9, 2C19, 2D6 and 3A4. Numerous *in vitro* assays to assess CYP inhibition have been developed and adapted for drug discovery. These systems differ in the sources and composition of the CYP enzyme assay (e.g., rhCYPs, HLMs, probe substrates) and in the detection method (e.g., radioactivity, fluorescence, luminescence and LC–MS) [16]. However, the two most popular approaches used to monitor DDIs through CYP inhibition are (i) rhCYPs with coumarin derivative probe substrates and fluorescence detection (rhCYP-fluorescent) [17] and (ii) HLM with drug probe substrates and LC–MS detection (HLM LC–MS) [18].

Since fluorescent-probe substrates lack specificity for each CYP enzyme, a single purified rhCYP enzyme is used in each assay and HLMs are not used. The most common approach used to assess inhibitory potency is to determine an IC_{50}, using a single probe substrate concentration, preferably equal to the K_m of the relevant CYP (Table 6.3).

Historically, a single LC–MS/MS method has been used for determination of CYP inhibitory potential using HLMs [18]. Over recent years, numerous laboratories have reported application cocktail incubations. This approach allows simultaneous determination of a NCE activity towards a number of DMEs [19–27]. However, numerous factors must be taken into consideration when establishing a cocktail assay. Firstly, probe substrates and their metabolites should exhibit minimal interference with each other. Inclusion of isotopically labelled internal standards wherever possible will help to correct for any suppression effects that may occur. Moreover, as alluded to earlier, probe specificity is an integral component. In this regard, the probe substrates shown in Table 6.3 are recognised by the Food and Drug Administration as being clinically relevant and/or specific to the CYPs being investigated.

While miniaturisation using advanced liquid handling technologies can achieve efficiency gains in conducting inhibition screens, increased throughput can be achieved by using single point determinations prior to the multipoint studies. Here a single concentration of NCE is investigated under the same conditions as described for a full IC_{50} study. Percent inhibition in this case is determined by the average percent of probe metabolite area in the presence of the NCE compared with that in the absence of NCE. The data from these studies can be used to classify compounds as high, moderate or low risk and subsequently flag their potential as inhibitors of CYPs. These single point data have also been used to calculate IC_{50} values because of the intrinsic relationship that exists between these values [28]. This has the effect of increasing the dynamic range of the single point assay and potentially increasing its resolution and so aiding differentiation between compounds.

The results for single point or full IC_{50} experiments should always be treated with some caution. While it is possible to sort compounds based on inhibitory potency from the *in vitro* experiment, whether or not this potency translates to an *in vivo* DDI requires a number of other factors such as dose etc. to be taken into account. Advances in modelling and simulation tools now make it possible to rank compounds based on parameters associated with the extent of a simulated *in vivo* DDI rather than just on *in vitro* potency. This will be discussed in more detail in a later section.

MECHANISM-BASED INHIBITION

The role played by mechanism-based inhibitors (irreversible inhibitors) is also a key focus during early discovery, as it can result in more profound and prolonged effect than that suggested by the therapeutic dose or exposure. Mechanism-based inhibition (MBI) occurs as a result of the CYP generating reactive intermediates that bind to the enzyme causing irreversible loss of activity. MBI chemistry has been comprehensively reviewed elsewhere [29–32] and will not be covered in detail here.

High-throughput assays employing the use of fluorescent substrates have been reported to access the reversible inhibitory potency of compounds, and with simple changes have been shown to be amenable to assessment of MBI potency. However, interference by fluorescent test inhibitors and a lack of probe sensitivity (requiring the use of costly recombinant systems) have limited this approach. In contrast, methods using human liver microsomes or rhCYPs, in conjunction with specific CYP probe, have been used routinely [33–36]. The most common approach is based on the enzyme inactivation constant (K_I) and the maximum rate of inactivation (k_{inact}). During early screening, an initial assay can be employed using human liver microsomes, normally at sufficiently high concentrations to provide catalytic activity following dilution into the activity mixture. A pre-incubation step is normally performed (to ensure identification of weak inhibitors, avoiding false negatives), after which an aliquot of the pre-incubation mix is added to the activity mix containing probe substrate, and further incubated (Figure 6.1A). This further incubation should be sufficiently long to observe measurable metabolite formation.

For these studies, the decrease in natural logarithm of the activity over time should be plotted for each inhibitor concentration. The negative slopes of each line represent the k_{obs} values (Figure 6.1B), and should be plotted against inhibitor concentration, from which the K_I and k_{inact} are generated using nonlinear regression of the data (Figure 6.1C). The values can then be used to assess the relative clinical risk of the inhibitor [32].

IN VITRO–IN VIVO EXTRAPOLATIONS (IVIVE) AND ASSOCIATED PREDICTION OF CLINICAL RISK

The previous sections have reviewed the various *in vitro* systems currently employed to assess drug metabolism, looking at NCEs as both victims and perpetrators of such interactions. As part of the drug discovery and development process, the *in vitro* data obtained from these experimental procedures require extrapolation to hepatic drug clearance (CL_H). Such IVIVEs were reported over 30 years ago [37], and have now become commonplace. The rationale behind this approach is that *in vitro* metabolic stability predicts *in vivo* clearance *reasonably* well. Historically, these extrapolations came from applying allometric principles to animal *in vitro* data. The application of allometric scaling will not be discussed here and the reader is referred to a recent publication that reviews this approach in detail [38]. Increasingly,

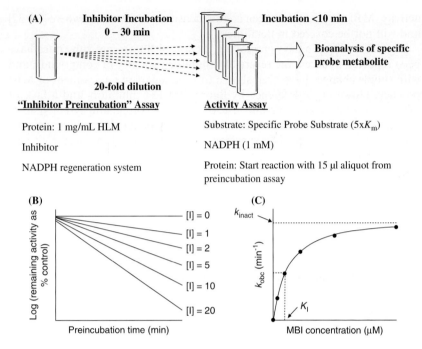

Fig. 6.1 Schematic representation of the steps involved in the experimental dilution approach used in the time-dependant inhibition assay (A) and graphical representations (B) and (C) used to determine enzyme inactivation constant (K_I) and the maximum rate of inactivation (k_{inact}).

human drug clearances are now being extrapolated from human *in vitro* data obtained from liver microsomes, hepatocytes (fresh or cryopreserved), liver slices or recombinant enzymes.

Hepatic drug clearance is the product of the hepatic blood flow (Q_L) and the extraction ratio (E) of the drug. Various hepatic clearance models have been developed to integrate the factors of flow, binding, transporters and enzymatic activities to predict the drug disposition in liver. Two conventionally used models are the "well-stirred" model, which views the liver as a well-stirred compartment with concentration of drug in the liver in equilibrium with that in the emergent blood [39], and the "parallel tube" model, which regards the liver as a series of parallel tubes with enzymes distributed evenly around the tubes and the concentration of drug declining along the length of the tube. Estimation of *in vivo* clearance from these different sources requires the application of appropriate scaling factors to correctly convert *in vitro* CL_{int} to *in vivo* CL_H (Figure 6.2). However, not surprisingly, different values have been used in the literature to extrapolate *in vitro* data, as reviewed by

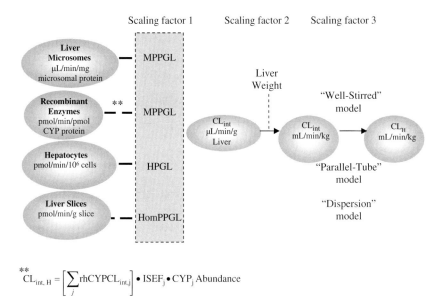

Fig. 6.2 Scaling approach used to estimate hepatic clearance (CL_H) from intrinsic clearances (CL_{int}) obtained using different in vitro systems.

Barter and co-workers [40], which may explain in part studies in which poor correlations were observed. Moreover, a recent study [41] has shown age-related changes in some of these scaling factors, suggesting a single generic value may not be suitable for CL_{int} extrapolations to CL_H.

In addition to the prediction of clearance, a growing emphasis is being placed on the prediction of adverse DDIs. Although this topic has been eloquently described in detail elsewhere [42–43], the current chapter will provide a brief overview of the application of *in vitro* data such as IC_{50}, K_i and K_I/K_{inact} generated in the experimental procedures described previously. The underlying principles for this approach have been known for many years [39], but demonstration that these early concepts could be used practically in the prediction of DDI has only been obtained over the past few years. The principles can be applied for new compounds as the potential cause of an interaction as well as the potential affected substrate. Equation (6.3) described by Rowland and Matin forms the basis of this approach:

$$\frac{CL_{control}}{CL_{inhibited}} = \frac{AUC_{inhibited}}{AUC_{control}} = \frac{1}{((f_{m\,(CYP)})/(1 + ([I]_{in\,vivo}/K_i))) + (1 - f_{m\,(CYP)})} \quad (6.3)$$

The terms $AUC_{inhibited}$ and $AUC_{control}$ refer to the exposures when co-administered in the presence and absence of a second drug, respectively. The term $f_{m\ CYP}$ refers to the fraction of clearance that is mediated by the enzyme that is affected. K_i is the reversible inhibition constant and $[I]_{in\ vivo}$ the concentration of the inhibitor available to bind to the affected enzyme *in vivo*. When dealing with an irreversible inactivator, the structure of the relationship remains the same but the $[I]_{in\ vivo}/K_i$ term is replaced with $([I]_{in\ vivo} \cdot k_{inact})/(k_{deg} \cdot ([I]_{in\ vivo} + K_I))$, where k_{inact} is the maximum inactivation rate constant, K_I the concentration of inactivator at which the observed inactivation rate is half of the maximum and k_{deg} the *in vivo* first order rate of degradation of the affected enzyme.

To determine an accurate prediction of $f_{m\ CYP}$ when analysing the effect of a given enzyme on a new compound, it is critical to collect *in vitro* data.

Bona-fide values for $f_{m\ CYP}$ are hard to obtain, even for well-established drugs. At best, values are estimates and inter-subject variability in this term probably overwhelms any investment in attempts to gain highly precise values. In order to predict whether a new compound will be a perpetrator of DDI, accurate definitions of $[I]_{in\ vivo}$ and K_i and/or K_I/K_{inact} are required.

CYP STRUCTURE–ACTIVITY RELATIONSHIPS

As previously described, the CYP family of enzymes is arguably the most important human DME system. However, despite the ubiquity of this enzyme system across nature, there are relatively few CYPs involved in human drug metabolism. All of the CYPs considered to be important in human drug metabolism have been studied to some degree with one or all of the techniques detailed above. Information produced from these studies has been used to gain an understanding of the high-level structural requirements, which will predispose a compound to interact with a particular member of this family, be that as a substrate or an inhibitor. Many of these structural requirements were described in the early 1990s prior to the solution of X-ray crystal structures for these enzymes [44–46].

Perhaps surprisingly, the subsequent solution of the crystal structures for many of these enzymes appears to have added little to the early analysis of structural requirements. The detailed structural knowledge has, however, helped in understanding which residues within the various active sites determine the nature of the structural requirements in the ligands.

CYP1A2

Initial studies on the CYP1A family characterised the substrates as being lipophilic planar polyaromatic/heteroaromatic molecules, with a small depth and a large area/depth ratio. Although to date there are no published X-ray crystal structures for CYP1A2, a number of homology models have been created which confirm the structural requirements detailed above and suggest amino acid residues within the protein which might define the binding site [47].

These homology models suggest that the active sites of the CYP1A enzymes are composed of several aromatic residues, which form a rectangular slot and restrict the size and shape of the cavity, so that only planar structures are able to occupy the binding site. This is in keeping with the initial observation and could explain the preference of CYP1A enzymes for hydrophobic, planar aromatic species that are able to partake in π–π interactions with these aromatic residues. In addition to the aromatic residues there are several residues able to form hydrogen bonds with substrate molecules. Such a model is able to rationalise that caffeine is N-demethylated at the 1, 3 and 7 positions by CYP1A2, of which the N-3-demethylation is the major pathway. Hence, it appears that binding to the active site of CYP1A2 requires certain molecular dimensions and hydrophobicity, together with defined hydrogen bonding and π–π interactions.

Unlike some of the other CYPs, CYP1A2 does not have a clear preference for acidic or basic molecules. It is able to metabolise basic compounds such as imipramine, but is inhibited by acidic compounds such as enoxacin.

CYP2C9

In terms of physicochemistry, the majority of the CYP2C9 substrates are anionic, including various nonsteroidal anti-inflammatory drugs (NSAIDs) such as flurbiprofen and diclofenac, or contain areas of hydrogen-bonding potential such as (*S*)-warfarin and tolbutamide. Therefore, it has been proposed that these groups are important in binding to the active site of CYP2C9. There are a number of substrate template models for CYP2C9, which typically produce templates where the hydrogen-bonding groups are positioned at a set distance and, because of the vectoral restrictions of hydrogen bonding, a set angle from the site of oxidation [48].

These SAR features were further defined through the construction of a protein homology model, which suggested that Arg-108, Phe-114 and Phe-476 are key residues involved in substrate binding and selectivity [49].

X-ray crystal structures of CYP2C9 are now published with warfarin [50] and flurbiprofen bound in the active site [51]. The warfarin-bound structure showed the warfarin binding to a site that was too distant from the haem for effective catalysis, but which may have been the site for allosteric activation of the enzyme. In the case of flurbiprofen, however, the substrate is bound in a conformation and distance from the haem that is consistent with its oxidation by the enzyme. This structure confirmed the important roles of Arg-108 and Phe-114 in the binding of this and potentially other anionic substrates to the enzyme.

CYP2C19

CYP2C19 is in the same sub-family of CYPs as CYP2C9 and therefore shares a high degree of sequence homology. Substrates for this enzyme include (R)-mephobarbital, moclobemide, proguanil, diazepam, omeprazole and imipramine, which do not show obvious structural or physicochemical similarities. Thus, it appears that CYP2C19 can bind compounds that are weakly basic such as diazepam, strongly basic such as imipramine or acidic compounds such as (R)-warfarin, which in physicochemical terms distinguishes it from CYP2C9. One possible explanation is that CYP2C19 binds substrates via hydrogen bonds, but uses a combination of a hydrogen bond donor and acceptor mechanisms.

While, as is the case for CYP1A2, there are currently no X-ray crystal structures for CYP2C19, the CYP2C9 structures have been used as the basis for a protein homology model which suggests that the differences in substrate specificity may be due in part to differences in lipophilicity between the active sites of the two enzymes [52]. This remains an area that requires further study.

CYP2D6

The overwhelming majority of CYP2D6 substrates are cationic and as such contain a basic nitrogen atom ($pK_a > 8$), which is ionised at physiological pH. It is the ionic interaction between this protonated nitrogen atom and an aspartic acid (Asp-301) residue that governs the majority of this binding. All the models of CYP2D6 show essentially the same characteristics, in which there is a 5–7 Å distance between this basic nitrogen atom and the site of metabolism [53].

The relative strength of this ionic interaction means that the affinity for substrates can be high, and that this CYP tends to have many examples of low-K_m and low-K_i interactions. There is also a suggestion from pharmacophore models that π–π interaction are also involved in the binding, based on the clustering of aromatic groups. Thus, although most of the substrates for CYP2D6 are basic, there are still marked differences in binding affinity, possibly attributable to other π–π, hydrophobic or steric interactions [54].

Following the crystallisation of a truncated form of this enzyme, many of the deductions made from pharmacophores, site-directed mutagenesis and/or homology models were confirmed [55]. The structure supported the role of Asp-301 in the ionic interaction, Phe-120 and Phe-483 in the π–π interactions, but suggested that Glu-213, a residue also suggested to have a role in substrate binding, might actually be involved in substrate recognition rather than binding, given its location in the enzyme.

CYP3A4

CYP3A4 appears to metabolise lipophilic drugs in positions largely dictated by the ease of hydrogen abstraction in the case of carbon hydroxylation, or electron abstraction in the case of N-dealkylation reactions. There are many drugs that are predominantly eliminated by CYP3A4 and many others where CYP3A4 is a secondary mechanism. The binding of substrates to CYP3A4 seems to be due essentially to lipophilic forces. Generally, such binding if based solely on hydrophilic interactions is relatively weak and without specific interactions, which allows motion of the substrate in the active site. Thus, a single substrate may be able to adopt more than one orientation in the active site, and there can be several products of the reaction. This is the case for the probe substrate midazolam, which forms 1- and 4-hydroxy metabolites via CYP3A4. There is considerable evidence for allosteric behaviour, due possibly to the simultaneous binding of two or more substrate molecules to the CYP3A4-active site [56]. Such binding can lead to atypical enzyme kinetics and inconsistent DDIs and is almost diagnostic of CYP3A4 involvement. By way of explanation, it has been suggested that the CYP3A4-active site may undergo substrate-dependent conformational changes [57], or there may be an alteration in the pool of active enzyme [58]. Whatever the case, it is not surprising that there is no useful template model for CYP3A4 substrates.

However, crystal structures are now available for this enzyme in its native state (no substrate or inhibitor bound) [59], with bound substrates (progesterone and erythromycin) and inhibitors (ketoconazole and metyrapone) [60, 61]. The structures show significant differences in the size of the active site, possibly reflecting the substrate-dependent conformational changes described above. The substrate-bound structures reveal relatively little about substrate binding since progesterone was located distal to the haem in what is postulated to be a substrate recognition site, and erythromycin was bound in a conformation which was not consistent with its known metabolism by this enzyme. Both ketoconazole and metyrapone are so-called "type II" inhibitors, which are typically characterised by a high-affinity-binding interaction between an aromatic nitrogen and the sixth co-ordination position of the haem iron. This interaction is evident in both crystal structures, as are various other lipophilic and polar interactions with the protein. Interestingly, the ketoconazole-bound structure revealed the presence of a second ketoconazole molecule in the active site. It is not clear whether this is an artefact of the crystallisation conditions or structural evidence to support the binding of multiple molecules in the active site of this enzyme.

OTHER CYP ENZYMES

Several other CYP enzymes are involved in the metabolism of pharmaceuticals, although they are still regarded as minor enzymes.

CYP2B6 is proposed as a major contributor to the metabolism of a number of drugs including bupropion [62], efavirenz [63] and cyclophosphamide [64] and has been implicated in the partial metabolism of many other drugs such as propofol where it contributes to 4-hydroxylation [65], although the major pathway of elimination is via glucuronidation.

To date no crystal structure for this enzyme has been published. Models based on the nature of CYP2B6 substrates and inhibitors have been produced [66, 67] which highlight the role of lipophilicity, hydrogen bonding and charge in binding to this enzyme.

CYP2C8 is an enzyme with a growing list of structurally diverse substrates including some major therapeutic agents including the glitazones, repaglinide, paclitaxel and cerivastatin. In addition CYP2C8 has been shown to further metabolise glucuronide conjugates of compounds such as gemfibrozil [68] and oestradiol [69]. Its ability to accommodate these potentially very polar metabolites differentiates it from other drug metabolising CYPs.

Sufficient identified substrates exist for this enzyme to build pharmacophore models [70]. One pharmacophore model describes substrates as requiring a hydrophobic chain with a polar or anionic group at the opposite end to the site of metabolism. Along the length of the hydrophobic chain were additional polar-binding sites as well as a hydrophilic/aromatic-binding site.

X-ray crystal structures of the native and substrate bound forms of the enzyme have been determined [71, 72]. From these structures it appears that the active site is large and "Y" or "T" shaped with the predominant interactions between substrate and protein being hydrophobic in nature, although anions also bind to Arg-241, which is broadly in keeping with the pharmacophore model.

MEDICINAL CHEMISTRY DESIGN CASE STUDIES

The complex and uncertain nature of SAR for CYP-mediated metabolism and inhibition remains an enduring design challenge to the medicinal chemist. Despite the advent of X-ray crystal structures for the majority of these enzymes, the broad substrate specificity continues to frustrate efforts to develop precise SARs within series.

An elegant example of designing out CYP2D6 inhibition has been reported at an early point in the Maraviroc CCR-5 programme (Table 6.4) [73].

The initial screening hit (1) showed very potent CYP2D6 inhibition ($IC_{50} = 40$ nM). Replacing the imidazopyridine with benzimidazole (2), resulted in a slight reduction in inhibition potency ($IC_{50} = 700$ nM). Potency was further reduced by replacing one of the benzhydryl phenyl groups with a secondary amide, to give (3) ($IC_{50} = 6,400$ nM). Using quantitative structure-activity relationship (QSAR) computational methods, a correlation was found between increased molecular volume of the secondary amide and decreased CYP2D6 inhibition with, for example *ortho*-chlorophenylamide (4) found to possess only very weak CYP2D6 inhibitory activity ($IC_{50} = 20$ µM).

An interesting example of medicinal chemistry design to attenuate CYP2C9 inhibition has recently been reported from within a series of progesterone ligands (Table 6.5) [74].

The difference in inhibitory potency between (5) and (6) has been rationalised by the increased proportion of ionisation of trifluoromethane sulphonamide (5) ($pK_a = 7.0$), relative to the methane sulphonamide (6) ($pK_a = 10.5$). This results in an enhanced ability of (5) to form interactions with key Arg-108 and Asn-204 residues within the active site of CYP2C9, relative to the less acidic analogue (6).

Balancing appropriate lipophilicity to ensure high-membrane permeability, with good CYP-mediated metabolic stability, is a continued area of focus for the medicinal chemist. It is generally accepted that moderating lipophilicity can play an important role in reducing the rate of metabolism

Table 6.4 CYP2D6 INHIBITION DATA FOR A SERIES OF CCR-5 ANTAGONISTS (1)–(4)

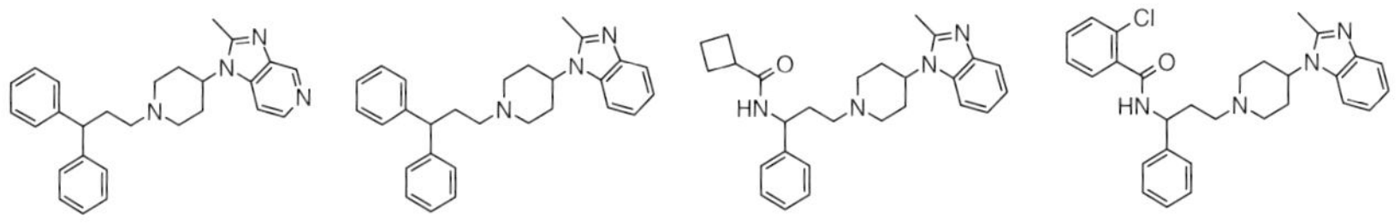

Compound	(1)	(2)	(3)	(4)
2D6 inhibition (IC_{50}, nM)	40	700	6,400	20,000

Table 6.5 CYP2C9 INHIBITION DATA FOR PROGESTERONE COMPOUNDS (5) AND (6)

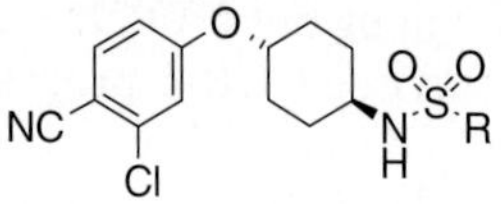

Compound	R	2C9 (IC_{50}, nM)	pK_a
(5)	-CF_3	750	7.0
(6)	-CH_3	>17,000	10.5

by certain CYPs. Sometimes, however, the SAR of the target of interest may require a degree of lipophilicity, or the presence of certain structural features to reach the required levels of potency, which results in an increased predisposition for CYP metabolism. In such circumstances, blocking metabolism has been reported as a successful strategy to retain both target pharmacology and stability. For example, NK-2 antagonist (7) [75, 76] (Table 6.6) was found to possess excellent metabolic stability, although only moderate binding potency ($pIC_{50} = 7.1$) against the target human NK-2 receptor.

Increasing the lipophilicity of the N-lactam substituent from cyclopropyl (7) to cyclohexyl (8) resulted in a 100-fold increase in potency. The attendant increase in $c\log P$ from 2.5 (7) to 4.2 (8), however, caused a perhaps not unexpected large drop in metabolic stability against CYP3A4. Modelling the CYP3A4 metabolism of (8) identified the cyclohexyl ring C-4 as a potentially vulnerable site for oxidation. As a strategy to block this oxidative route, 4, 4-difluorocyclohexyl analogue (9) was prepared. Analogue (9) showed significantly enhanced metabolic stability relative to (8) in human liver microsomal preparations ($t_{1/2}$ 80 min vs. 15 min) despite possessing similar lipophilicity ($c\log P$ 3.9) to (8).

The subtle, complex and often unpredictable SAR surrounding CYP substrate specificity is nicely illustrated by the series of selective serotonin re-uptake inhibitors shown in Table 6.7 [77–79].

The carboxamido-substituted diphenylether (10) was found to be metabolised solely by CYP2D6. The closely analagous sulphonamide (11) was, however, found to be predominantly metabolised by CYP2C9, together with lesser contributions from CYP2D6 and CYP3A4. A further interesting pairwise comparison is provided by pyridine analogues (12) and (13). It may be of significance that both analogue pairs show the same structural switch between *meta*-fluoro (10 and 12) and *meta*-methyl (11 and 13) in the distal

Table 6.6 HUMAN MICROSOMAL STABILITY DATA FOR NK-2 ANTAGONISTS (7)–(9)

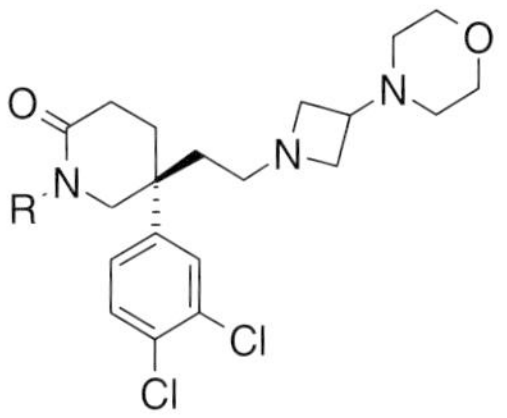

Compound	R	CHO pIC_{50}	HLM ($t_{1/2}$, min)	c $\log$ P
(7)	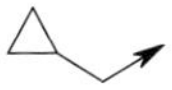 cyclopropylmethyl	7.1	>120	2.5
(8)	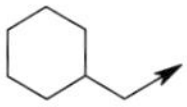 cyclohexylmethyl	9.1	15	4.2
(9)	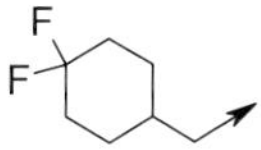 (4,4-difluorocyclohexyl)methyl	8.1	80	3.9

Table 6.7 CYP CONTRIBUTIONS TO THE METABOLISM OF A SERIES OF SELECTIVE SEROTONIN REUPTAKE INHIBITORS (10)–(13)

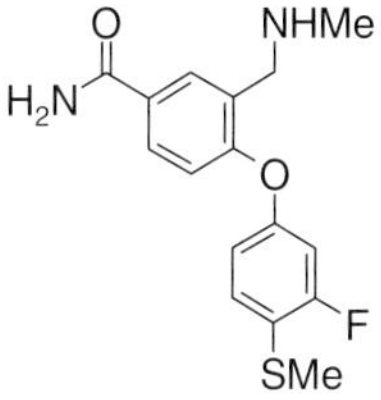

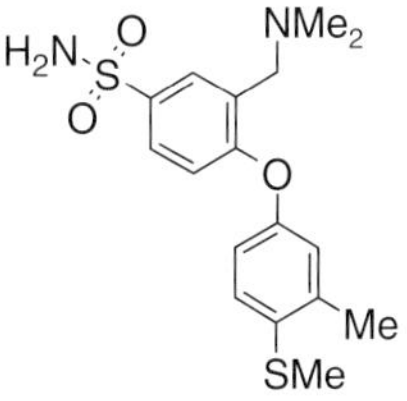

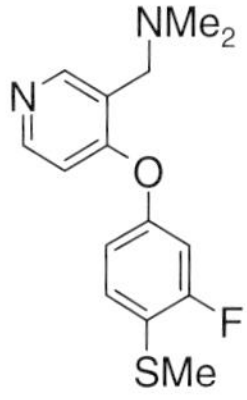

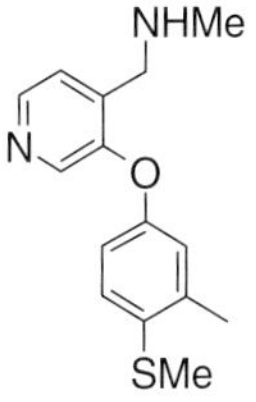

Compound		(10)	(11)	(12)	(13)
SRI (IC_{50}, nM)		6	6	9	9
% P450 contribution to metabolism	2D6	100	11	60–70	34
	3A4	–	14	–	41
	2C9	–	75	–	6

phenyl ring, although no compelling rationalisation has emerged to date to support these experimental observations.

CONCLUSIONS

There are a number of experimental techniques available to determine whether compounds are substrates or inhibitors of various CYP enzymes. During the early 1990s, these techniques were used to develop high-level understanding of the structural requirements that predisposed molecules to bind to these enzymes as substrates or inhibitors.

The solution of the X-ray crystal structure for many of these enzymes during the intervening years has apparently added very little to this understanding of structural requirements. Indeed, the case study examples presented in this manuscript show that while these high-level requirements hold true, the broad substrate specificity of these enzymes continues to frustrate efforts to develop series-specific metabolic SARs.

REFERENCES

[1] Gomez-Lechon, M.J., Donato, T., Ponsoda, X. and Castell, J.V. (2003) *Altern. Lab. Anim.* **31**, 257–265.
[2] Lamb, D.C., Waterman, M.R., Kelly, S.L. and Guengerich, F.P. (2007) *Curr. Opin. Biotechnol.* **18**, 504–512.
[3] Krueger, S.K. and Williams, D.E. (2005) *Pharmacol. Ther.* **106**, 357–387.
[4] Olinga, P., Elferink, M.G., Draaisma, A.L., Merema, M.T., Castell, J.V., Perez, G and Groothuis, G.M. (2008) *Eur. J. Pharm. Sci.* **33**, 380–389.
[5] Persson, K.P., Ekehed, S., Otter, C., Lutz, E.S., McPheat, J., Masimirembwa, C.M. and Andersson, T.B. (2006) *Pharm. Res.* **23**, 56–69.
[6] Lloyd, T.D., Orr, S., Skett, P., Berry, D.P. and Dennison, A.R. (2003) *Cell Tissue Bank* **4**, 3–15.
[7] Nussler, A.K., Wang, A., Neuhaus, P., Fischer, J., Yuan, J., Liu, L., Zeilinger, K., Gerlach, J., Arnold, P.J. and Albrecht, W. (2001) *Altex* **18**, 91–101.
[8] Bjornsson, T.D., Callaghan, J.T., Einolf, H.J., Fischer, V., Gan, L., Grimm, S., et al. (2003) *J. Clin. Pharmacol.* **43**, 443–469.
[9] McGinnity, D.F. and Riley, R.J. (2001) *Biochem. Soc. Trans.* **29**, 135–139.
[10] Grime, K. and Riley, R.J. (2006) *Curr. Drug Metab.* **7**, 251–264.
[11] Masimirembwa, C.M., Bredberg, U. and Andersson, T.B. (2003) *Clin. Pharmacokinet.* **42**, 515–528.
[12] Kumar, V., Rock, D.A., Warren, C.J., Tracy, T.S. and Wahlstrom, J.L. (2006) *Drug Metab. Dispos.* **34**, 1903–1908.
[13] Dickins, M., Galetin, A. and Proctor, N. (2007) *In* "Comprehensive Medicinal Chemistry II". Taylor, J.B. and Triggle, D.J. (eds), pp. 827–846. Elsevier, Oxford.
[14] Proctor, N.J., Tucker, G.T. and Rostami-Hodjegan, A. (2004) *Xenobiotica* **34**, 151–178.

[15] Shou, M., Lin, Y., Lu, P., Tang, C., Mei, Q., Cui, D., et al. (2001) *Curr. Drug Metab.* **2**, 17–36.
[16] Zlokarnik, G., Grootenhuis, P.D. and Watson, J.B. (2005) *Drug Discov. Today* **10**, 1443–1450.
[17] Crespi, C.L. and Stresser, D.M. (2000) *J. Pharmacol. Toxicol. Methods* **44**, 325–331.
[18] Walsky, R.L. and Obach, R.S. (2004) *Drug Metab. Dispos.* **32**, 647–660.
[19] Gao, Z.W., Shi, X.J., Yu, C., Li, S.J. and Zhong, M.K. (2007) *Yao Xue Xue Bao* **42**, 589–594.
[20] Tolonen, A., Petsalo, A., Turpeinen, M., Uusitalo, J. and Pelkonen, O. (2007) *J. Mass Spectrom.* **42**, 960–966.
[21] Di, L., Kerns, E.H., Li, S.Q. and Carter, G.T. (2007) *Int. J. Pharm.* **335**, 1–11.
[22] Testino, S.A., Jr. and Patonay, G. (2003) *J. Pharm. Biomed. Anal.* **30**, 1459–1467.
[23] Adedoyin, A., Frye, R.F., Mauro, K. and Branch, R.A. (1998) *Br. J. Clin. Pharmacol.* **46**, 215–219.
[24] Dixit, V., Hariparsad, N., Desai, P. and Unadkat, J.D. (2007) *Biopharm. Drug Dispos.* **28**, 257–262.
[25] Dierks, E.A., Stams, K.R., Lim, H.K., Cornelius, G., Zhang, H. and Ball, S.E. (2001) *Drug Metab. Dispos.* **29**, 23–29.
[26] Bu, H.Z., Magis, L., Knuth, K. and Teitelbaum, P. (2001) *Rapid Commun. Mass Spectrom.* **15**, 741–748.
[27] Smith, D., Sadagopan, N., Zientek, M., Reddy, A. and Cohen, L. (2007) *J. Chromatogr. B Analyt. Technol. Biomed. Life Sci.* **850**, 455–463.
[28] Gao, F., Johnson, D.L., Ekins, S., Janiszewski, J., Kelly, K.G., Meyer, R.D. and West, M. (2002) *J. Biomol. Screen* **7**, 373–382.
[29] Fontana, E., Dansette, P.M. and Poli, S.M. (2005) *Curr. Drug Metab.* **6**, 413–454.
[30] Hollenberg, P.F., Kent, U.M. and Bumpus, N.N. (2008) *Chem. Res. Toxicol.* **21**, 189–205.
[31] Kalgutkar, A.S., Obach, R.S and Maurer, T.S. (2007) *Curr. Drug Metab.* **8**, 407–447.
[32] Venkatakrishnan, K., Obach, R.S. and Rostami-Hodjegan, A. (2007) *Xenobiotica* **37**, 1225–1256.
[33] Polasek, T.M. and Miners, J.O. (2007) *Expert Opin. Drug Metab. Toxicol.* **3**, 321–329.
[34] Ghanbari, F., Rowland-Yeo, K., Bloomer, J.C., Clarke, S.E., Lennard, M.S., Tucker, G.T. and Rostami-Hodjegan, A. (2006) *Curr. Drug Metab.* **7**, 315–334.
[35] Van, L.M., Heydari, A., Yang, J., Hargreaves, J., Rowland-Yeo, K., Lennard, M.S., Tucker, G.T. and Rostami-Hodjegan, A. (2006) *J. Psychopharmacol.* **20**, 834–841.
[36] Riley, R.J., Grime, K. and Weaver, R. (2007) *Expert Opin. Drug Metab. Toxicol.* **3**, 51–66.
[37] Rane, A., Wilkinson, G.R. and Shand, D.G. (1977) *J. Pharmacol. Exp. Ther.* **200**, 420–424.
[38] Lave, T., Coassolo, P. and Reigner, B. (1999) *Clin. Pharmacokinet.* **36**, 211–231.
[39] Rowland, M. and Matin, S.B. (1973) *J. Pharmacokinet. Biopharm.* **1**, 553–567.
[40] Barter, Z.E., Bayliss, M.K., Beaune, P.H., Boobis, A.R., Carlile, D.J., Edwards, R.J., et al. (2007) *Curr. Drug Metab.* **8**, 33–45.
[41] Johnson, T.N., Rostami-Hodjegan, A. and Tucker, G.T. (2006) *Clin. Pharmacokinet.* **45**, 931–956.
[42] Obach, R.S. (2003) *Drugs Today (Barc.)* **39**, 301–338.
[43] Venkatakrishnan, K., von Moltke, L.L., Obach, R.S. and Greenblatt, D.J. (2003) *Curr. Drug Metab.* **4**, 423–459.

[44] Obach, R.S. and Reed-Hagen, A.E. (2002) *Drug Metab. Dispos.* **30**, 831–837.
[45] Smith, D.A. and Jones, B.C. (1992) *Biochem. Pharmacol.* **44**, 2089–2098.
[46] Smith, D.A., Ackland, M.J. and Jones, B.C. (1997) *Drug Disc. Today* **2**, 479–486.
[47] Lewis, D.F.V and Lake, B.G. (1996) *Xenobiotica* **26**, 723–753.
[48] Jones, B.C., Hawksworth, G., Horne, V.A., Newlands, A., Morsman, J., Tute, M.S. and Smith, D.A. (1996) *Drug Metab. Dispos.* **24**, 1–7.
[49] de Groot, M.J., Alex, A.A. and Jones, B.C. (2002) *J. Med. Chem.* **45**, 1983–1993.
[50] Williams, P.A., Cosme, J., Ward, A., Angrove, H.C., Vinkovic, D.M. and Jhoti, H. (2003) *Nature (London)* **424**, 464–468.
[51] Wester, M.R., Yano, J.K., Schoch, G.A., Yang, C., Griffin, K.J., Stout, C.D. and Johnson, E.F. (2004) *J. Biol. Chem.* **279**, 35630–35637.
[52] Wang, J.-F., Wei, D.-Q., Zheng, S.-Y., Li, Y.-X. and Chou, K.-C. (2007) *Biochem. Biophys. Res. Commun.* **355**, 513–519.
[53] de Groot, M.J., Ackland, M.J., Horne, V.A., Alex, A.A. and Jones, B.C. (1999) *J. Med. Chem.* **42**, 1515–1524.
[54] Halliday, R.C., Jones, B.C., Park, B.K. and Smith, D.A. (1997) *Eur. J. Drug Metab. Pharmacokinet.* **22**, 291–294.
[55] Rowland, P., Blany, F.E., Smyth, M.G., Jones, J.J., Leydon, V.R., Oxbrow, A.K., et al. (2006) *J. Biol. Chem.* **281**, 7614–7622.
[56] Niwa, T., Murayama, N. and Yamazaki, H. (2008) *Curr. Drug Metab.* **9**, 453–462.
[57] Koley, A.P., Robinson, R.C. and Friedman, F.K. (1996) *Biochimie* **78**, 706–713.
[58] Koley, A.P., Buters, J.T.M., Robinson, R.C., Markowitz, A. and Friedman, F.K. (1997) *J. Biol. Chem.* **272**, 3149–3152.
[59] Yano, J.K., Wester, M.R., Schoch, G.A., Griffin, K.J., Stout, C.D. and Johnson, E.F. (2004) *J. Biol. Chem.* **279**, 38091–38094.
[60] Ekroos, M. and Sjoegren, T. (2006) *Proc. Natl. Acad. Sci. USA* **103**, 13682–13687.
[61] Williams, P.A, Cosme, J., Vinkovic, D.M., Ward, A., Angove, H.C., Day, P.J., Vonrhein, C., Tickle, I.J. and Jhoti, H. (2004) *Science (Washington, DC)* **305**, 683–686.
[62] Hesse, L.M., Venkatakrishnan, K., Court, M.H., von Moltke, L.L., Duan, S.X., Shader, R.I. and Greenblatt, D.J. (2000) *Drug Metab. Dispos.* **28**, 1176–1183.
[63] Ward, B.A., Gorski, J.C., Jones, D.R., Hall, S.D., Flockhart, D.A. and Desta, Z. (2003) *J. Pharmacol. Exp. Ther.* **306**, 287–300.
[64] Huang, Z., Roy, P. and Waxman, D.J. (2000) *Biochem. Pharmacol.* **59**, 961–972.
[65] Oda, Y., Hamaoka, N., Hiroi, T., Imaoka, S., Hase, I., Tanaka, K., Funae, Y., Ishizaki, T. and Asada, A. (2001) *Br. J. Clin. Pharmacol.* **51**, 281–285.
[66] Ekins, S., Iyer, M., Krasowski, M.D. and Kharasch, E.D. (2008) *Curr. Drug Metab.* **9**, 363–373.
[67] Korhonen, L.E., Turpeinen, M., Rahnasto, M., Wittekindt, C., Poso, A., Pelkonen, O., Rauniol, H. and Juvonen, R.O. (2007) *Br. J. Pharmacol.* **150**, 932–942.
[68] Ogilvie, B.W., Zhang, D., Li, W., Rodrigues, A.D., Gipson, A.E., Holsapple, J., Toren, P. and Parkinson, A. (2006) *Drug Metab. Dispos.* **34**, 191–197.
[69] Delaforge, M., Pruvost, A., Perrin, L. and Andre, F. (2005) *Drug Metab. Dispos.* **33**, 466–473.
[70] Melet, A., Marques-Soares, C., Schoch, G.A., Macherey, A.-C., Jaouen, M., Dansette, P.M., Sari, M.-A., Johnson, E.F. and Mansuy, D. (2004) *Biochemistry* **43**, 15379–15392.
[71] Schoch, G.A., Yano, J.K., Wester, M.R, Griffin, K.J., Stout, C.D. and Johnson, E.F. (2004) *J. Biol. Chem.* **279**, 9497–9503.

[72] Schoch, G.A., Yano, J.K., Sansen, S., Dansette, P.M., Stout, C.D. and Johnson, E.F. (2008) *J. Biol. Chem.* **2889**, 17227–17237.
[73] Armou, D., de Groot, M.J., Edwards, M., Perros, M., Pric, D.A., Stammen, B.L. and Wood, A. (2006) *Chem. Med. Chem.* **1**, 706–709.
[74] Skerratt, S. (2008) "19th Symposium on Medicinal Chemistry in Eastern England". University of Hertfordshire, Hatfield, UK.
[75] Middleton, D.S., MacKenzie, A.R., Newman, S.D., Corless, C., Warren, A., Marchington, A.P. and Jones, B. (2005) *Bioorg. Med. Chem. Lett.* 3957–3961.
[76] Middleton, D.S. (2002) 223rd American Chemical Society Meeting, Orlando, FL, USA.
[77] Middleton, D.S., Andrews, M., Glossop, P., Gymer, G., Hepworth, D., Jessiman, A., et al. (2008) *Bioorg. Med. Chem. Lett.* **18**, 4018–4021.
[78] Middleton, D.S. (2005) 230th American Chemical Society Meeting. Washington, DC, USA.
[79] Middleton, D.S. (2006) "17th Symposium on Medicinal Chemistry in Eastern England". University of Hertfordshire, Hatfield, UK.

Subject Index

A80915A, 147
Acid pump antagonists (APAs), 78
Acid-related diseases (ARD)
  defined, 76. See also H^+/K^+ ATPase proton pump inhibitor; Potassium-competitive inhibitors; Proton pump inhibitors (PPIs)
ACS. See Acute coronary syndrome (ACS)
Acute coronary syndrome (ACS), 167
3-acyl-4-anilinoquinoline derivatives, 126–135
Acyloxyalkyl, 86
Adenocard™, 164
Adenosine A_1 agonists
  synthetic non-purinergic, 190–194
Adenosine A_1 receptor
  binding affinities, 183, 184
  chemistry
    adenosine-derived A_1 agonists, 171–189
    agonists with non-adenosine-derived structures, 189–194
  overview, 163–164
  pharmacology
    function, 165–167
    species differences, 167
    structure, 165–167
    therapeutic application, 167–171
  selectivities, 183, 184
Adenosine derivatives, 185, 189
Adenosine-derived A_1 agonists
  adenosine derivatives, 189
  C^2-substituted adenosine derivatives, 182–183
  N^6-aryl- and arylalkyl-substituted adenosine derivatives, 180–182
  N^6-cycloalkyl-substituted adenosine derivatives, 172–178
  N^6-heterocyclic adenosine derivatives, 178–180
  purine moieties variation, 184–189
  ribose variation, 184–189
  structure activity relationship, 189

Adenosine triphosphatase activity, of the protein, 77
Adipocytes, and adenosine A_1 receptor, 171
AGN 201904-Z, 86
Agonistic activity, non-purinergic agonists, 192–194
Agonists, with non-adenosine-derived structures
  non-purinergic natural products, 189–190
  synthetic non-purinergic adenosine A_1 agonists, 190–194
AHR-9294, 126
Ala335, 150
ALE-36, 144
Alkoxyalkyl groups, 86
Alkoxycarbonyl, 86
Aminoethyl, 86
Amino-halogeno pyridinylmethylsulfinyl-imidazo[4,5- b]pyridine, 85
AMP-579, 185, 186
Angina pectoris, and adenosine A_1 receptor, 168
Anilinomethylsulfinylbenzimidazoles, 85
AR-H47108, 110
AR-H047108, 98
AR-H047116, 98
Arofylline (LAS-31025), 56
Asp-301, 255
Aspirin, 83
AstraZeneca, 223
Atrasentan for cancer, 207, 209, 229
Atrial fibrillation, and adenosine A_1 receptor, 168–169
Atrio-ventricular (A-V) node, 164
AU-1421, 140–141
AU-2064, 140–141
AVE-8112, 58
A-V nodal conduction block, 167
A-V node. See Atrio-ventricular (A-V) node
Avosentan, for diabetic nephropathy, 208, 216, 217
AWD-12-281, 52

SUBJECT INDEX

AZD-0865, 135
AZD-5745, 98
AZD-9139, 98

Banyu pyridine derivatives, 141
Benzimidazoles
  5-aza-, 116
  benzimidazole NH, 86
  chromanyl-4-yl substituent in, 114
  deuterated analogues, 116
  inhibition of gastric acid secretion, 114–116
  isomeric, 114, 115
  N1-methyl substitution, 113
  ortho substitution in, 113
  tricyclic, 117–119
Benzothiazine-based acids, 228
8-benzylamino imidazo[1,2-a]pyridines, 97
Benzyloxy-imidazopyridines, 91
Bethanechol, 83
BIBN4096BS, 3, 12–15
Binding affinities, 182, 183, 184, 185
  adenosine A_1 receptor, 177–178, 183, 184
  AMP-579, 186
  in DDT_1 MF-2 cells, 177
  of GR-79236, 175
  in GTPγS binding assay, 179
  N^6-bicycloalkyladenosines, 174
  N^6-cycloalkyladenosines, 175
  and selectivities of AMP-579, 186
  and selectivities of ring-constrained (N)-methanocarba nucleosides, 187
  and selectivities of VCP102 and analogues, 181
  VCP102, 181
Bis-anilinopyrimidine derivatives, 137
BMS series, heteroaryl sulfonamides, 218, 220–221
Bosentan
  for digital ulcer disease, 207, 214, 216
  for PAH, 207, 208, 214, 216
BQ-123, peptide antagonists, 211–212
Bradycardia, 168
Brain haemorrhage, clazosentan for, 208, 209, 216
Butenolides, 228
BY-359, 104
BY-841. *See* Pumaprazole

Byk99, 150
BYK61359, 104

Ca^{2+} ATPase, 77
CABG. *See* Coronary artery bypass graft (CABG)
Caffeine, 2
Calcitonin gene-related peptide (CGRP), 3
^{14}C-aminopyrine accumulation, 88
Cancer, atrasentan for, 207, 209, 229
Capsaicin-induced dermal blood flow (CIDV) model, 30–31
Carboxylic acid antagonists
  benzothiazine-based acids, 228
  butenolides, 228
  diarylpropionic acids, 224–225
  indane-2-carboxylic acids, 225, 227
  pyrrolidine-3-carboxylic acid, 229–230
Cardioprotection, and adenosine A_1 receptor, 169–170
Catalytic cycle, 77
CC-1088, 56
C3-cyanomethyl group, 92
CDC-801, 56
CDP-840, 56
Central nervous system, and adenosine A_1 receptor, 170
CEREP Bioprint database, 223
C7-(4-fluorobenzyloxy) group, 125, 126
CGRP receptor antagonists, 8–9
  azabenzimidazolinone derivative, 25
  benzodiazepinone-based, 22–24
  BIBN4096BS, 12–14, 12–15
  C-6 aryl substituents, 25–26
  2′-(4-chlorophenyl)dihydroquinine, 11
  2,3-difluorophenyl analogue, 26
  methyl and ethyl analogues, 27
  MK-0974, 16, 28–31
  N-(4-piperidinyl)piperazine, 12
  orally active, 15–16
  pyridinone, 17–18
  (3*R*)-amino-(6*S*)-phenyl caprolactam, 24
  SB-211973, 11–12
  SB-273779, 11–12
  spirohydantoin-based, 16–17
  spirohydantoin replacements, 19–22
CHD. *See* Coronary heart disease (CHD)
Chinese hamster ovary cells, 166

2′-(4-chlorophenyl)dihydroquinine, 11
Cilomilast, 49, 50
Clazosentan, for brain haemorrhage, 208, 209, 216
C6-N-hydroxyethyl amide (AZD-0865), 98–100
Coronary artery bypass graft (CABG), 167
Coronary heart disease (CHD), 168
C^2-substituted adenosine derivatives, 182–183
C7-THIQ group, 124
CVT-3619, 178
Cyclic sulfenamide, 79
Cyclopropane carboxylic acid, 59
CYP1A, 252–253
CYP3A4, 255–256
CyP450 3A4, 82, 83
CYP2B6, 256
CYP2C8, 256
CYP2C9, 253–254
 inhibition, 257, 258
CYP2C19, 254
CyP450 2C19, 80, 82, 84
CYP2D6, 254–255
 inhibition, 257
Cys813, 84
Cys815, 149
Cys822, 84
Cys813Thr, 148
Cytochrome P450 (CYP), 240
 chemical inhibitors, 245–246
 CYP1A2, 252–253
 CYP3A4, 255–256
 CYP2B6, 256
 CYP2C8, 256
 CYP2C9, 253–254, 257, 258
 CYP2C19, 254
 CYP2D6, 254–255, 257
 and medicinal chemistry, 256–260
 reaction phenotyping, 245
Cytosol, DME, 243–244

Darusentan, for hypertension, 208, 225
DBM-819, 132
DDT_1 MF-2 cells, 177
Diabetic nephropathy, avosentan, 208, 216, 217
Diarylpropionic acids, 224–225

Digital ulcer disease, bosentan for, 207, 214, 216
Dihydroergotamine (DHT), 2
Dithiothreitol, 88
DME. *See* Drug metabolising enzymes (DME)
Drug metabolising enzymes (DME)
 CYP. *See* Cytochrome P450 (CYP)
 defined, 240
 hepatocytes, 242–243
 liver slices, 242
 MBI. *See* Mechanism-based inhibition (MBI)
 Phase I metabolism, 240, 241
 Phase II metabolism, 240, 241
 reversible inhibition, 247–248
 sources, 241–242
 subcellular fractions. *See* Subcellular fractions, DME
 in vitro studies
   chemical inhibitors, 245–246
   metabolic stability, 244–245
   phenotyping, 245

Ebselen, 89
ELB-353, 58
Endothelin. *See also* Endothelin receptor antagonists
 amino acid structure, 205
 biological effects, 205–206
Endothelin converting enzymes (ECE), 204
Endothelin-1 (ET-1), 204
Endothelin-2 (ET-2), 204
Endothelin-3 (ET-3), 204
Endothelin receptor antagonists
 carboxylic acids
   benzothiazine-based, 228
   butenolide, 228
   diarylpropionic acids, 224–225
   indane-2-carboxylic acids, 225, 227
   pyrrolidine-3-carboxylic acid, 229–230
 clinical status, 207–209
 N-acylsulfonamide, 230–231
 overview, 210–211
 peptide based. *See* Peptide antagonists
 side effects, 209
 structural biology, 207

sulfonamides
  heteroaryl. *See* Heteroaryl sulfonamides
  pyrimidine. *See* Pyrimidine sulfonamides
Enterochromaffin (ECL) cells, 90
Eosinophils, 39–40
Ergotamine (ET), 2
Erythromycin, 255
Esomeprazole, 80, 83

Flavin mono-oxygenase (FMO), 240
Flurbiprofen, 253–254
FR 139317, peptide antagonists, 212, 213

Gastroesophageal reflux disease (GERD), 76
  AZD-0865 therapy, 99–100
  esomeprazole for, 83
Glu-213, 255
Glu795, 149
Gly795, 150
GPD-1116, 58
GRC-4039, 57
GTPγS binding assay, 179
GW-493838, 180
GW842470. *See* AWD-12-281

H. pylori infection, 84, 95, 120
H 335/25 compound, 130
Hepatic drug clearance (CL_H), 250–251
  models for, 250
Hepatocytes, DME, 242–243
Heteroaryl sulfonamides, 218–224
  BMS series, 218, 220–221
  properties of, 219
  PS-433540, 221
  pyrazine, 223
  sitaxsentan, 222
  TBC-3711, 222
  UK-419106, 223, 224
  UK-434643, 223–224
  ZD-1611, 223
  ZD-4054, 223
Histamine-stimulated acid secretion, 83
H^+/K^+ ATPase IC_{50} values, 121–127
H^+/K^+ ATPase proton pump inhibitor, 76
  alternative types, 142–147
  Cys822 of, 85
  and *de novo* protein synthesis, 87

gastric acid secretion by, 77
  inhibition of, 79
  purification, 78
  reversible inhibition, 91
H_2 receptor antagonists (H_2RAs), 76
Human recombinant A_1 receptors, 166
4-hydroxyderricin, 146

IDDB, 223
Ile816, 149
Ile820, 149
Imidazo[1,2a]pyrano[2,3-c]pyridine
  analogue, 93
Imidazo[1,2-a]thieno[3,2-c]pyridine SPI-447,
  142–143
Imidazole, 139, 142
Imidazopyridines, 139
  alternatives to, 110–112
  aryl, heteroaryl and heterocyclyl
    substituents at the C6-position, 103
  BY-841, 93–96
  Byk Gulden group studies, 94
  carbamate derivative of, 94
  6-carboxamide substituted N-tricyclic
    derivatives, 108
  C3-CH$_2$CN compounds, 103
  chromane moiety of, 102
  C6-N-hydroxyethyl amide (AZD-0865),
    98–100
  compound binding results, 92
  deuterated analogues of compounds, 101
  2,6-dialkyl substitution of, 96
  2-hydroxypropionylamino and ureido
    substituents of, 94
  and inhibition of gastrin-induced gastric
    acid secretion, 96
  modifications in C8 substituent, 100
  SCH-28080, 91–92, 94
  with substitution at both 7- and
    8-positions, 103–107
  tricyclic analogues, 103–109
  and two-atom linker with phenyl group, 93
  Yamanouchi studies, 93
  YM-020, 93
Indane-2-carboxylic acids, 225, 227
Indomethacin, 83
INO-8875, 188
Inpharmatica's Drugstore and Drugbank, 223

Intersystem extrapolation factor (ISEF), 247
IPL-455903, 55
I-Pr analogue AHR-9294, 126
Ischemic preconditioning, and adenosine A_1 receptor, 169–170
ISEF. See Intersystem extrapolation factor (ISEF)
IY81149, 81

K-8974, 217, 218
Ketoconazole, 255–256

L-454560, 58–59
Lansoprazole, 80, 81, 83–84, 87
Leminoprazole, 82
Leu141, 150
Leu811, 149
Ligand efficiency (LE), 210–211
Ligand lipophilicity efficiency (LLE), 211
Lipid metabolism, and adenosine A_1 receptor, 171
Lirimilast (BAY-19-8004), 56
Liver S9 fractions, DME, 243
Liver slices, DME, 242
LLE. See Ligand lipophilicity efficiency (LLE)

MDPQ, 132
ME3407, 82
Mechanism-based inhibition (MBI), 248–249
MEM 1414, 55
MEM 1917, 55
β-mercaptoethanol, 144
Met334, 150
Met336, 149
Metabolic stability, DME, 244–245
Metyrapone, 255–256
Microsomes, DME, 244
Migraine headache
 CGRP and, 4–5
  affinity antagonist interactions, 9–10
  antagonists of receptors. See CGRP receptor antagonists
  receptor, 5–7
  selectivity vs adrenomedullin (AM) receptors, 9
  in vitro assays, 7–8

 in vivo potency, 10–11
 early treatments, 2
 epidemiology and clinical manifestations, 4
MK-0873, 56
MK-0974, 3, 16, 28–31
ML3000, 147
Myocardial infarction (MI), and adenosine A_1 receptor, 169–170

N-acylsulfonamide antagonists, 230–231
Na^+/K^+ ATPase, 77, 82
2-naphthyl moiety, 124
1,8-naphthyridine N-oxide, 59
N^6-Aryl- and arylalkyl-substituted adenosine derivatives, 180–182
N^6-bicycloalkyladenosines, adenosine receptor binding, 174
N^6-cyclic adenosine analogues, 173
N^6-cycloalkyladenosines, 175
N^6-cycloalkyl-substituted adenosine derivatives, 172–178
NERD patients, 76
Neutrophils, 40
N^6-Heterocyclic adenosine derivatives, 178–180
N-methanocarba nucleosides, 187
Non-purinergic natural products, 189–190
N^6-substituted-5′-modified adenosine synthesis, 176

Oglemilast (GRC-3886), 52
Omeprazole, 79, 80, 82, 90, 91, 99, 135
 (S)-enantiomer of, 83
Omeprazole with MDR2 inhibitor MK-571, 87
OPC22575, 82

PAH. See Pulmonary arterial hypertension (PAH)
Pantoprazole, 80, 81, 83, 84
Parallel tube model, 250
Paroxysmal supraventricular tachyarrhythmias (PSVT), 169
PD-145065, peptide antagonists, 212, 213
PDE4 inhibitors. See Phosphodiesterase (PDE) enzyme family
Pentagastrin-induced acid secretion, 94

SUBJECT INDEX

Pentagastrin-stimulated acid secretion, 108, 111
Peptide antagonists, 211–213
  BQ-123, 211–212
  FR 139317, 212, 213
  PD-145065, 212, 213
  TAK-044, 212, 213
Phe332, 150
Phenotyping, DME, 245
Phosphodiesterase (PDE) enzyme family
  compounds in phase I clinical trials, 56–58
  compounds in phase II clinical trials, 52–56
  compounds in phase III clinical trials, 49–51
  compounds in pre-clinical evaluation, 58–60
  consequences of inhibition, 39–44, 47–48
  marketed compounds, 48–49
  molecular cloning, localisation and regulation, 38
  patent applications submitted, 61–63
  PDE1–11, 37
  potential indications for, 47
  in vitro pharmacology, 38–44
    in airway epithelial cells, 42
    in endothelial cells, 42–43
    in eosinophils, 39–40
    in fibroblasts, 44
    in lymphocytes, 41–42
    in mast cells and basophils, 43
    in monocytes and macrophages, 40–41
    in neutrophils, 40
    in pulmonary artery smooth muscle cells, 44
    in smooth muscle cells, 43
    in vascular smooth muscle cells (VSMCs), 44
  in vivo pharmacology, 44–47
    asthma, 44–45
    COPD, 45
    cough, 45
    depression, 46–47
    inflammatory bowel disease, 46
    memory, 46
    multiple sclerosis, 45
    pulmonary artery hypertension, 47
    renal diseases, 47
    rheumatoid arthritis, 46

Phospholipase C modulation, adenosine A_1 receptor effect, 166
Piclamilast (RPR-73401), 56
Potassium-competitive inhibitors, 148–150
Pro798, 150
Progesterone, 255
Proline-rich sequence binding (SH3) proteins, 38
Proton pump inhibitors (PPIs), 76
  acid suppression, 89–91
  combinations, 87
  derivatives, 87
  irreversible, duration of pH, 78
  mechanism of inhibition of H^+/K^+ ATPase by, 79
  modifications, 85–89
  N-sulphonyl analogues of, 86
  off-label use of, 90
  pyridine-2-ylmethylsulphinyl-1 H-benzimidazole, 80
  shortcomings, 80
Prous Integrity, 223
PS-433540, 221
PSVT. *See* Paroxysmal supraventricular tachyarrhythmias (PSVT)
Pulmonary arterial hypertension (PAH)
  ambrisentan for, 225
  bosentan for, 207, 208, 214, 216
  sitaxsentan for, 222
Pumaprazole, 93, 94, 96, 105
Purine moieties variation in adenosine A_1 receptor agonists, 184–189
Pyrazine, 223
Pyridine-2-ylmethylsulfinyl-1 H-benzimidazoles, 85
Pyridoimidazole, 85
Pyridopyrimidine derivatives, 138
Pyrimidine sulfonamides
  avosentan, 216, 217
  bosentan, 214, 216
  clazosentan, 216
  K-8974, 217, 218
  properties of, 215
  TA-0201, 217, 218
  tezosentan, 216
  YM-598, 217, 218
Pyrrolidine-3-carboxylic acid, 229–230
Pyrrolo[3,2-c]pyridine derivatives, 121

Pyrrolo[2,3-d]pyridazines, 120–121
Pyrrolo or 2,3-dihydropyrroloquinoline
 template, 132–133
Pyrrolopyridine isomers, 121–126
Pyrrolopyridine systems, 112
Pyrroloquinolines, 134

Rabeprazole, 80, 81, 84, 90
 pK_a of, 80
RAF. See Relative activity factor (RAF)
Receptor desensitization, 166
Recombinant enzymes, 244
Relative activity factor (RAF), 246
Revaprazan hydrochloride, 135
Reversible inhibition, DME, 247–248
Ribose variation in adenosine A_1 receptor
 agonists, 184–189
Right ventricular failure, tezosentan for,
 208, 216
Roflumilast, 50–51
RS13232A, 89

S-0139, 232
S-3337, 82
Saviprazole, 81
SB-211973, 11–12
SB-273779, 11–12
SB-641257, 135
SCH-28080, 87, 91, 92, 110, 126, 129, 142,
 146, 148
SCH-32651, 110
SCH35191, 60
SCH-351591, 57
SCH365351, 60
Scopadulcic acid B, 146
Scopadulciol, 146
Seledenoson, 177
Sepracor, 84
Sitaxsentan, 222
SKF 96067, 126, 127
SKF 97574, 128
SKF quinazoline derivatives, 138
Sofalcone, 145–146
Sophoradin, 146
Soraprazan, 104, 105, 106
SPI-447, 149
Spiro-tricyclicbenzimidazole
 analogues, 118

Subcellular fractions, DME
 cytosol, 243–244
 liver S9 fractions, 243
 microsomes, 244
 recombinant enzymes, 244
Sulfenamide, 87
Sulfenic acid, 87
Sulfonamide antagonists. See Heteroaryl
 sulfonamides
 See Pyrimidine sulfonamides
Sulphonamide, 258
Synthetic non-purinergic adenosine A_1
 agonists, 190–194

T330, 87
TA-0201, 217, 218
TAK-044, peptide antagonists,
 212, 213
TBC-3711, 222
Tenatoprazole (TU-199), 81, 85
Tetomilast (OPC-6535), 51
Tetracyclic sulfenamide, 80
Tetrahydroisoquinoline (THIQ)
 moiety, 121
Tezosentan, for right ventricular failure,
 208, 216
Therapeutic application, adenosine A_1
 receptor
 angina pectoris, 168
 atrial fibrillation, 168–169
 cardioprotection, 169–170
 central nervous system, 170
 ischemic preconditioning, 169–170
 lipid metabolism, 171
1,2,4-thiadiazolo[4, 5-a]benzimidazole
 and imidazo[1,2-d]-1,2,
 4-thiadiazole, 88
Thiazolo-guanidine derivatives, 139
Thiocyanate, 92
Thr929, 149
TMPFPIP, 148
Tofimilast (CP-325366), 56
Trans vinyl compound, 93
Triazolo[1,5-a]pyridines, 119
Triazolo[4,3-a]pyridines, 119
Triazolopyridines, 119–120
Trifluoperazine, 144
Triptans, 2

Tyr799, 150
Tyr802, 149
Tyr928, 149
Tyr801Phe, 149

UK-419106, 223, 224
UK-434643, 223–224
UK-500,001, 53

VCP102, 181, 182
Verapamil, 144

Warfarin, 253
Well-stirred model for drug disposition in liver, 250

YH-1885, 135
YJA20379-1, 88
YJA-20379-2, 89
YJA20379-8, 129
YM-020, 93
YM-598, 217, 218
YM-976, 56, 57
Yuhan pyrimidine derivatives, 136–137

ZD-1611, 223
ZD-4054, 223–224
Zegerid, 85–86
Zibotentan 25, 223
Zolimidine, 110

Cumulative Index of Authors for Volumes 1–47

The volume number, (year of publication) and page number are given in that order.

Aboul-Ela, F., 39 (2002) 73
Adam, J., 44 (2006) 209
Adams, J.L., 38 (2001) 1
Adams, S.S., 5 (1967) 59
Afshar, M., 39 (2002) 73
Agrawal, K.C., 15 (1978) 321
Albrecht, W.J., 18 (1981) 135
Albrecht-Küpper, B., 47 (2009) 163
Allain, H., 34 (1997) 1
Allen, M.J., 44 (2006) 335
Allen, N.A., 32 (1995) 157
Allender, C.J., 36 (1999) 235
Altmann, K.-H., 42 (2004) 171
Andrews, P.R., 23 (1986) 91
Ankersen, M., 39 (2002) 173
Ankier, S.I., 23 (1986) 121
Appendino, G., 44 (2006) 145
Arrang, J.-M., 38 (2001) 279
Armour, D., 43 (2005) 239
Aubart, K., 44 (2006) 109

Badger, A.M., 38 (2001) 1
Bailey, E., 11 (1975) 193
Ballesta, J.P.G., 23 (1986) 219
Bamford, M., 47 (2009) 75
Banner, K.H., 47 (2009) 37
Banting, L., 26 (1989) 253; 33 (1996) 147
Barbier, A.J., 44 (2006) 181
Barker, G., 9 (1973) 65
Barnes, J.M., 4 (1965) 18
Barnett, M.I., 28 (1991) 175
Batt, D.G., 29 (1992) 1
Beaumont, D., 18 (1981) 45
Beckett, A.H., 2 (1962) 43; 4 (1965) 171
Beckman, M.J., 35 (1998) 1
Beddell, C.R., 17 (1980) 1
Beedham, C., 24 (1987) 85
Beeley, L.J., 37 (2000) 1

Beher, D., 41 (2003) 99
Beisler, J.A., 19 (1975) 247
Bell, J.A., 29 (1992) 239
Belliard, S., 34 (1997) 1
Benfey, B.G., 12 (1975) 293
Bentué-Ferrer, D., 34 (1997) 1
Bernstein, P.R., 31 (1994) 59
Besra, G.S., 45 (2007) 169
Bhowruth, V., 45 (2007) 169
Binnie, A., 37 (2000) 83
Bischoff, E., 41 (2003) 249
Biswas, K., 46 (2008) 173
Black, M.E., 11 (1975) 67
Blandina, P., 22 (1985) 267
Bond, P.A., 11 (1975) 193
Bonta, I.L., 17 (1980) 185
Booth, A.G., 26 (1989) 323
Boreham, P.F.I., 13 (1976) 159
Bös, M., 44 (2006) 65
Bowman, W.C., 2 (1962) 88
Bradner, W.T., 24 (1987) 129
Bragt, P.C., 17 (1980) 185
Brain, K.R., 36 (1999) 235
Branch, S.K., 26 (1989) 355
Braquet, P., 27 (1990) 325
Brezina, M., 12 (1975) 247
Brooks, B.A., 11 (1975) 193
Brown, J.R., 15 (1978) 125
Brunelleschi, S., 22 (1985) 267
Bruni, A., 19 (1982) 111
Buckingham, J.C., 15 (1978) 165
Bulman, R.A., 20 (1983) 225
Burgey, C.S., 47 (2009) 1

Camaioni, E., 42 (2004) 125
Carman-Krzan, M., 23 (1986) 41
Carruthers, N.I., 44 (2006) 181
Cassells, A.C., 20 (1983) 119

Casy, A.F., 2 (1962) 43; 4 (1965) 171; 7 (1970) 229; 11 (1975) 1; 26 (1989) 355
Casy, G., 34 (1997) 203
Caton, M.P.L., 8 (1971) 217; 15 (1978) 357
Chambers, M.S., 37 (2000) 45
Chang, J., 22 (1985) 293
Chappel, C.I., 3 (1963) 89
Chatterjee, S., 28 (1991) 1
Chawla, A.S., 17 (1980) 151; 22 (1985) 243
Chen, C., 45 (2007) 111
Chen, J.J., 46 (2008) 173
Cheng, C.C., 6 (1969) 67; 7 (1970) 285; 8 (1971) 61; 13 (1976) 303; 19 (1982) 269; 20 (1983) 83; 25 (1988) 35
Cherry, M., 44 (2006) 1
Clark, R.D., 23 (1986) 1
Clitherow, J.W., 41 (2003) 129
Cobb, R., 5 (1967) 59
Cochrane, D.E., 27 (1990) 143
Corbett, J.W., 40 (2002) 63
Costantino, G., 42 (2004) 125
Coulton, S., 31 (1994) 297; 33 (1996) 99
Cowley, P.M., 44 (2006) 209
Cox, B., 37 (2000) 83
Crossland, J., 5 (1967) 251
Crowshaw, K., 15 (1978) 357
Cushman, D.W., 17 (1980) 41
Cuthbert, A.W., 14 (1977) 1

Dabrowiak, J.C., 24 (1987) 129
Daly, M.J., 20 (1983) 337
D'Arcy, P.F., 1 (1961) 220
Daves, G.D., 13 (1976) 303; 22 (1985) 1
Davies, G.E., 2 (1962) 176
Davies, R.V., 32 (1995) 115
De Clercq, E., 23 (1986) 187
De Gregorio, M., 21 (1984) 111
De Luca, H.F., 35 (1998) 1
De, A., 18 (1981) 117
Deaton, D.N., 42 (2004) 245
Demeter, D.A., 36 (1999) 169
Denyer, J.C., 37 (2000) 83
Derouesné, C., 34 (1997) 1
Dimitrakoudi, M., 11 (1975) 193
Donnelly, M.C., 37 (2000) 83
Dover, L.G., 45 (2007) 169
Draffan, G.H., 12 (1975) 1

Drewe, J.A., 33 (1996) 233
Drysdale, M.J., 39 (2002) 73
Dubinsky, B., 36 (1999) 169
Duckworth, D.M., 37 (2000) 1
Duffield, J.R., 28 (1991) 175
Durant, G.J., 7 (1970) 124
Dvorak, C.A., 44 (2006) 181

Eccleston, J.F., 43 (2005) 19
Edwards, D.I., 18 (1981) 87
Edwards, P.D., 31 (1994) 59
Eglen, R.M., 43 (2005) 105
Eldred, C.D., 36 (1999) 29
Ellis, G.P., 6 (1969) 266; 9 (1973) 65; 10 (1974) 245
Evans, B., 37 (2000) 83
Evans, J.M., 31 (1994) 409

Falch, E., 22 (1985) 67
Fantozzi, R., 22 (1985) 267
Feigenbaum, J.J., 24 (1987) 159
Ferguson, D.M., 40 (2002) 107
Feuer, G., 10 (1974) 85
Finberg, J.P.M., 21 (1984) 137
Fletcher, S.R., 37 (2000) 45
Flörsheimer, A., 42 (2004) 171
Floyd, C.D., 36 (1999) 91
François, I., 31 (1994) 297
Frank, H., 27 (1990) 1
Freeman, S., 34 (1997) 111
Fride, E., 35 (1998) 199

Gale, J.B., 30 (1993) 1
Ganellin, C.R., 38 (2001) 279
Garbarg, M., 38 (2001) 279
Garratt, C.J., 17 (1980) 105
Gerspacher, M., 43 (2005) 49
Gill, E.W., 4 (1965) 39
Gillespie, P., 45 (2007) 1
Ginsburg, M., 1 (1961) 132
Glennon, R.A., 42 (2004) 55
Goldberg, D.M., 13 (1976) 1
Goodnow, Jr. R.A., 45 (2007) 1
Gould, J., 24 (1987) 1
Graczyk, P.P., 39 (2002) 1
Graham, J.D.P., 2 (1962) 132
Green, A.L., 7 (1970) 124
Green, D.V.S., 37 (2000) 83; 41 (2003) 61

Greenhill, J.V., 27 (1990) 51; 30 (1993) 206
Griffin, R.J., 31 (1994) 121
Griffiths, D., 24 (1987) 1
Griffiths, K., 26 (1989) 299
Groenewegen, W.A., 29 (1992) 217
Groundwater, P.W., 33 (1996) 233
Guile, S.D., 38 (2001) 115
Gunda, E.T., 12 (1975) 395; 14 (1977) 181
Gylys, J.A., 27 (1990) 297

Hacksell, U., 22 (1985) 1
Haefner, B., 43 (2005) 137
Hall, A.D., 28 (1991) 41
Hall, S.B., 28 (1991) 175
Halldin, C., 38 (2001) 189
Halliday, D., 15 (1978) 1
Hammond, S.M., 14 (1977) 105; 16 (1979) 223
Hamor, T.A., 20 (1983) 157
Haning, H., 41 (2003) 249
Hanson, P.J., 28 (1991) 201
Hanus, L., 35 (1998) 199
Hargreaves, R.B., 31 (1994) 369
Harris, J.B., 21 (1984) 63
Harrison, T., 41 (2003) 99
Hartley, A.J., 10 (1974) 1
Hartog, J., 15 (1978) 261
Heacock, R.A., 9 (1973) 275; 11 (1975) 91
Heard, C.M., 36 (1999) 235
Heinisch, G., 27 (1990) 1; 29 (1992) 141
Heller, H., 1 (1961) 132
Henke, B.R., 42 (2004) 1
Heptinstall, S., 29 (1992) 217
Herling, A.W., 31 (1994) 233
Hider, R.C., 28 (1991) 41
Hill, S.J., 24 (1987) 30
Hillen, F.C., 15 (1978) 261
Hino, K., 27 (1990) 123
Hjeds, H., 22 (1985) 67
Holdgate, G.A., 38 (2001) 309
Hooper, M., 20 (1983) 1
Hopwood, D., 13 (1976) 271
Hosford, D., 27 (1990) 325
Hu, B., 41 (2003) 167
Hubbard, R.E., 17 (1980) 105
Hudkins, R.L., 40 (2002) 23
Hughes, R.E., 14 (1977) 285

Hugo, W.B., 31 (1994) 349
Hulin, B., 31 (1994) 1
Humber, L.G., 24 (1987) 299
Hunt, E., 33 (1996) 99
Hutchinson, J.P., 43 (2005) 19

Ijzerman, A.P., 38 (2001) 61
Imam, S.H., 21 (1984) 169
Ince, F., 38 (2001) 115
Ingall, A.H., 38 (2001) 115
Ireland, S.J., 29 (1992) 239

Jacques, L.B., 5 (1967) 139
James, K.C., 10 (1974) 203
Jameson, D.M., 43 (2005) 19
Jászberényi, J.C., 12 (1975) 395; 14 (1977) 181
Jenner, F.D., 11 (1975) 193
Jennings, L.L., 41 (2003) 167
Jewers, K., 9 (1973) 1
Jindal, D.P., 28 (1991) 233
Jones, B.C., 41 (2003) 1; 47 (2009) 239
Jones, D.W., 10 (1974) 159
Jorvig, E., 40 (2002) 107
Judd, A., 11 (1975) 193
Judkins, B.D., 36 (1999) 29

Kadow, J.F., 32 (1995) 289
Kapoor, V.K., 16 (1979) 35; 17 (1980) 151; 22 (1985) 243; 43 (2005) 189
Kawato, Y., 34 (1997) 69
Kelly, M.J., 25 (1988) 249
Kendall, H.E., 24 (1987) 249
Kennett, G.A., 46 (2008) 281
Kennis, L.E.J., 33 (1996) 185
Kew, J.N.C., 46 (2008) 131
Khan, M.A., 9 (1973) 117
Kiefel, M.J., 36 (1999) 1
Kilpatrick, G.J., 29 (1992) 239
Kindon, N.D., 38, (2001) 115
King, F.D., 41 (2003) 129
Kirst, H.A., 30 (1993) 57; 31 (1994) 265
Kitteringham, G.R., 6 (1969) 1
Kiyoi, T., 44 (2006) 209
Knight, D.W., 29 (1992) 217
Körner, M., 46 (2008) 205
Kobayashi, Y., 9 (1973) 133
Koch, H.P., 22 (1985) 165
Kopelent-Frank, H., 29 (1992) 141

CUMULATIVE AUTHOR INDEX

Kramer, M.J., 18 (1981) 1
Krause, B.R., 39 (2002) 121
KrogsgaardLarsen, P., 22 (1985) 67
Kulkarni, S.K., 37 (2000) 135
Kumar, K., 43 (2005) 189
Kumar, M., 28 (1991) 233
Kumar, S., 38 (2001) 1; 42 (2004) 245
Kwong, A.D., 39 (2002) 215

Lambert, P.A., 15 (1978) 87
Launchbury, A.P., 7 (1970) 1
Law, H.D., 4 (1965) 86
Lawen, A., 33 (1996) 53
Lawson, A.M., 12 (1975) 1
Leblanc, C., 36 (1999) 91
Lee, C.R., 11 (1975) 193
Lee, J.C., 38 (2001) 1
Lenton, E.A., 11 (1975) 193
Lentzen, G., 39 (2002) 73
Letavic, M.A., 44 (2006) 181
Levin, R.H., 18 (1981) 135
Lewis, A.J., 19 (1982) 1; 22 (1985) 293
Lewis, D.A., 28 (1991) 201
Lewis, J.A., 37 (2000) 83
Li, Y., 43 (2005) 1
Lien, E.L., 24 (1987) 209
Lightfoot, A.P., 46 (2008) 131
Ligneau, X., 38 (2001) 279
Lin, T.-S., 32 (1995) 1
Liu, M.-C., 32 (1995) 1
Livermore, D.G.H., 44 (2006) 335
Llinas-Brunet, M., 44 (2006) 65
Lloyd, E.J., 23 (1986) 91
Lockhart, I.M., 15 (1978) 1
Lord, J.M., 24 (1987) 1
Lowe, I.A., 17 (1980) 1
Lucas, R.A., 3 (1963) 146
Lue, P., 30 (1993) 206
Luscombe, D.K., 24 (1987) 249

Mackay, D., 5 (1967) 199
Main, B.G., 22 (1985) 121
Malhotra, R.K., 17 (1980) 151
Malmström, R.E., 42 (2004) 207
Manchanda, A.H., 9 (1973) 1
Mander, T.H., 37 (2000) 83
Mannaioni, P.F., 22 (1985) 267

Maroney, A.C., 40 (2002) 23
Martin, I.L., 20 (1983) 157
Martin, J.A., 32 (1995) 239
Masini, F., 22 (1985) 267
Matassova, N., 39 (2002) 73
Matsumoto, J., 27 (1990) 123
Matthews, R.S., 10 (1974) 159
Maudsley, D.V., 9 (1973) 133
May, P.M., 20 (1983) 225
McCague, R., 34 (1997) 203
McFadyen, I., 40 (2002) 107
McLelland, M.A., 27 (1990) 51
McNeil, S., 11 (1975) 193
Mechoulam, R., 24 (1987) 159; 35 (1998) 199
Meggens, A.A.H.P., 33 (1996) 185
Megges, R., 30 (1993) 135
Meghani, P., 38 (2001) 115
Merritt, A.T., 37 (2000) 83
Metzger, T., 40 (2002) 107
Michel, A.D., 23 (1986) 1
Middlemiss, D.N., 41 (2003) 129
Middleton, D.S., 47 (2009) 239
Miura, K., 5 (1967) 320
Moncada, S., 21 (1984) 237
Monck, N.J.T., 46 (2008) 281
Monkovic, I., 27 (1990) 297
Montgomery, J.A., 7 (1970) 69
Moody, G.J., 14 (1977) 51
Mordaunt, J.E., 44 (2006) 335
Morris, A., 8 (1971) 39; 12 (1975) 333
Morrison, A.J., 44 (2006) 209
Mort, C.J.W., 44 (2006) 209
Mortimore, M.P., 38 (2001) 115
Munawar, M.A., 33 (1996) 233
Murchie, A.I.H., 39 (2002) 73
Murphy, F., 2 (1962) 1; 16 (1979) 1
Musallan, H.A., 28 (1991) 1
Musser, J.H., 22 (1985) 293

Natoff, I.L., 8 (1971) 1
Neidle, S., 16 (1979) 151
Nell, P.G., 47 (2009) 163
Nicholls, P.J., 26 (1989) 253
Niewöhner, U., 41 (2003) 249
Nodiff, E.A., 28 (1991) 1
Nordlind, K., 27 (1990) 189
Nortey, S.O., 36 (1999) 169

O'Hare, M., 24 (1987) 1
O'Reilly, T., 42 (2004) 171
Ondetti, M.A., 17 (1980) 41
Ottenheijm, H.C.J., 23 (1986) 219
Oxford, A.W., 29 (1992) 239

Paget, G.E., 4 (1965) 18
Palatini, P., 19 (1982) 111
Palazzo, G., 21 (1984) 111
Palfreyman, M.N., 33 (1996) 1
Palmer, D.C., 25 (1988) 85
Palmer, M.J., 47 (2009) 203
Parkes, M.W., 1 (1961) 72
Parnham, M.J., 17 (1980) 185
Parratt, J.R., 6 (1969) 11
Patel, A., 30 (1993) 327
Paul, D., 16 (1979) 35; 17 (1980) 151
Pearce, F.L., 19 (1982) 59
Peart, W.S., 7 (1970) 215
Pellicciari, R., 42 (2004) 125
Perni, R.B., 39 (2002) 215
Petrow, V., 8 (1971) 171
Picard, J.A., 39 (2002) 121
Pike, V.W., 38 (2001) 189
Pinder, R.M., 8 (1971) 231; 9 (1973) 191
Poda, G., 40 (2002) 107
Ponnudurai, T.B., 17 (1980) 105
Potter, B.V.L., 46 (2008) 29
Powell, W.S., 9 (1973) 275
Power, E.G.M., 34 (1997) 149
Press, N.J., 47 (2009) 37
Price, B.J., 20 (1983) 337
Prior, B., 24 (1987) 1
Procopiou, P.A., 33 (1996) 331
Purohit, M.G., 20 (1983) 1

Ram, S., 25 (1988) 233
Rampe, D., 43 (2005) 1
Reader, J., 44 (2006) 1
Reckendorf, H.K., 5 (1967) 320
Reddy, D.S., 37 (2000) 135
Redshaw, S., 32 (1995) 239
Rees, D.C., 29 (1992) 109
Reitz, A.B., 36 (1999) 169
Repke, K.R.H., 30 (1993) 135
Richards, W.G., 11 (1975) 67
Richardson, P.T., 24 (1987) 1
Roberts, L.M., 24 (1987) 1

Rodgers, J.D., 40 (2002) 63
Roe, A.M., 7 (1970) 124
Rose, H.M., 9 (1973) 1
Rosen, T., 27 (1990) 235
Rosenberg, S.H., 32 (1995) 37
Ross, K.C., 34 (1997) 111
Roth, B., 7 (1970) 285; 8 (1971) 61;
  19 (1982) 269
Roth, B.D., 40 (2002) 1
Rowley, M., 46 (2008) 1
Russell, A.D., 6 (1969) 135; 8 (1971) 39;
  13 (1976) 271; 31 (1994) 349;
  35 (1998) 133
Ruthven, C.R.J., 6 (1969) 200

Sadler, P.J., 12 (1975) 159
Salvatore, C.A., 47 (2009) 1
Sampson, G.A., 11 (1975) 193
Sandler, M., 6 (1969) 200
Saporito, M.S., 40 (2002) 23
Sarges, R., 18 (1981) 191
Sartorelli, A.C., 15 (1978) 321;
  32.(1995) 1
Saunders, J., 41 (2003) 195
Schiller, P.W., 28 (1991) 301
Schmidhammer, H., 35 (1998) 83
Schön, R., 30 (1993) 135
Schunack, W., 38 (2001) 279
Schwartz, J.-C., 38 (2001) 279
Schwartz, M.A., 29 (1992) 271
Scott, M.K., 36 (1999) 169
Sewell, R.D.E., 14 (1977) 249;
  30 (1993) 327
Shank, R.P., 36 (1999) 169
Shaw, M.A., 26 (1989) 253
Sheard, P., 21 (1984) 1
Shepherd, D.M., 5 (1967) 199
Silver, P.J., 22 (1985) 293
Silvestrini, B., 21 (1984) 111
Singh, H., 16 (1979) 35; 17 (1980) 151;
  22 (1985) 243; 28 (1991) 233
Skidmore, J., 46 (2008) 131
Skotnicki, J.S., 25 (1988) 85
Slater, J.D.H., 1 (1961) 187
Sliskovic, D.R., 39 (2002) 121
Smith, H.J., 26 (1989) 253; 30 (1993) 327
Smith, R.C., 12 (1975) 105
Smith, W.G., 1 (1961) 1; 10 (1974) 11

Solomons, K.R.H., 33 (1996) 233
Sorenson, J.R.J., 15 (1978) 211;
  26 (1989) 437
Souness, J.E., 33 (1996) 1
Southan, C., 37 (2000) 1
Spencer, P.S.J., 4 (1965) 1; 14 (1977) 249
Spinks, A., 3 (1963) 261
Ståhle, L., 25 (1988) 291
Stark, H., 38 (2001) 279
Steiner, K.E., 24 (1987) 209
Stenlake, J.B., 3 (1963) 1; 16 (1979) 257
Stevens, M.F.G., 13 (1976) 205
Stewart, G.A., 3 (1963) 187
Studer, R.O., 5 (1963) 1
Su, X., 46 (2008) 29
Subramanian, G., 40 (2002) 107
Sullivan, M.E., 29 (1992) 65
Suschitzky, J.L., 21 (1984) 1
Swain, C.J., 35 (1998) 57
Swallow, D.L., 8 (1971) 119
Sykes, R.B., 12 (1975) 333
Szallasi, A., 44 (2006) 145

Talley, J.J., 36 (1999) 201
Taylor, E.C., 25 (1988) 85
Taylor, E.P., 1 (1961) 220
Taylor, S.G., 31 (1994) 409
Tegnér, C., 3 (1963) 332
Terasawa, H., 34 (1997) 69
Thomas, G.J., 32 (1995) 239
Thomas, I.L., 10 (1974) 245
Thomas, J.D.R., 14 (1977) 51
Thompson, E.A., 11 (1975) 193
Thompson, M., 37 (2000) 177
Tibes, U., 46 (2008) 205
Tilley, J.W., 18 (1981) 1
Timmerman, H., 38 (2001) 61
Traber, R., 25 (1988) 1
Tucker, H., 22 (1985) 121
Tyers, M.B., 29 (1992) 239

Upton, N., 37 (2000) 177

Valler, M.J., 37 (2000) 83
Van de Waterbeemd, H., 41 (2003) 1
Van den Broek, L.A.G.M., 23 (1986) 219
Van Dijk, J., 15 (1978) 261
Van Muijlwijk-Koezen, J.E., 38 (2001) 61

Van Wart, H.E., 29 (1992) 271
Vaz, R.J., 43 (2005) 1
Vicker, N., 46 (2008) 29
Vincent, J.E., 17 (1980) 185
Volke, J., 12 (1975) 247
Von Itzstein, M., 36 (1999) 1
Von Seeman, C., 3 (1963) 89
Von Wartburg, A., 25 (1988) 1
Vyas, D.M., 32 (1995) 289

Waigh, R.D., 18 (1981) 45
Wajsbort, J., 21 (1984) 137
Walker, R.T., 23 (1986) 187
Walls, L.P., 3 (1963) 52
Walz, D.T., 19 (1982) 1
Ward, W.H.J., 38 (2001) 309
Waring, W.S., 3 (1963) 261
Wartmann, M., 42 (2004) 171
Watson, N.S., 33 (1996) 331
Watson, S.P., 37 (2000) 83
Wedler, F.C., 30 (1993) 89
Weidmann, K., 31 (1994) 233
Weiland, J., 30 (1993) 135
West, G.B., 4 (1965) 1
White, P.W., 44 (2006) 65
Whiting, R.L., 23 (1986) 1
Whittaker, M., 36 (1999) 91
Whittle, B.J.R., 21 (1984) 237
Wiedling, S., 3 (1963) 332
Wiedeman, P.E., 45 (2007) 63
Wien, R., 1 (1961) 34
Williams, T.M., 47 (2009) 1
Wikström, H., 29 (1992) 185
Wikström, H.V., 38 (2001) 189
Wilkinson, S., 17 (1980) 1
Williams, D., 44 (2006) 1
Williams, D.R., 28 (1991) 175
Williams, J., 41 (2003) 195
Williams, J.C., 31 (1994) 59
Williams, K.W., 12 (1975) 105
Williams-Smith, D.L., 12 (1975) 191
Wilson, C., 31 (1994) 369
Wilson, H.K., 14 (1977) 285
Witte, E.C., 11 (1975) 119
Wold, S., 25 (1989) 291
Wood, A., 43 (2005) 239
Wood, E.J., 26 (1989) 323
Wright, I.G., 13 (1976) 159

Wyard, S.J., 12 (1975) 191
Wyman, P.A., 41 (2003) 129

Yadav, M.R., 28 (1991) 233
Yates, D.B., 32 (1995) 115
Youdim, K., 47 (2009) 239
Youdim, M.B.H., 21 (1984) 137

Young, P.A., 3 (1963) 187
Young, R.N., 38 (2001) 249

Zalacain, M., 44 (2006) 109
Zee-Cheng, R.K.Y., 20 (1983) 83
Zon, G., 19 (1982) 205
Zylicz, Z., 23 (1986) 219

Cumulative Index of Subjects for Volumes 1–47

The volume number, (year of publication) and page number are given in that order.

ACAT inhibitors, 39 (2002) 121
Adamantane, amino derivatives, 18 (1981) 1
Adenosine A_1 receptor ligands, 47 (2009) 163
Adenosine A_3 receptor ligands, 38 (2001) 61
Adenosine triphosphate, 16 (1979) 223
Adenylate cyclase, 12 (1975) 293
Adipose tissue, 17 (1980) 105
Adrenergic agonists, β_3-, 41 (2003) 167
Adrenergic blockers, α-, 23 (1986) 1
  ß-, 22 (1985) 121
α_2-Adrenoceptors, antagonists, 23 (1986) 1
Adrenochrome derivatives, 9 (1973) 275
Adriamycin, 15 (1978) 125; 21 (1984) 169
AIDS, drugs for, 31 (1994) 121
Aldehyde thiosemicarbazones as antitumour agents, 15 (1978) 321; 32 (1995) 1
Aldehydes as biocides, 34 (1997) 149
Aldose reductase inhibitors, 24 (1987) 299
Allergy, chemotherapy of, 21 (1984) 1; 22 (1985) 293
Alzheimer's disease, chemotherapy of, 34 (1997) 1; 36 (1999) 201
  M1 agonists in, 43 (2005) 113
Amidines and guanidines, 30 (1993) 203
Aminoadamantane derivatives, 18 (1981) 1
Aminopterins as antitumour agents, 25 (1988) 85
8-Aminoquinolines as antimalarial drugs, 28 (1991) 1; 43 (2005) 220
Analgesic drugs, 2 (1962) 43; 4 (1965) 171; 7 (1970) 229; 14 (1977) 249
Anaphylactic reactions, 2 (1962) 176
Angiotensin, 17 (1980) 41; 32 (1995) 37
Anthraquinones, antineoplastic, 20 (1983) 83
Antiallergic drugs, 21 (1984) 1; 22 (1985) 293; 27 (1990) 34

Antiapoptotic agents, 39 (2002) 1
Antiarrhythmic drugs, 29 (1992) 65
Antiarthritic agents, 15 (1978) 211; 19 (1982) 1; 36 (1999) 201
Anti-atherosclerotic agents, 39 (2002) 121
Antibacterial agents, 6 (1969) 135; 12 (1975) 333; 19 (1982) 269; 27 (1990) 235; 30 (1993) 203; 31 (1994) 349; 34 (1997)
  resistance to, 32 (1995) 157; 35 (1998) 133
Antibiotics, antitumour, 19 (1982) 247; 23 (1986) 219
  carbapenem, 33 (1996) 99
  ß-lactam, 12 (1975) 395; 14 (1977) 181; 31 (1994) 297; 33 (1996) 99
  macrolide, 30 (1993) 57; 32 (1995) 157
  mechanisms of resistance, 35 (1998) 133
  polyene, 14 (1977) 105; 32 (1995) 157
  resistance to, 31 (1994) 297; 32 (1995) 157; 35 (1998) 133
Anticancer agents — *see* Antibiotics, Antitumour agents
Anticonvulsant drugs, 3 (1963) 261; 37 (2000) 177
Antidepressant drugs, 15 (1978) 261; 23 (1986) 121
Antidiabetic agents, 41 (2003) 167; 42 (2004) 1
Antiemetic action of 5-HT_3 antagonists, 27 (1990) 297; 29 (1992) 239
Antiemetic drugs, 27 (1990) 297; 29 (1992) 239
Antiepileptic drugs, 37 (2000) 177
Antifilarial benzimidazoles, 25 (1988) 233
Antifolates as anticancer agents, 25 (1988) 85; 26 (1989) 1

Flavonoids, physiological and nutritional aspects, 14 (1977) 285
Fluorescence-based assays, 43 (2005) 19
Fluoroquinolone antibacterial agents, 27 (1990) 235
mechanism of resistance to, 32 (1995) 157
Folic acid and analogues, 25 (1988) 85; 26 (1989) 1
Formaldehyde, biocidal action, 34 (1997) 149
Free energy, biological action and linear, 10 (1974) 205

GABA, heterocyclic analogues, 22 (1985) 67
$GABA_A$ receptor ligands, 36 (1999) 169
Gas-liquid chromatography and mass spectrometry, 12 (1975) 1
Gastric $H+/K+$-ATPase inhibitors, 31 (1994) 233
Genomics, impact on drug discovery, 37 (2000) 1
Glutaraldehyde, biological uses, 13 (1976) 271
as sterilizing agent, 34 (1997) 149
Gold, immunopharmacology of, 19 (1982) 1
Growth hormone secretagogues 39 (2002) 173
Guanidines, 7 (1970) 124; 30 (1993) 203

Halogenoalkylamines, 2 (1962) 132
Heparin and heparinoids, 5 (1967) 139
Hepatitis C virus NS3-4 protease, inhibitors of, 39 (2002) 215
Hepatitis C virus NS3/NS4A protease inhibitors, 44 (2006) 65
Herpes virus, chemotherapy, 23 (1985) 67
Heterocyclic analogues of GABA, 22 (1985) 67
Heterocyclic carboxaldehyde thiosemicarba- zones, 16 (1979) 35; 32 (1995) 1
Heterosteroids, 16 (1979) 35; 28 (1991) 233
H^+/K^+ ATPase inhibitors, 47 (2009) 75
High-throughput screening techniques, 37 (2000) 83; 43 (2005) 43
Histamine, H_3 ligands, 38 (2001) 279; 44 (2006) 181
Hit identification, 45 (2007) 1

H_2-antagonists, 20 (1983) 337
receptors, 24 (1987) 30; 38 (2001) 279
release, 22 (1985) 26
secretion, calcium and, 19 (1982) 59
$5-HT_{1A}$ receptors, radioligands for *in vivo* studies, 38 (2001) 189
$5-HT_{2C}$ ligands, 46 (2008) 281
Histidine decarboxylases, 5 (1967) 199
Histone deacetylase inhibitors, 46 (2008) 205
HIV CCR5 antagonists in, 43 (2005) 239
proteinase inhibitors, 32 (1995) 239
HIV integrase inhibitors, 46 (2008) 1
HMG-CoA reductase inhibitors, 40 (2002) 1
Human Ether-a-go-go (HERG), 43 (2005) 1
Hydrocarbons, carcinogenicity of, 10 (1974) 159
11β-Hydroxysteroid dehydrogenase inhibitors, 46 (2008) 29
Hypersensitivity reactions, 4 (1965) 1
Hypocholesterolemic agents, 39 (2002) 121; 40 (2002) 1
Hypoglycaemic drugs, 1 (1961) 187; 18 (1981) 191; 24 (1987) 209; 30 (1993) 203; 31(1994) 1
Hypolipidemic agents, 40 (2002) 1
Hypotensive agents, 1 (1961) 34; 30 (1993) 203; 31 (1994) 409; 32 (1995) 37, 115

Immunopharmacology of gold, 19 (1982) 1
Immunosuppressant cyclosporins, 25 (1988) 1
India, medicinal research in, 22 (1985) 243
Influenza virus sialidase, inhibitors of, 36 (1999) 1
Information retrieval, 10 (1974) 1
Inotropic steroids, design of, 30 (1993) 135
Insulin, obesity and, 17 (1980) 105
Ion-selective membrane electrodes, 14 (1977) 51
Ion transfer, 14 (1977) 1
Irinotecan, anticancer agent, 34 (1997) 68
Isothermal titration calorimetry, in drug design, 38 (2001) 309
Isotopes, in drug metabolism, 9 (1973) 133
stable, 15 (1978) 1

Kappa opioid non-peptide ligands, 29 (1992) 109; 35 (1998) 83